Methods in Computational Biology

Methods in Computational Biology

Special Issue Editors

Ross P. Carlson
Herbert M. Sauro

MDPI • Basel • Beijing • Wuhan • Barcelona • Belgrade

MDPI

Special Issue Editors

Ross P. Carlson
Montana State University Bozeman
USA

Herbert M. Sauro
University of Washington
USA

Editorial Office
MDPI
St. Alban-Anlage 66
4052 Basel, Switzerland

This is a reprint of articles from the Special Issue published online in the open access journal *Processes* (ISSN 2227-9717) from 2018 to 2019 (available at: https://www.mdpi.com/journal/processes/special_issues/methods_biology)

For citation purposes, cite each article independently as indicated on the article page online and as indicated below:

LastName, A.A.; LastName, B.B.; LastName, C.C. Article Title. *Journal Name* **Year**, *Article Number*, Page Range.

ISBN 978-3-03921-163-0 (Pbk)
ISBN 978-3-03921-164-7 (PDF)

Cover image courtesy of S. Lee McGill and Ross P. Carlson.

Contents

About the Special Issue Editors vii

Ross P. Carlson and Herbert M. Sauro
Special Issue: Methods in Computational Biology
Reprinted from: *Processes* **2019**, *7*, 205, doi:10.3390/pr7040205 1

Veronica L. Porubsky and Herbert M. Sauro
Application of Parameter Optimization to Search for Oscillatory Mass-Action Networks Using
Python
Reprinted from: *Processes* **2019**, *7*, 163, doi:10.3390/pr7030163 4

Poonam Phalak and Michael A. Henson
Metabolic Modeling of *Clostridium difficile* Associated Dysbiosis of the Gut Microbiota
Reprinted from: *Processes* **2019**, *7*, 97, doi:10.3390/pr7020097 21

Kerri-Ann Norton, Chang Gong, Samira Jamalian and Aleksander S. Popel
Multiscale Agent-Based and Hybrid Modeling of the Tumor Immune Microenvironment
Reprinted from: *Processes* **2019**, *7*, 37, doi:10.3390/pr7010037 41

André H. Erhardt
Early Afterdepolarisations Induced by an Enhancement in the Calcium Current
Reprinted from: *Processes* **2019**, *7*, 20, doi:10.3390/pr7010020 64

Frances Pool, Peter K. Sweby and Marcus J. Tindall
An Integrated Mathematical Model of Cellular Cholesterol Biosynthesis and
Lipoprotein Metabolism
Reprinted from: *Processes* **2018**, *6*, 134, doi:10.3390/pr6080134 80

Parham Farzan and Marianthi G. Ierapetritou
A Framework for the Development of Integrated and Computationally Feasible Models of
Large-Scale Mammalian Cell Bioreactors
Reprinted from: *Processes* **2018**, *6*, 82, doi:10.3390/pr6070082 114

**Justin T. Roberts, Dillon G. Patterson, Valeria M. King, Shivam V. Amin, Caroline J. Polska,
Dominika Houserova, Aline Crucello, Emmaline C. Barnhill, Molly M. Miller,
Timothy D. Sherman and Glen M. Borchert**
ADAR Mediated RNA Editing Modulates MicroRNA Targeting in Human Breast Cancer
Reprinted from: *Processes* **2018**, *6*, 42, doi:10.3390/pr6050042 130

Tim Daniel Rose and Jean-Pierre Mazat
FluxVisualizer, a Software to Visualize Fluxes through Metabolic Networks
Reprinted from: *Processes* **2018**, *6*, 39, doi:10.3390/pr6050039 144

Ashley E. Beck, Kristopher A. Hunt and Ross P. Carlson
Measuring Cellular Biomass Composition for Computational Biology Applications
Reprinted from: *Processes* **2018**, *6*, 38, doi:10.3390/pr6050038 154

**C. Anthony Hunt, Ahmet Erdemir, William W. Lytton, Feilim Mac Gabhann,
Edward A. Sander, Mark K. Transtrum and Lealem Mulugeta**
The Spectrum of Mechanism-Oriented Models and Methods for Explanations of
Biological Phenomena
Reprinted from: *Processes* **2018**, *6*, 56, doi:10.3390/pr6050056 181

About the Special Issue Editors

Ross P. Carlson is a Professor at the Department of Chemical and Biological Engineering at Montana State University, Bozeman. He has an interdisciplinary education, including a Ph.D. in chemical engineering, an M.Sc in microbial engineering, and a B.Sc. in biochemistry. His research focuses on combining in silico systems biology with both in vitro and in situ experimental systems ranging from medical infections to biofuel production and nutrient cycling in Yellowstone National Park hot spring mats. His research aims to identify design principles for mass and energy transfer within metabolic systems with the ultimate goal of system control.

Herbert M. Sauro is an Associate Professor at the Department of Bioengineering in the University of Washington, Seattle. His work focuses on (1) developing credible and reliable computational models of protein signaling pathways; (2) the use of mathematics and computation to help understand the dynamics and operation of cellular processes, particularly through the use of metabolic control analysis; (3) disseminating best practices in systems biology modeling by developing standards such as SBML and SBOL; and (4) supporting journals to help disseminate reproducible models. He is currently Director of the NIH Center of Reproducible Biomedical Modeling and a member of the Cancer Systems Biology Consortium. He has written numerous papers, including a number of textbooks on modeling in systems biology.

processes

MDPI

Editorial

Special Issue: Methods in Computational Biology

Ross P. Carlson [1],* and Herbert M. Sauro [2],*

[1]   Department of Chemical and Biological Engineering, Montana State University, Bozeman, MT 59717, USA
[2]   Department of Bioengineering, University of Washington, Seattle, WA 98195-5061, USA
*   Correspondence: rossc@montana.edu (R.P.C.); hsauro@uw.edu (H.M.S.); Tel.: +406-994-3631 (R.P.C.);
     +206-685-2119 (H.M.S.)

Received: 8 April 2019; Accepted: 8 April 2019; Published: 11 April 2019

check for
updates

Biological systems are multiscale with respect to time and space, exist at the interface of biological and physical constraints, and their interactions with the environment are often nonlinear. These systems are being quantified in ever increasing detail using rapidly developing omics technologies; yet, it is difficult to predict the dynamic and spatial behavior of even the simplest model systems. Computational biology approaches are essential for leveraging the omics data to develop and test new theories on biological organization. This is a major challenge for the life sciences, including the medical, environmental, and bioprocess fields.

A primary goal of this Special Issue "Methods in Computational Biology" is the communication of computational biology methods, which can extract biological design principles from complex data, described in enough detail to permit reproduction of the results. This issue integrates highly interdisciplinary researchers such as biologists, computer scientists, engineers and mathematicians to advance biological systems analysis. A summary of the contributions to the Special Issue are provided in the following section; many of the contributions are mentioned more than once because their content includes themes that fall under multiple categories.

Reviews of Computational Methods

The Special Issue includes two contributions which review and synthesize important aspects of computational analysis. In Hunt et al. [1], the authors summarize, organize and provide examples of seven different 'mechanism-oriented' model types and discuss how they can be employed to analyze biological phenomena. Coverage includes not only a mathematical description, but also solvers and simulation considerations. Norton et al. [2] provide a thorough review of agent-based modeling of tumor cells, tumor cell heterogeneity as well as tumor interactions with host immune system components and local physical environments.

Computational Analysis of Biological Dynamics: From Molecular to Cellular to Tissue/Consortia Level

Life is an inherently dynamic process. The Special Issue includes analysis of dynamic processes on molecular, cellular, tissue and microbial consortia size scales. A comparison of the different size scales identifies mathematical and computational approaches that span scales. Porubsky and Sauro [3] examine molecular level processes—for instance, gene networks—and present methodologies for optimizing parameters necessary to obtain models that exhibit oscillating behavior. Erhardt [4] examines cellular scale systems and the role of calcium-induced oscillations in cardiac cells and its role in cardiac arrhythmis. The study applies a number of theories and methods including bifurcation theory, numerical bifurcation analysis, and geometric singular perturbation theory to study nonlinear multi time scale systems. Pool et al. [5] studies intra- and extracellular processes associated with cholesterol and lipoprotein metabolism and how intervention strategies such as statins or diet can influence metabolism. Farzan and Ierapetritou [6] report on multicellular scale systems and analyze interactions between mammalian cells and the bioreactor environment with the ultimate goal of optimizing bioprocess applications. Norton et al. [2] study systems on the cellular and tissue scales and examine the interactions between tumor cells, host immune cells and local microenvironments. Phalak and Henson [7] study a multicellular scale system quantifying the dynamic interactions between multiple microorganisms, including the exchange of metabolites, and the role of time and space on microbial infections.

The Interface of Biotic and Abiotic Processes

Life occurs at the interface of biological and physical constraints. Biological processes, including metabolism, are constrained by physical processes such as chemical transport to and from the cell. Phalak and Henson [7] analyze how assemblages of different microorganisms can organize along chemical gradients established by an imbalance between biological reaction rates and abiotic diffusion rates. These gradients lead to spatial distributions of cell types and often enhanced system robustness. Farzan and Ierapetritou [6] consider the interface of mammalian cells and convective transport processes which ultimately influence the local chemical, thermal and mechanical environments. The work also discusses computational optimization and selection of solvers for these types of modeling applications.

Processing of Large Data Sets for Enhanced Analysis

Modern biology is rapidly becoming a study of large sets of data. Roberts et al [8] analyze tools for extracting additional information from microRNA extracted from breast cancer by measuring 50,000 recurrent editing sites. The data identifies the presence of additional levels of complexity in microRNAs which influences how the molecules interact with target mRNA.

Representing the output from computational biology efficiently, in a manner that facilitates communication, is often difficult. Rose and Mazat [9] present software that enables the visualization of metabolic flux data using a graphical user interface that permits rapid and simplified formatting of data.

Parameters Optimization and Measurements

Computational representations of life require parameters. Parameter identification is a major challenge and a focus of many studies. The Special Issue includes contributions which focus on optimizing parameters required to represent biphasic systems, including generalized mass action networks, relevant to gene signaling and metabolite networks [3], as well as calcium-induced oscillation in cardiac cells [4]. Beck et al. [10] provide detailed methods for experimentally measuring key parameters required for genome-scale metabolic models, including the biomass synthesis reaction. The authors then demonstrate how different biomass parameters produce very different results based on the interaction of electron balances and metabolism.

This Special Issue is coordinated with the Metabolic Pathway Analysis 2017 conference held in Bozeman, MT and Interagency Modeling and Analysis Group (IMAG) MultiScale Modeling (MSM) working groups (https://www.imagwiki.nibib.nih.gov/).

References

1. Hunt, C.; Erdemir, A.; Lytton, W.; Mac Gabhann, F.; Sander, E.; Transtrum, M.; Mulugeta, L. The Spectrum of Mechanism-Oriented Models and Methods for Explanations of Biological Phenomena. *Processes* **2018**, *6*, 56. [CrossRef]
2. Norton, K.A.; Gong, C.; Jamalian, S.; Popel, A.S. Multiscale Agent-Based and Hybrid Modeling of the Tumor Immune Microenvironment. *Processes* **2019**, *7*, 37. [CrossRef] [PubMed]
3. Porubsky, V.L.; Sauro, H.M. Application of Parameter Optimization to Search for Oscillatory Mass-Action Networks Using Python. *Processes* **2019**, *7*, 163. [CrossRef]

4. Erhardt, A. Early After depolarisations Induced by an Enhancement in the Calcium Current. *Processes* **2019**, *7*, 20. [CrossRef]
5. Pool, F.; Sweby, P.; Tindall, M. An Integrated Mathematical Model of Cellular Cholesterol Biosynthesis and Lipoprotein Metabolism. *Processes* **2018**, *6*, 134. [CrossRef]
6. Farzan, P.; Ierapetritou, M. A Framework for the Development of Integrated and Computationally Feasible Models of Large-Scale Mammalian Cell Bioreactors. *Processes* **2018**, *6*, 82. [CrossRef]
7. Phalak, P.; Henson, M.A. Metabolic Modeling of Clostridium difficile Associated Dysbiosis of the Gut Microbiota. *Processes* **2019**, *7*, 97. [CrossRef]
8. Roberts, J.; Patterson, D.; King, V.; Amin, S.; Polska, C.; Houserova, D.; Crucello, A.; Barnhill, E.; Miller, M.; Sherman, T.; et al. ADAR Mediated RNA Editing Modulates MicroRNA Targeting in Human Breast Cancer. *Processes* **2018**, *6*, 42. [CrossRef] [PubMed]
9. Rose, T.; Mazat, J.P. FluxVisualizer, a Software to Visualize Fluxes through Metabolic Networks. *Processes* **2018**, *6*, 39. [CrossRef]
10. Beck, A.; Hunt, K.; Carlson, R. Measuring Cellular Biomass Composition for Computational Biology Applications. *Processes* **2018**, *6*, 38. [CrossRef]

processes

MDPI

Article

Application of Parameter Optimization to Search for Oscillatory Mass-Action Networks Using Python

Veronica L. Porubsky * and Herbert M. Sauro

Department of Bioengineering, University of Washington, Seattle, WA 98105, USA; hsauro@uw.edu
* Correspondence: verosky@uw.edu; Tel.: +1-206-685-2119

Received: 21 February 2019; Accepted: 13 March 2019; Published: 18 March 2019

check for
updates

Abstract: Biological systems can be described mathematically to model the dynamics of metabolic, protein, or gene-regulatory networks, but locating parameter regimes that induce a particular dynamic behavior can be challenging due to the vast parameter landscape, particularly in large models. In the current work, a Pythonic implementation of existing bifurcation objective functions, which reward systems that achieve a desired bifurcation behavior, is implemented to search for parameter regimes that permit oscillations or bistability. A differential evolution algorithm progressively approximates the specified bifurcation type while performing a global search of parameter space for a candidate with the best fitness. The user-friendly format facilitates integration with systems biology tools, as Python is a ubiquitous programming language. The bifurcation–evolution software is validated on published models from the BioModels Database and used to search populations of randomly-generated mass-action networks for oscillatory dynamics. Results of this search demonstrate the importance of reaction enrichment to provide flexibility and enable complex dynamic behaviors, and illustrate the role of negative feedback and time delays in generating oscillatory dynamics.

Keywords: parameter optimization; differential evolution; evolutionary algorithm; bistable switch; oscillator; turning point bifurcation; Hopf bifurcation; biological networks; mass-action networks; BioModels Database

1. Introduction

Biological systems exhibit dynamic behaviors due to the regulation of metabolites, proteins, or genetic components, and these dynamics are frequently represented by a series of nonlinear equations for the purpose of computational modeling. Dynamical behaviors in biological systems are dependent on motifs within the network, defined by the species interactions and rate laws which construct the overall network topology. However, the behavior of the system is also heavily influenced by the parameter values attributed to rate constants, regulatory elements, and initial concentrations of floating and boundary species in the network, such that the behavior may shift depending on the current parameter regime. When modeling these biological systems, it may be desirable to obtain a particular dynamic behavior to approximate a physiologically-relevant result. Cell cycle oscillations have been studied for decades but underlying mechanisms remain a topic of interest to systems biologists [1]. Neuroscientists are constructing computational models that exhibit complex oscillatory dynamics to explore the effects of parameter variation, which enriches their understanding of disorders like Parkinson's and could have implications for treatment [2]. Developing such models requires knowledge of the parameter regimes that permit complex dynamic behaviors, and this knowledge is not always available from experimental data. Searching the landscape which defines parameter space can be a computationally-intensive task, as this landscape is N-dimensional, where N represents the number of parameters in the model, causing the search space to expand dramatically as the number of parameters defining the system increases. Algorithms to scan high-dimensional parameter spaces have been developed and extensively researched, using combinations of global and local searches to define the landscape of computational models and estimate model parameters [3–5]. Still, there is a need for efficient parameter optimization tools to search for hallmark dynamic behaviors in systems biology.

A tool implemented in C# was previously developed by Chickarmane et al. to optimize parameter values of biological network models defined by systems of nonlinear equations for bifurcation behavior [6]. Using information about the eigenvalues of bifurcated systems, the authors developed objective functions to independently optimize parameters for either Hopf bifurcations, characteristic of oscillatory systems, or for turning point bifurcations, which can lead to bistability [6]. Such functionality would be desirable in a Pythonic computing environment for those interested in modeling biological systems, as Python is a more ubiquitous computer language in the biological sciences, implemented by expert and novice computer scientists alike, is easily-interpreted by the user, and facilitates integration with existing software for modeling and simulation in systems biology. In the current work, bifurcation–evolution software (evolveBifurcation v1.0.0, Seattle, WA, USA, 2019) is developed in which these objective functions are adapted from C# into user-friendly Python code, and global and local optimization algorithms are implemented for parameter evolution in computational models available through the BioModels Database [7]. The bifurcation–evolution software is then employed to search for oscillatory dynamics in populations of randomly-generated mass-action kinetic models of variable size, and oscillatory models discovered during this search are analyzed to understand how a reduced network topology generates oscillations.

2. Materials and Methods

This bifurcation–evolution software relies on standard biological network manipulation and analysis tools available through Tellurium, a Python environment for dynamical modeling of biological networks, and the associated library for simulation of biological models, libRoadRunner [8,9]. The algorithm implemented relies on progressively approximating an acceptable solution to the bifurcation-specific objective function by evolving a population of parameter value vectors. Each parameter vector represents a single point in the landscape of available parameter space that the network can occupy, and vectors which minimize the objective function approximate the global minimum of parameter space, where the desired bifurcation is achieved.

2.1. Objective Function

The objective functions introduced by Chickarmane et al. are re-implemented in the current work, and enable optimization for either switch-like or oscillatory behavior, depending on the bifurcation type selected by the user [6]. Both objective functions rely on intrinsic properties of eigenvalues corresponding to the parameter set governing a system of nonlinear equations at steady state.

2.1.1. Optimization for Turning Point Bifurcations

Turning point bifurcations, capable of introducing bistability and switch-like behavior, can be discovered by minimizing the following objective function as previously described [6]:

$$\epsilon = \frac{\prod \lambda_i}{\left(1 - 0.99 \times e^{-|\prod \lambda_{Min}|}\right)}. \tag{1}$$

A turning point bifurcation requires that one eigenvalue is zero. This objective function is effective for evolving turning point bifurcations because the numerator, which is the product of all eigenvalues of the system, will force the system to assume eigenvalues that approximate zero during minimization. The denominator introduces a penalty for systems in which all eigenvalues are becoming very small, suggesting they are all moving towards the imaginary axis [6]. λ_{Min} includes all eigenvalues except the smallest eigenvalue, so that no penalty results from the system achieving one zero-valued eigenvalue.

It is not guaranteed that a turning point bifurcation will be reached. Pitchfork and transcritical bifurcations could also result.

2.1.2. Optimization for Oscillatory Systems

For a Hopf Bifurcation bifurcation, which occurs in an oscillatory system, the following objective function is minimized as previously described [6]:

$$\epsilon = \frac{\prod \lambda_i^R}{\prod (1 - 0.99 \times e^{-|\lambda_i^I|})}. \tag{2}$$

A Hopf bifurcation requires the real part of one of the complex conjugate eigenvalues to approach zero, which is accounted for in the numerator of the objective function, where λ_R corresponds to all real components of eigenvalues that have a non-zero complex component. The denominator enhances optimization for systems that have complex conjugate eigenvalues by awarding a penalty to systems with no imaginary component.

2.1.3. Steady State Solver

Optimizing for either bifurcation requires that the model is at steady state before performing the eigenvalue analysis. Steady state represents the solution to the system of differential equations comprising the model when the rates of change of all species equal zero. In order to bring the system to steady state, the Newton-based solver implemented in this work iterates through all independent floating species in the system and takes a step defined by the following equation:

$$s^i = -\alpha (J^{-1} \cdot v)^i. \tag{3}$$

Boldface denotes matrix and vector quantities. In this equation, the dot product of the inverted Jacobian, J^{-1}, and the rates of change, v, define the direction of the step, and the step size, α, is selected to gradually approximate the steady state value for each floating species in the network. s represents a vector of all independent floating species in the network, and s^i represents a single species in the vector. The step size scalar multiplier is adjusted to ensure that the floating species maintains a positive concentration during the steady state approximation. To ensure that the steady state is reached, the Frobenius norm of the rates of change vector is computed and compared to a predefined tolerance level which approximates zero. If the norm is less than the tolerance level, indicating that the concentrations of floating species are not changing significantly, the steady state is reached.

2.2. Parameter Selection and Value Assignment

Global parameter values, floating species initial concentrations, and boundary species concentrations are optimized in the bifurcation–evolution software. Conserved sum parameters, which arise in biological models due to moiety conservation through reversible cycles, are removed from the optimization routine, enabling flexibility in the selection of species concentrations [10–12].

Parameter ranges can be specified by the user or automatically specified within the function by referencing initial values contained in the model when it is passed to the function. If the user specifies the bounds, they must submit a sequence defining the upper and lower bounds for each parameter, such that the length of the sequence is equal to the number of parameters undergoing optimization, N. The sequence is thus specified as follows: $[(bound_{min}^1, bound_{max}^1), ..., (bound_{min}^N, bound_{max}^N)]$. Alternatively, the user can specify that all parameters should fall within a uniform range by setting the parameter range argument equal to $[(bound_{min}, bound_{max})]$.

If the model submitted for optimization is known to permit the desired bifurcation under an optimal parameter regime, and has been assigned parameter values that are a good approximation for the bifurcation type, the user can choose to omit the parameter assignment. Differential evolution, the global optimization algorithm implemented in the bifurcation evolution tool, performs poorly with large parameter value ranges, so selecting appropriate bounds is critical. To accommodate for this, the automatic selection, which is the default setting in the tool, attempts to narrow the ranges in a generalizable manner. First, the algorithm checks the current parameter value, p^i, in the loaded model, and, if the value is zero, the algorithm creates a range of parameter values from 1×10^{-25} to 10, approximating zero but prohibiting possible failure of the algorithm if the parameter appears in the denominator of a rate law. For parameter values less than or equal to 10, the assigned parameter range is from $\frac{p^i}{10}$ to 10. All parameter values greater than 10 receive a range from $\frac{p^i}{10}$ to $2p^i$. The automatic assignment allows appropriate flexibility if the range of suitable parameter values for bifurcation is unknown, but relies on the initial parameter value to provide a suitable estimate. If an appropriate range is available for a given parameter through reliable experimental data, manual assignment may be preferable, particularly if this assignment further narrows the range.

Within the differential evolution global optimization algorithm, parameter values are selected from the assigned ranges using a random uniform distribution, such that it is equally likely to choose any value within the assigned range. As this would greatly reduce the frequency of assigned parameter values less than 1, and possibly prevent a parameter from occupying the ideal parameter space to achieve the desired bifurcation behavior, the algorithm selects from a random uniform log distribution of the parameter ranges for all parameters with an upper bound of 10. This ensures that values across multiple orders of magnitude are equally likely to be selected.

2.3. Differential Evolution Algorithm

In the Pythonic approach to this tool, a simple implementation of the differential evolution algorithm developed by Storn and Price was integrated within the bifurcation module, to perform a search in parameter space for global minima of the objective function [13].

2.3.1. Initializing a Population

The algorithm begins by initializing a population of parameter vectors which represent solutions to the objective function. Parameter space in this multi-dimensional optimization problem contains N dimensions, where N is the number of parameters being optimized. These vectors are populated with elements assigned randomly selected values within the predefined bounds specified for N_i, a single parameter, such that the members in the initial population occupy diverse regions of parameter space. While Pythonic versions of this algorithm have been developed, the version available through the scipy.optimize package, frequently used for similar optimization problems, does not allow unfit members to be discarded from the starting population [14]. This makes optimization inefficient, slows convergence and increases the likelihood that the algorithm will terminate before a sufficient minima is reached. The current implementation of the algorithm discards all members with an objective function evaluation above a predetermined threshold before evolving the population. This threshold value coincides with penalty functions included within the bifurcation objective function so that parameter vectors which do not reach steady state, or which have multiple eigenvalues approaching zero in both the real and complex component, are discarded from the solution.

2.3.2. Recombination

During a single round of differential evolution, each member of the population undergoes recombination to construct a trial vector. While iterating through each element, the trial vector is populated with parameter values taken from the member at the current population index or from a mutant vector. If a random number chosen from between zero and one is smaller than the crossover probability, the trial vector receives the parameter value from the mutant vector, as long as the parameter value remains within the acceptable range. Otherwise, the trial vector receives the element from the current population member.

2.3.3. Mutation

Mutated parameter values are generated using the following algebraic expression where boldface denotes vector quantities:

$$\boldsymbol{v}^i = \boldsymbol{m_{best_1}}^i + F(\boldsymbol{m_{best_2}}^i - \boldsymbol{m_{best_3}}^i). \tag{4}$$

The expression shows that the trial vector element v^i, where i designates the index of the parameter value being mutated, is the sum of a population member element $m_{best_1}^i$ and the scaled difference between two additional population member elements, $m_{best_{2,3}}^i$. The three m_{best} vectors correspond to randomly selected members of the population that won a single round of tournament selection. The winner of tournament selection is the parameter value vector that has a lower objective function evaluation. While each tournament selection is between population members sampled without replacement, the selection of m_{best} vectors for mutation between rounds of tournament selection allows sampling with replacement. As a result, m_{best} vectors may be identical. F is the mutation constant, and can be specified by the user.

2.3.4. Selection

Once the trial member is constructed, the fitness of the member is evaluated. If the trial member has a lower objective function evaluation than the original member at the current population index, the trial member is more fit and selected to replace the original member in the population.

2.3.5. Termination

After the entire population of parameter vectors has undergone recombination, the stopping criteria are assessed to determine if the population has converged on a solution. Termination of differential evolution is achieved when the maximum number of generations has been reached or when the threshold value is met. The threshold value can be selected to consider the smallest eigenvalue of the system, such that the eigenvalue must be sufficiently close to zero to have reached the bifurcation behavior, or the threshold value will correspond to the best objective function fitness from all members in the population. The fitness threshold is the default stopping criteria.

2.3.6. Conditions for Optimal Convergence

There are several input parameters to the differential evolution algorithm that can be manipulated to shift the balance between fast and accurate convergence. Generally, increasing the population size and mutation constant while decreasing the recombination constant will improve the chance that the algorithm converges on a global minimum. However, this will result in computational costs that slow convergence. A population size of 50, and mutation and recombination constants of 0.5, are assigned as default values for the algorithm and typically enable rapid convergence on a suitable solution.

2.4. Local Optimization Algorithm

Following the differential evolution routine, the objective function can be minimized further using an optional bounded Broyden–Fletcher–Goldfarb–Shanno algorithm to provide a final local optimization step [15–18]. The algorithm uses approximated Hessian updates that are dependent on the approximate gradient at the point in parameter space where the current parameter vector rests, such that it minimizes in the direction of steepest descent. A one-dimensional line search is implemented to determine the step size. This local optimization dramatically reduces the final objective function evaluation for both oscillators and turning points, often yielding a fitness that is minimized by multiple orders of magnitude. However, this step is not recommended for most Hopf bifurcation optimization problems, as fitness values smaller than 1×10^{-3} frequently correspond to damped oscillatory models.

2.5. Random Network Generation

To determine the frequency of oscillators in random networks, a network generator was used which permitted four types of reactions and variable numbers of floating and boundary species. Floating species are state variables, and therefore the concentrations of these species are variable in time during the course of a simulation [19]. Boundary species are fixed and independent of the model state, and are therefore either constant sources to the system or sinks, constant outputs [19]. The random networks were assigned simple mass-action kinetic rate laws and included the reactions summarized in Figure 1. Mass-action kinetic rate laws are proportional to the concentration of the reactant species in the biochemical reaction. Figure 1 therefore defines the mass-action rate laws, v, used in the random network generator as the product of a rate constant, k, and the concentration of the reactants involved. Species concentrations are represented by placing brackets around the species name. The generator excludes reactions that violate moiety conservation, and requires that at least one species is a boundary species. For the purpose of analysis of networks with a specified number of species and reactions, which are discussed in the frequency analysis, only networks which did not have orphaned species and which had at least three floating species were passed to the final populations. Three floating species was selected as the minimum cutoff because the smallest system exhibiting a Hopf bifurcation contained three floating species [20]. For parameter value assignment, the random network generator assigned concentrations and rate constants with arbitrary units (a.u.). Initial concentrations ranged from 1 to 10 a.u., and rate constants ranged from 1×10^{-3} to 2 a.u. The random network generator was used to create populations of random networks that could be sent to the bifurcation–evolution software to evolve oscillatory dynamics. The default settings in the tool were used for optimization, and, as a result, optimized species initial concentrations ranged from 0.1 to 10.0 a.u., and rate constant value assignments ranged from 1×10^{-4} to 10.0 a.u. Models that could not reach steady state, or which contained negative concentrations, were omitted from analysis. Models that obtained a sufficiently low fitness value after optimization were reset and underwent two additional rounds of optimization to increase the probability of achieving sustained oscillations given an appropriate network architecture, accounting for stochasticity in the algorithm. Following parameter optimization of all networks, populations of a minimum of 1100 randomly-generated networks for each network size were manually assessed for oscillatory dynamics by simulating the model with optimized parameters and inspecting the time-course of all floating species concentrations.

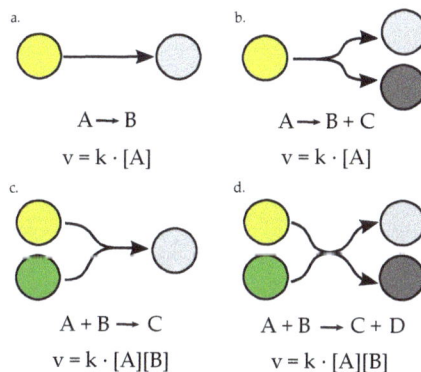

Figure 1. Types of reactions permitted in randomly-generated networks, governed by laws of mass-action. Reactions are depicted visually and written with standard biochemical reaction formatting. Mass-action kinetic rate laws, v, for each reaction are defined by the product of rate constant, k, and the concentrations of reactant species. (**a**) unimolecular–unimolecular reaction; (**b**) unimolecular–bimolecular reaction; (**c**) bimolecular–unimolecular reaction; (**d**) bimolecular–bimolecular reaction.

2.6. Machine Specifications

All computations with runtime calculations were performed on an Intel(R) Core(TM) i5-6300HQ CPU (Intel Corporation, Santa Clara, CA, USA) at 2.30 GHz with 8.00 GB RAM.

2.7. Data Repository

All data required to construct the figures in the main text and the bifurcation evolution algorithm sourcecode, evolveBifurcation.py, are publically available on Github. The repository location is provided in the Supplementary Materials.

3. Results

3.1. Testing Bifurcation–Evolution Software on Models from the BioModels Database

To demonstrate the efficacy of the bifurcation–evolution software, models from the BioModels Database underwent parameter optimization for turning point or Hopf bifurcations, depending on the dynamic properties described for each model in the referenced publications. The rate laws describing the models tested are not limited to mass-action kinetics, and demonstrate that the algorithm is effective for optimization of more complex systems. The results of these test cases are shown in Table 1. Graphical output of the optimized networks and model files are available in the supplementary data. Most of these models were tested in the previous work, so we have demonstrated that the Pythonic implementation maintains functionality for all previous test cases [6]. All models were evolved using the default settings in the bifurcation–evolution software, with the exception of the threshold value, which was set independently for turning point and Hopf bifurcations. All turning point models were evolved with a fitness threshold of 10^{-3} to induce bistability. All models capable of a Hopf bifurcation were evolved with a fitness threshold of 5 to optimize for oscillatory dynamics. These threshold values were chosen empirically. The runtime is the average number of seconds to complete a single optimization, taken over 100 attempts.

The largest model tested, a negative feedback and bi-rhythmic oscillator containing 10 state variables and 46 global parameters, could achieve oscillatory dynamics after optimization [21]. However, a single run with the default parameters in the bifurcation–evolution software lasted 22 minutes, with the majority of this time allocated to initializing a population of 50 members that could achieve steady state before the differential evolution routine could begin. In some of the simulations of the optimized model, chaotic oscillatory behavior described by the authors arose, which was not observed in any of the other models tested.

Figure 2 shows the result of optimizing for a turning point bifurcation in the Hervagault bistable switch model [22]. Before parameter optimization, the concentrations of S1 and S2 reach a single steady state despite a parameter sweep in the global parameter, J1_k. After parameter optimization, both S1 and S2 achieve two distinct steady states, demonstrating the bistable dynamics of the model in a parameter regime with eigenvalues that satisfy the condition for a turning point bifurcation. Figure 3 shows the result of optimizing for a Hopf bifurcation in the modified Edelstein relaxation oscillator model [23]. Before parameter optimization, species A reaches a steady state upon simulation. However, once optimized, species A achieves sustained oscillatory dynamics.

3.2. Oscillation Discovery in Randomly-Generated Networks

The bifurcation–evolution software was used to search populations of randomly-generated networks for models exhibiting oscillatory dynamics. Table 2 shows the percentage of sustained oscillators in populations of randomly-generated networks with variable network sizes. Networks either have an equal ratio (1:1) of species to reactions, or are enriched with 50% more reactions than species (1:1.5). Figure 4 shows the frequencies of oscillatory dynamics in networks of variable size. The top row of Figure 4 contains frequency data of all oscillating systems, including those with sustained and damped dynamics. The bottom row contains only the frequency of sustained oscillatory dynamics. The frequency of sustained oscillators increases for larger networks in both the 1:1 and 1:1.5 network sizes. However, networks with a 1:1.5 ratio have a higher frequency of oscillatory dynamics than networks with a 1:1 ratio when comparing networks with the same number of species, therefore increasing the ratio of reactions to species increases the frequency of oscillators in randomly-generated populations.

Table 1. Test cases from the BioModels Database. Table model names and descriptions adapted from Chickarmane et al. with average runtime values generated by optimizing model parameter values for all test cases using the Pythonic bifurcation–evolution software [6].

Turning Point Model	Description	Runtime (s)
Tyson et al., 2003 (Figure 1e) [24]	Irreversible bistable switch	9.10
Tyson et al., 2003 (Figure 1f) [24]	Reversible bistable switch	8.88
Edelstein, 1970 [25]	Autocatalytic bistable switch	17.70
Hervagault and Canu, 1987 [22]	Bistable switch	6.72
Angeli et al., 2004 [26]	Bistable switch cdc2/wee1	8.16
Oscillatory Model	**Description**	**Runtime (s)**
Tyson et al., 2003 (Figure 2a) [24]	Negative feedback oscillator	1.24
Tyson et al., 2003 (Figure 2b) [24]	Activator-inhibitor oscillator	1.07
Tyson et al., 2003 (Figure 2c) [24]	Substrate-depletion oscillator	0.99
Nicolis and Prigogine, 1977 [27]	Autocatalytic oscillator	0.16
Heinrich et al., 1977 [28]	Positive feedback oscillator	0.16
Seno et al., 1978 [23]	Modified Edelstein relaxation oscillator	1.54
Kholodenko, 2000 [29]	Mitogen-activated protein kinase feedback oscillator	42.57
Goldbeter, 1991 [30]	Mitotic oscillator	1.20
Francois et al., 2005 [31]	Mixed feedback loop oscillator	6.18
Lavrentovich and Hemkin, 2008 [32]	Spontaneous Ca^{2+} oscillator	1.70

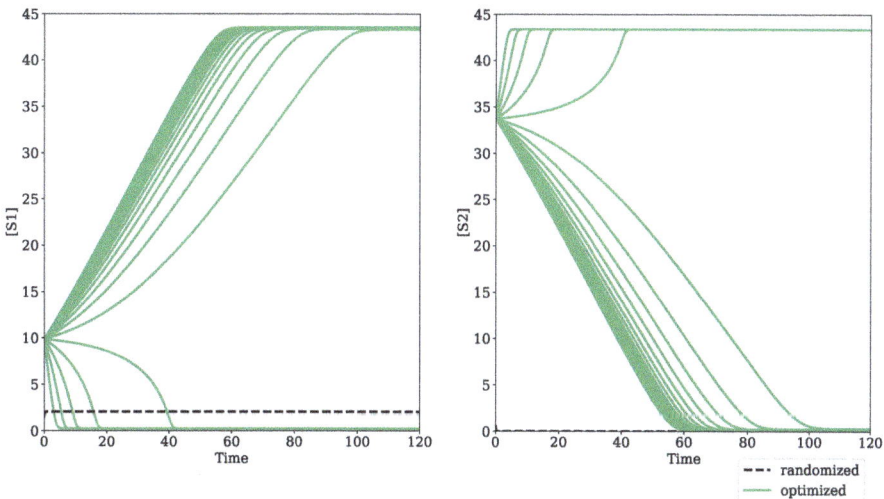

Figure 2. Optimized dynamic concentration changes in a bistable over time. The dashed black trace is the result randomizing all parameter values in the bistable switch model from Hervagault and Canu, 1987, using Tellurium and libRoadRunner functionalities by selecting from a uniform distribution between 0.001 and 15.0 and performing a parameter sweep of parameter J1_k from 0.0 to 2.0. The green trace is the optimized bistable output for both floating species in the model in response to an identical parameter sweep after the randomized parameters were evolved with default settings in the bifurcation–evolution software.

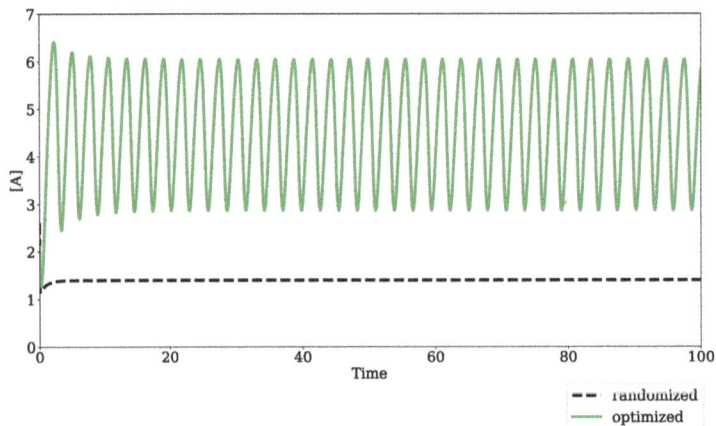

Figure 3. Optimized dynamic output of an oscillatory model. The dashed black trace is the result of simulating the modified Edelstein relaxation oscillator from Seno et al., 1978, using Tellurium and libRoadRunner functionalities after randomizing parameter values from a uniform distribution between log(0.01) and log(10.0). The green trace is the optimized oscillatory output after the randomized model underwent parameter evolution with default settings in the bifurcation–evolution software.

In Figures 5 and 6, oscillator frequency within a given network size is binned by the number of floating species in the network. All networks tested which exhibited sustained oscillations had a minimum of four floating species. In networks which have an equal number of reactions and species, sustained oscillatory dynamics were only achieved in networks that had fewer than $n - 2$ floating species, where n is the total number of species in the network, as shown in Figure 5. As demonstrated by the histograms in Figure 6, for populations with a 50% enrichment in the number of reactions, networks could achieve sustained oscillations with the maximum number of floating species. In most cases, an intermediate number of floating species achieved sustained oscillatory dynamics with the highest frequency for networks with either a 1:1 or 1:1.5 ratio. Enriching the number of reactions shifted the histogram towards higher numbers of floating species. For all network sizes, the spread of the histogram increased when damped oscillators were considered.

Table 2. Percentage of systems exhibiting sustained oscillatory dynamics in populations containing 1100 randomly-generated networks. Each population is defined by a characteristic network size, or number of species and reactions. Networks containing orphaned species were not included in these populations.

Species	Reactions	Oscillators
5	8	0.7%
6	6	0.0%
6	9	2.0%
7	7	0.1%
7	11	4.2%
8	8	0.3%
8	12	6.1%
9	9	1.1%
9	14	9.0%
10	10	1.3%

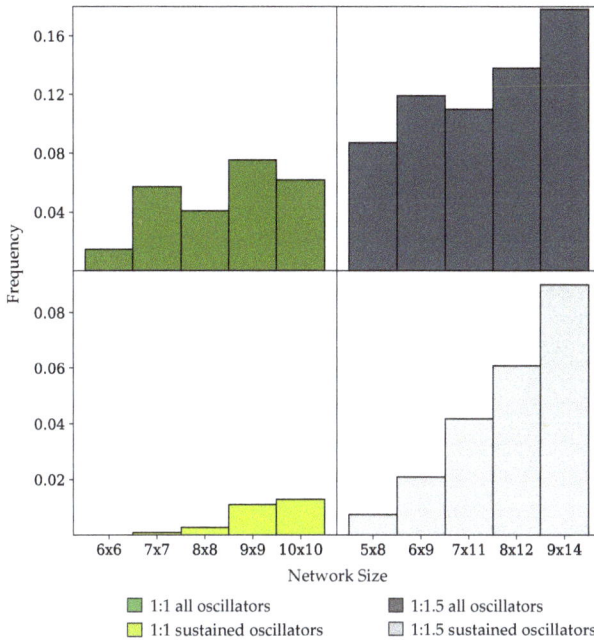

Figure 4. Frequency of occurrence of oscillatory dynamics in populations of randomly-generated networks of variable size, disallowing orphan species. Network sizes are labeled in the form, number of species × number of reactions. Networks either contain an equal ratio of species to reactions (1:1) or a 50% enrichment in the number of reactions (1:1.5). (**Top row**) frequency of networks exhibiting sustained oscillators and damped oscillators; (**Bottom row**) frequency of sustained oscillating networks.

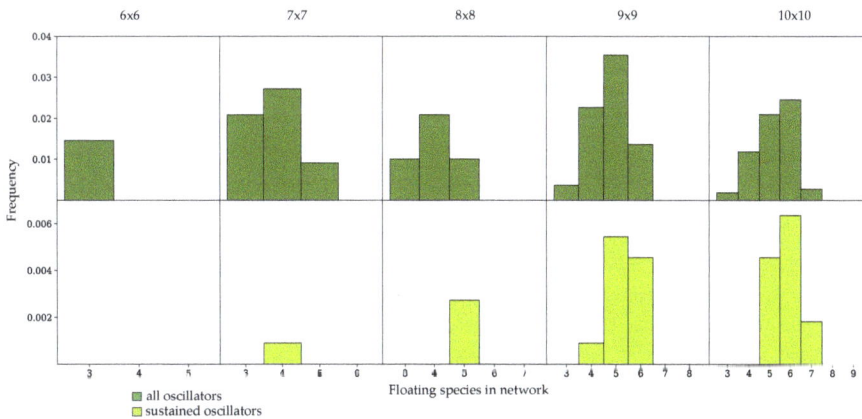

Figure 5. Frequency of occurrence of oscillatory dynamics in randomly-generated networks with a 1:1, number of species to number of reactions ratio, binned by the number of floating species in the network. (**Top row**) frequency of oscillatory dynamics, including sustained and damped systems; (**Bottom row**) frequency of sustained oscillatory dynamics.

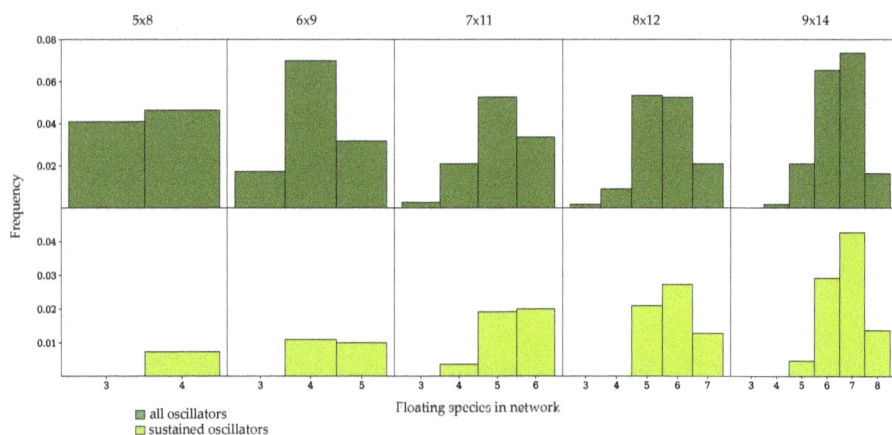

Figure 6. Frequency of occurrence of oscillatory dynamics in randomly-generated networks with a 1:1.5, number of species to number of reactions ratio, binned by number of floating species in the network. (**Top row**) Frequency of oscillatory dynamics, including sustained and damped systems; (**Bottom row**) Frequency of sustained oscillatory dynamics.

3.3. Components of Randomly-Generated Networks Responsible for Oscillation

Next, a population of 10,000 randomly-generated 10 species, 10 reaction networks were generated, allowing for the existence of orphaned species, which are disconnected from the network. In addition, 42.6% of these networks had at least one orphaned species, and the frequency of oscillatory dynamics was greatest in the subset of networks containing orphaned species. Figures 7 and 8 show two randomly-generated networks which were pulled from the population to study the components of the network topology, or species connectivity, responsible for oscillatory dynamics. Both networks have nine total species that participate in reactions, indicating that one of the species was orphaned, and therefore does not contribute to the network dynamics. The reduced forms of these networks, which maintain oscillatory behavior, were generated by removing unnecessary and redundant nodes or reactions, and compounding constant parameters within the rate law. These reduced forms show the species and reactions responsible for oscillatory dynamics.

The reduced network in Figure 7 shows feedback loops that contribute to oscillatory behavior. The first pathway of interest is the sequence of unimolecular–unimolecular reactions from species S1 to S6, with two intermediate nodes, S3 and S5. This pathway shows that, as the concentration of species S1 increases or decreases, there is an impact on species S6 in the same direction of change, accompanied with a time delay in the signal due to the intermediate nodes. Continuing the cycle, an increase in species S6 contributes to a rise in species S2 and subsequent production of S5, feeding back positively into the pathway that produces species S6 and sustaining the cycle. However, negative feedback of S6 on S1 is key for enabling oscillations. While species S6 increases, the bi-molecular reaction between S6 and S1 which produces S2 simultaneously reduces the concentration of species S1 available for the uni-molecular reaction producing species S3. Together, the feedback and time delay promote oscillation. This cycle repeats without dampening given appropriate global parameter values.

The reduced network in Figure 8 also contains feedback loops, involving the reversible reactions between species S2, S9,and S4, as well as the reactions between species S2, S7 and S3. The optimized model drives the conversions of S2 to S9 and S9 to S4 at a fast rate, while the conversion of S4 to S9 occurs much more slowly, as determined by the rate constants for these reactions. Since species S9 feeds back to S4 when generating S2, negative feedback arises. It is important to note that reactions involving S8 effectively promote a time delay due to the involvement of an intermediate node in the production of S2, which impacts the phase of the oscillations. However, removing species S8 allows for a simplified reversible reaction between S4 and S9 in which the time delay created by the intermediate node can be partially compensated for by adjusting the rate constants in the reversible reaction. In the second loop, S3 exerts negative feedback on S2—as the concentration of S3 increases due to greater conversion of S2 to S7, more of species S2 is consumed in the bi-molecular reaction with S3, reducing the production of species S7. Again, species S7 serves to promote a delay in the signal.

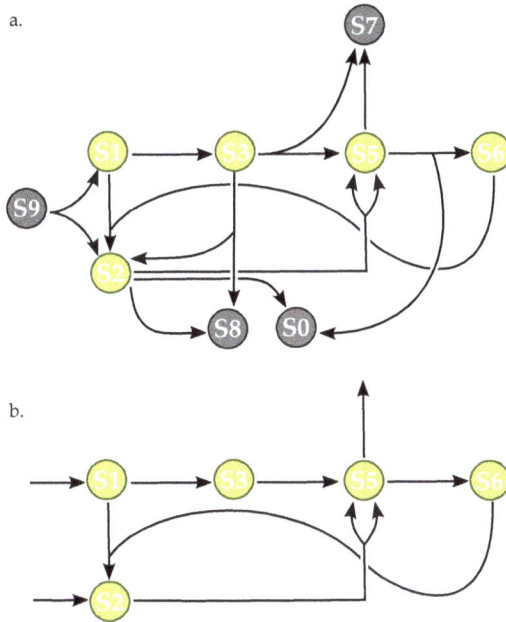

Figure 7. Randomly-generated network with nine species and 10 reactions. Dark grey circles represent boundary nodes, and are replaced with constant production and decay rates in the reduced network. Green circles represent floating species. The network corresponds to Network 428 in the supplementary data of 10 × 10 networks allowing orphaned species. (**a**) complete network capable of oscillatory dynamics; (**b**) reduced network with essential species and reactions necessary for oscillation.

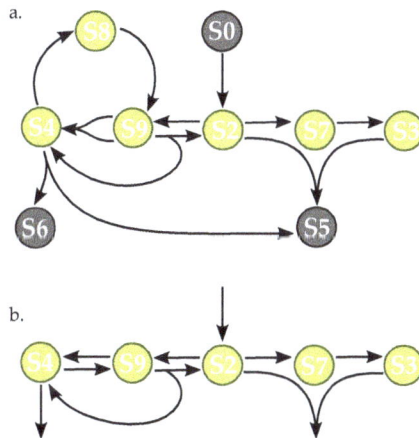

Figure 8. Randomly-generated network with nine species and 10 reactions. The network corresponds to Network 7814 in the supplementary data of 10 × 10 networks allowing orphaned species. (**a**) complete network capable of oscillatory dynamics; (**b**) reduced network with essential species and reactions necessary for oscillation.

4. Discussion

4.1. Algorithm Evaluation

In this paper, a Pythonic interpretation of existing objective functions to evaluate the fitness of system parameters for bifurcation behavior is presented, and a differential evolution algorithm is implemented for progressive parameter optimization. The bifurcation–evolution software relies on iterative minimization of an objective function that rewards networks with eigenvalues approximating the specified bifurcation behavior. Several published models from the BioModels Database were tested, and demonstrated that the algorithm performs well using default conditions for models with few parameters, evolving the desired bifurcation behavior with a short runtime and within a few generations. The bifurcation–evolution software could produce oscillatory dynamics in an mitogen-activated protein kinase feedback oscillator model with 30 parameters (including 22 global parameters and eight state variables) in less than 45 seconds. However, the largest model tested, which contained 56 total parameters for optimization with a Hopf bifurcation objective, required a 22 minute runtime, suggesting that the bifurcation–evolution software may not be efficient for larger models. This extended runtime was attributed to the task of generating an initial population of suitable members, indicating that the randomized parameter value selection created many individuals which could not reach steady state, excluding these parameter sets from the population. It is likely that this model is also sensitive, only reaching steady state and achieving oscillatory dynamics for a small subset of parameter space, such that small changes in parameter values dramatically alter the model dynamics.

The bifurcation–evolution software presents several optional arguments that the user can alter to increase the probability of finding a suitable solution or to expedite convergence, as the stiffness of the model will affect the efficiency of the algorithm converging on a solution. Increasing the mutation constant and population size, or lowering the recombination constant will improve the chance of converging on a global minima. However, these actions will also increase the time required for convergence. For larger models which complicate the steady state solver computation, it may be desirable to decrease the population size and increase the maximum number of iterations so that computation time is shifted from initializing the population to performing the evolution routine. Additionally, differential evolution is sensitive to parameter ranges. The default settings for parameter range selection attempts to create narrow, parameter-specific ranges while still providing sufficient space for non-trivial randomization. Providing broad ranges will generally reduce the efficiency of the differential evolution algorithm and greatly reduce the chance of converging on a suitable solution. However, the user can further restrict or expand these ranges when appropriate to improve the chance of convergence.

While the results of testing the bifurcation–evolution software on multiple published biological models and on randomly-generated mass-action networks suggest that the algorithm previously implemented by Chickarmane et al. is suitable for detecting turning point and Hopf bifurcations, there are some limitations. The objective function for turning point bifurcations cannot detect bistable systems automatically. Instead, the user must pass the optimized model to external software that permits bifurcation analysis or perform manual parameter scans to search for bistability, as in Figure 1. The Pythonic implementation presented does not distinguish between systems with turning point bifurcations and pitchfork or transcritical bifurcations, so it would be desirable to add an additional step to eliminate parameter sets that correspond to these bifurcation types as in the previous implementation in C#. Similarly, the objective function for Hopf bifurcations still permits damped oscillators to be represented as solutions with good fitness values as determined by the eigenvalues of the optimized parameter set, depending on the stringency of the threshold value. Chaotic oscillatory dynamics also resulted following parameter optimization in a subset of the tested cases in the model of drosophila circadian rhythms by Leloup et al. [21]. While chaos and birhythmicity are known to arise in this model, the presence of these dynamics in the parameter-optimized model suggests that the algorithm does not exclude global bifurcations from the evolving population. While a system with a zero real component of one member of a complex conjugate pair would be a true Hopf bifurcation that produces sustained oscillations, the bifurcation–evolution software only approximates this behavior. Additionally, since the bifurcation–evolution software introduces stochasticity and depends on appropriate parameter ranges for convergence, it is not guaranteed to produce the bifurcation behavior selected on every run. This may require that the function is called multiple times before a satisfactory solution is reached.

4.2. Oscillator Frequency in Randomly-Generated Network Populations

The frequency data presented shows the intuitive result that larger random networks are more likely to contain components that permit oscillatory dynamics. All randomly-generated networks which exhibited sustained oscillations contained at least four floating species, although a system with three floating species could conceivably achieve a Hopf bifurcation with the appropriate network topology [20]. This suggests that oscillatory networks with three floating species are rare in randomly-generated populations and sensitive to the parameter regime, such that oscillatory networks occupy only a small region of parameter space. Oscillatory dynamics arise in networks that have negative feedback and a time delay, as described for the networks in Figures 7 and 8. These are more likely to arise in larger networks because there are more nodes to participate in complex loops and engage in multiple reactions, increasing the likelihood of randomly generating a motif that provides the feedback or time delay architecture. Figure 8 shows that it is advantageous to have a large number of floating species to produce complex dynamics, as the reduced network has two three-species loops that incorporate negative feedback, but the individual loops could not sustain oscillatory dynamics when the second loop was removed. However, the data also suggests that the number of reactions available to floating species in the network exerts greater control over the frequency of oscillatory dynamics than does the total number of species.

Intentionally increasing the ratio of reactions to species greatly increases the frequency of oscillatory dynamics following parameter optimization, as shown by the oscillator enrichment in the 1:1.5 networks shown in Figure 5 compared to the 1:1 networks shown in Figure 4. In the case where the number of reactions and species are equal, as in Figure 4, oscillatory dynamics only appear in networks with fewer than $n-2$ floating species, where n is the total number of boundary and floating species in the network. The networks which produce oscillators tend to have the reaction density shifted onto the floating species, with minimal reaction density between boundary species. This shift creates a local enrichment of reaction density on the floating species, permitting greater control over the dynamic interactions between these species by providing flexibility to the network. This is not necessary in the networks that have a 1:1.5 ratio, in which networks with the maximal number of floating species can give rise to oscillatory dynamics because sufficient flexibility is conferred by the intentional enrichment in the number of reactions. As a result, networks with a 1:1 ratio of total species to reactions can approximate the behavior seen in networks with a 1:1.5 ratio by decreasing the number of floating species in the network and minimizing the number of trivial reactions between boundary species that contribute to network dynamics. The importance of reaction enrichment is further confirmed by the study of the 10 species, 10 reactions randomly-generated network populations in which orphaned species were permitted. The majority of oscillatory networks in this population included at least one orphaned species, which enriches the reaction density between the remaining species in the network and provides additional flexibility.

Together, these data demonstrate the importance of providing a mass-action network with sufficient connectivity to enable dynamic bifurcation behaviors to arise. True biological systems, which may incorporate complex rate laws involving cooperativity and enzyme kinetics, could circumvent such limitations in network size and reaction enrichment. In addition to the studies presented describing the frequency of oscillatory dynamics, further studies should explore the sizes of the parameter landscapes which permit desired bifurcation behaviors. This could better explain the sensitivity of some optimized networks to variation in the parameter regime, and could provide researchers with a greater understanding of the impact of small perturbations on overall system dynamics. The presented bifurcation evolution tool would facilitate such a study. With this tool, modeling of complex biological systems which exhibit oscillatory dynamics or turning point bifurcations can be readily optimized to achieve such behavior given the appropriate network topology. The tool also facilitates exploration of parameter space in models for which the full range of dynamic behavior is unknown, which could enable detection of rare dynamic behavior inside a small subspace of the parameter landscape and inform experimental studies with possible implications for understanding disease and the molecular underpinnings of life.

5. Conclusions

Biological systems are often described with systems of nonlinear equations to model the dynamics of metabolic, protein, or gene-regulatory networks. While the topology of such models constrains the range of dynamic behavior that can be achieved, parameter regimes that fully define the reaction rates and interactions, as well as the state variables of the system, also determine whether a given dynamic behavior will arise. Locating parameter regimes that induce bifurcations can be challenging due to the size of the parameter landscape. In this work, a Pythonic bifurcation–evolution software is presented which employs existing bifurcation objective functions and a differential evolution algorithm that minimizes these objectives to improve the fitness of the best candidate parameter regime, progressively optimizing the system for either a turning point or Hopf bifurcation. The objective functions reward systems with steady state eigenvalues approximating those characteristic of the desired bifurcation. The bifurcation–evolution software is validated using published models from the BioModels Database, confirming that the algorithm performs well for evolving bistability and oscillations. The bifurcation-evolution software was subsequently used to determine the frequency of oscillators in randomly-generated mass-action network populations, and the results of this search indicate that populations of large random networks with 50% reaction enrichment achieve oscillatory dynamics more frequently than networks with fewer species or with an equivalent number of species and reactions. This demonstrates the importance of reaction enrichment for flexibility that enables complex dynamic behaviors. An analysis of selected randomly-generated mass-action networks from these populations shows that negative feedback and time delays are involved in the generation of oscillatory dynamics. Ultimately, the studies presented in the current work demonstrate the utility of the bifurcation-evolution software for efficiently exploring parameter space for a solution that satisfies the bifurcation objective. Due to the ubiquity of Python in computational biology, this Pythonic bifurcation-evolution software can be readily understood and easily-integrated with existing modeling software, providing the modeling community with a tool that could help detect of rare dynamics and inform experimental studies of disease and the molecular mechanisms underlying a range of biological phenomena.

Supplementary Materials: All supplementary materials required to generate the data and figures in the main text are available at https://github.com/vporubsky/evolve-bifurcation. evolveBifurcation.py contains the sourcecode for the bifurcation evolution algorithm described in the text. Optimized bifurcated BioModels represented in an Antimony string format and time-course simulation output is included for all test cases. All randomly-generated networks are provided in Antimony formats and time-course simulation output for each network is included.

Author Contributions: V.L.P. was responsible for data curation, formal analysis, investigation, methodology, software, validation, visualization, and writing the original manuscript draft. H.M.S. was responsible for conceptualization, funding acquisition, project administration, resources, and assisted in formal analysis and visualization. Both V.L.P. and H.M.S. reviewed and edited the manuscript.

Funding: H.M.S. is very grateful for the support provided by National Institutes of Health grants GM123032-01, U01HL122199-02, and P41EB023912. V.L.P. is grateful for the support of National Institutes of Health grant P41EB023912.

Acknowledgments: V.L.P. would like to thank Kiri Choi for valuable discussions about integrating steady state solvers and the functionality provided by Tellurium and libRoadRunner for implementing the bifurcation evolution algorithm. V.L.P. would also like to thank J. Kyle Medley for providing insight into the utility and shortcomings of several global and local optimization algorithms.

Conflicts of Interest: The authors declare no conflict of interest. The funders had no role in the design of the study; in the collection, analyses, or interpretation of data; in the writing of the manuscript, or in the decision to publish the results.

References

1. Csikasz-Nagy, A. Computational systems biology of the cell cycle. *Brief. Bioinform.* **2009**, *10*, 424–434. [CrossRef] [PubMed]
2. Pavlides, A.; Hogan, S.J.; Bogacz, R. Computational Models Describing Possible Mechanisms for Generation of Excessive Beta Oscillations in Parkinson's Disease. *PLoS Comput. Biol.* **2015**, *1*, e1004609. [CrossRef] [PubMed]
3. Zamora-Sillero, E.; Hafner, M.; Ibig, A.; Stelling, J.; Wagner, A. Efficient characterization of high-dimensional parameter spaces for systems biology. *BMC Syst. Biol.* **2011**, *5*, 142. [CrossRef] [PubMed]
4. Liu, Y.; Gunawan, R. Parameter estimation of dynamic biological network models using integrated fluxes. *BMC Syst. Biol.* **2014**, *8*, 127. [CrossRef]
5. Ashyraliyev, M.; Fomekong-Nanfack, Y.; Kaandorp, J.A.; Blom, J.G. Systems biology: Parameter estimation for biochemical models. *FEBS J.* **2009**, *276*, 886–902. [CrossRef] [PubMed]
6. Chickarmane, V.; Paladugu, S.R.; Bergmann, F.; Sauro, H.M. Bifurcation discovery tool. *Bioinformatics* **2005**, *21*, 3688–3690. [CrossRef]
7. Li, C.; Donizelli, M.; Rodriguez, N.; Dharuri, H.; Endler, L.; Chelliah, V.; Li, L.; He, E.; Henry, A.; Stefan, M.I.; et al. BioModels Database: An enhanced, curated and annotated resource for published quantitative kinetic models. *BMC Syst. Biol.* **2010**, *4*, 92. [CrossRef]
8. Choi, K.; Medley, J.K.; Cannistra, C.; Konig, M.; Smith, L.; Stocking, K.; Sauro, H.M. Tellurium: A Python Based Modeling and Reproducibility Platform for Systems Biology. *Biosystems* **2018**, *171*, 74–79. [CrossRef]
9. Somogyi, E.T.; Bouteiller, J.M.; Glazier, J.A.; König, M.; Medley, J.K.; Swat, M.H.; Sauro, H.M. libRoadRunner: A high performance SBML simulation and analysis library. *Bioinformatics* **2015**, *31*, 3315–3321. [CrossRef]
10. Reich, J.G.; Selkov, E.E. *Energy Metabolism of the Cell: A Theoretical Treatise*; Academic Press: London, UK, 1981; p. 345.
11. Sauro, H.M.; Ingalls, B. Conservation analysis in biochemical networks: Computational issues for software writers. *Biophys. Chem.* **2004**, *109*, 1–15. [CrossRef]
12. Sauro, H.M. *Systems Biology: An Introduction to Metabolic Control Analysis*; Ambrosius Publishing: Seattle, WA, USA, 2018.
13. Storn, R.; Price, K. Differential Evolution—A Simple and Efficient Heuristic for global Optimization over Continuous Spaces. *J. Glob. Optim.* **1997**, *11*, 341–359. [CrossRef]
14. scipy.optimize.differential_evolution—SciPy v1.1.0 Reference Guide. Available online: https://docs.scipy.org/doc/scipy/reference/generated/scipy.optimize.differential_evolution.html (accessed on 5 November 2018).
15. Broyden, C.G. The Convergence of a Class of Double-rank Minimization Algorithms 1. General Considerations. *IMA J. Appl. Math.* **1970**, *6*, 76–90. [CrossRef]
16. Fletcher, R. A new approach to variable metric algorithms. *Comput. J.* **1970**, *13*, 317–322. [CrossRef]
17. Goldfarb, D. A Family of Variable-Metric Methods Derived by Variational Means. *Math. Comput.* **1970**, *24*, 23. [CrossRef]
18. Shanno, D.F. Conditioning of Quasi-Newton Methods for Function Minimization. *Math. Comput.* **1970**, *24*, 647. [CrossRef]
19. Sauro, H.M. *Systems Biology: Introduction to Pathway Modeling*, 1st ed.; Ambrosius Publishing: Seattle, WA, USA, 2017.
20. Wilhelm, T.; Heinrich, R. Smallest chemical reaction system with Hopfbifurcation. *J. Math. Chem.* **1995**, *17*, 1–14. [CrossRef]
21. Leloup, J.; Goldbeter, A. Chaos and birhythmicity in a model for circadian oscillations of the PER and TIM proteins in drosophila. *J. Theor. Biol.* **1999**, *198*, 445–459. [CrossRef]
22. Hervagault, O.; Canu, P. *Bistability and Irreversible Transitions in a Simple Substrate Cyclet*; Technical Report; University of Technology of Compiègne: Compiègne, France, 1987.

23. Seno, M.; Iwamoto, K.; Sawada, K. *Instability and Oscillatory Behavior of Membrane-Chemical Reaction Systems*; Technical Report; University of Tokyo: Tokyo, Japan, 1978.

24. Tyson, J.J.; Chen, K.C.; Novak, B. Sniffers, buzzers, toggles and blinkers: Dynamics of regulatory and signaling pathways in the cell. *Curr. Opin. Cell Biol.* **2003**, *15*, 221–231. [CrossRef]

25. Edelstein, B.B. Biochemical Model With Multiple Steady States and Hysteresis. *J. Theor. Biol.* **1970**, *29*, 57–62. [CrossRef]

26. Angeli, D.; Ferrell, J.E.; Sontag, E.D. Detection of multistability, bifurcations, and hysteresis in a large class of biological positive-feedback systems. *Proc. Natl. Acad. Sci. USA* **2004**, *101*, 1822–1827. [CrossRef]

27. Nicolis, G.; Prigogine, I. *Self-Organization in Nonequilibrium Systems*; Wiley-Blackwell: Hoboken, NJ, USA, 1977; Volume 82, p. 672.

28. Heinrich, R.; Rapoport, S.M.; Rapoport, T.A. Metabolic regulation and mathematical models. *Proq. Biophys. Mol. Biol.* **1977**, *32*, 1–82.

29. Kholodenko, B.N. Negative feedback and ultrasensitivity can bring about oscillations in the mitogen-activated protein kinase cascades. *Eur. J. Biochem.* **2000**, *267*, 1583–1588. [CrossRef] [PubMed]

30. Goldbeter, A. A minimal cascade model for the mitotic oscillator involving cyclin and cdc2 kinase. *Proc. Natl. Acad. Sci. USA* **1991**, *88*, 9107–9111. [CrossRef] [PubMed]

31. François, P.; Hakim, V. Core genetic module: The mixed feedback loop. *Phys. Rev. E* **2005**, *72*, 031908. [CrossRef] [PubMed]

32. Lavrentovich, M.; Hemkin, S. A mathematical model of spontaneous calcium(II) oscillations in astrocytes. *J. Theor. Biol.* **2008**, *251*, 553–560. [CrossRef] [PubMed]

processes

MDPI

Article

Metabolic Modeling of *Clostridium difficile* Associated Dysbiosis of the Gut Microbiota

Poonam Phalak and Michael A. Henson *

Department of Chemical Engineering and Institute of Applied Life Science, University of Massachusetts, Amherst, MA 01003, USA; pphalak@umass.edu
* Correspondence: mhenson@umass.edu; Tel.: +1-413-545-3481

Received: 16 December 2018; Accepted: 6 February 2019; Published: 15 February 2019

check for updates

Abstract: Recent in vitro experiments have demonstrated the ability of the pathogen *Clostridium difficile* and commensal gut bacteria to form biofilms on surfaces, and biofilm development in vivo is likely. Various studies have reported that 3%–15% of healthy adults are asymptomatically colonized with *C. difficile*, with commensal species providing resistance against *C. difficile* pathogenic colonization. *C. difficile* infection (CDI) is observed at a higher rate in immunocompromised patients previously treated with broad spectrum antibiotics that disrupt the commensal microbiota and reduce competition for available nutrients, resulting in imbalance among commensal species and dysbiosis conducive to *C. difficile* propagation. To investigate the metabolic interactions of *C. difficile* with commensal species from the three dominant phyla in the human gut, we developed a multispecies biofilm model by combining genome-scale metabolic reconstructions of *C. difficile*, *Bacteroides thetaiotaomicron* from the phylum Bacteroidetes, *Faecalibacterium prausnitzii* from the phylum Firmicutes, and *Escherichia coli* from the phylum Proteobacteria. The biofilm model was used to identify gut nutrient conditions that resulted in *C. difficile*-associated dysbiosis characterized by large increases in *C. difficile* and *E. coli* abundances and large decreases in *F. prausnitzii* abundance. We tuned the model to produce species abundances and short-chain fatty acid levels consistent with available data for healthy individuals. The model predicted that experimentally-observed host-microbiota perturbations resulting in decreased carbohydrate/increased amino acid levels and/or increased primary bile acid levels would induce large increases in *C. difficile* abundance and decreases in *F. prausnitzii* abundance. By adding the experimentally-observed perturbation of increased host nitrate secretion, the model also was able to predict increased *E. coli* abundance associated with *C. difficile* dysbiosis. In addition to rationalizing known connections between nutrient levels and disease progression, the model generated hypotheses for future testing and has the capability to support the development of new treatment strategies for *C. difficile* gut infections.

Keywords: gut microbiota dysbiosis; *Clostridium difficile* infection; bacterial biofilms; metabolic modeling

1. Introduction

The gut microbiota comprise a complex ecological system that maintains a critical symbiotic relationship with the human host [1,2]. The microbiota provide essential nutrients such as short-chain fatty acids (SCFAs; acetate, butyrate, and propionate), support colonization resistance to pathogens, participate in the degradation of toxic compounds, and regulate the immune responses [3–7]. Bacteroidetes and Firmicutes are the two dominant phyla in the healthy gut, comprising approximately 90% of the community. Other important but less abundant phyla are Proteobacteria, Actinobacteria, Euryarchaeota and Verrucomicrobia, as well as Eukaryota such as fungi [8,9]. The gut microbiota composition can be altered by numerous factors including diet, antibiotic treatment, stress, and lifestyle [10,11]. Dietary components including carbohydrates, protein, fat, and host secretions such as primary bile acids and nitrate play a particularly important role in shaping microbiota abundances [12–17]. Unhealthy alterations of the gut microbiota are termed as dysbiosis and represent imbalances in species abundances associated with diseases such as inflammatory bowel diseases, Crohn's disease, obesity, and diabetes [18–20].

The anaerobic bacterium *Clostridium difficile* is an opportunistic human pathogen responsible for infections in the colon of the human gastrointestinal tract [21]. Various studies have reported that 3%–15% of healthy adults are asymptomatically colonized with *C. difficile* [22–28]. Commensal species in healthy gut usually provide resistance against *C. difficile* pathogenic colonization. *C. difficile* infection (CDI) is most common in patients previously treated with broad spectrum antibiotics that disrupt the healthy gut microbiota and reduce competition for available nutrients [29], resulting in dysbiosis conducive to *C. difficile* propagation [30–33]. CDI symptoms can range from mild diarrhea to severe and life-threatening colitis [21,34]. *C. difficile* virulence is attributable to the secretion of the high molecular weight toxins A and B that promote epithelial tissue damage and rapid fluid loss. Some *C. difficile* strains have developed resistance to common antibiotics while also exhibiting more severe pathogenicity [35]. CDI has become particularly common in hospital settings due to the ability of *C. difficile* to form spores that adhere to surfaces and resist common disinfectant protocols. Studies estimate that almost 500,000 CDI cases occur within the U.S. annually [36], resulting in 29,000 deaths and over $4.8 billion in associated costs in acute care facilities alone [37].

Numerous experimental studies have demonstrated that *C. difficile* [38–41] can form biofilms in vitro. The other commensal bacteria [42,43] can form biofilms in vivo, which are well known to exhibit phenotypes distinct from planktonic cultures. For example, bacteria in biofilms can tolerate antimicrobial concentrations 10,000-times higher than the same bacteria grown planktonically, making the development of effective treatment strategies a major challenge [44,45]. This difficulty is partially attributable to the spatially-varying biofilm environment, which has profound effects on biofilm development and function [46–48]. Mechanistic understanding of the relationships between biofilm spatial variations, species–species interactions, and host–species interactions remains inadequate to systematically analyze and rationally treat CDI [49]. To address these challenges, we added *C. difficile* to our previous multispecies biofilm model [50,51] consisting of three representative species from the phyla Bacteroidetes (*Bacteroides thetaiotaomicron*), Firmicutes (*Faecalibacterium prausnitzii*), and Proteobacteria (*Escherichia coli*). Model simulations were performed to connect host-induced nutrient changes in the gut environment with observed alternations of species abundances and SCFA levels [52–54] to unravel the metabolic determinants of CDI.

2. Results

2.1. Discovery of Putative Byproduct Cross-Feeding Relationships

Our previous modeling study [50] without *C. difficile* generated three byproduct cross-feeding relationships that were predicted to be necessary and sufficient for the coexistence of the three species: *B. thetaiotaomicron* consumption of ethanol secreted by *E. coli* and *F. prausnitzii* consumption of acetate and succinate secreted by *B. thetaiotaomicron* and *E. coli*. Preliminary flux balance analysis (FBA) with the *C. difficile* reconstruction showed that acetate, butyrate, and propionate were the major byproducts, and succinate and formate could be uptaken as carbon sources in the presence of glucose. With this knowledge, the four-species biofilm model was analyzed to discover additional cross-feeding relationships that support *C. difficile* coexistence with the three commensal species. Each species was allowed to consume glucose, the eight amino acids, and any available byproduct (acetate, CO_2, ethanol, formate, lactate, and succinate), assuming no differences in uptake kinetics across species and byproducts (see Materials and Methods). Simulations with a biofilm thickness of 40 microns and bulk concentrations of 8 mmol/L glucose and 0.5 mmol/L each amino acid at the biofilm-stool interface corresponding to the healthy case (Table 1) were run for 300 h to ensure a steady-state solution consistent with obtaining a mature biofilm. A particular cross-feeding relationship was deemed significant if at least one uptake or secretion flux exceeded 1 mmol/gDW·h.

Table 1. Nutrient concentrations used for healthy and three dysbiosis simulation cases in mmol/L.

Nutrient	Healthy	High Amino Acids, Low Glucose	High Primary Bile Acids	High Nitrate
Glucose	8.0	4.0	8.0	4.0
Cysteine	0.5	1.0	0.5	1.0
Isoleucine	0.5	1.0	0.5	1.0
Leucine	0.5	1.0	0.5	1.0
Methionine	0.5	1.0	0.5	1.0
Proline	0.5	1.0	0.5	1.0
Serine	0.5	1.0	0.5	1.0
Tryptophan	0.5	1.0	0.5	1.0
Valine	0.5	1.0	0.5	1.0
Nitrate	0	0	0	0.4
Taurocholate	0	0	1.5	1.5

The biofilm model predicted significant cross-feeding of acetate, ethanol, formate, and succinate between the four species (Figure 1A). Lactate and CO_2 cross-feeding were insignificant. Importantly for this study, *C. difficile* was predicted to: (1) consume formate secreted by *F. prausnitzii* and *E. coli*; (2) compete with *F. prausnitzii* for succinate secreted by *B. thetaiotaomicron*; and (3) synthesize acetate for consumption by *F. prausnitzii* (Figure 1B). Experimentally, *C. difficile* has been shown to uptake succinate and produce butyrate [55] and to produce acetate by consuming formate directly or indirectly by uptaking CO_2 and H_2 [56]. Consequently, we hypothesized that formate and succinate cross-feeding could play a role in *C. difficile* propagation in vivo.

To test community stability and robustness in the absence of *C. difficile*, the same simulation was performed with the initial *C. difficile* biomass concentration set to zero. The resulting three-species community remained stable with *B. thetaiotaomicron*:*F. prausnitzii*:*E. coli* abundances of 66%:27%:7%, consistent with a healthy gut community (Supplementary Materials Figure S1). These predictions were aligned with our previous study [50].

2.2. Characterization of Healthy Gut Microbiota

With the putative cross-feeding relationships (Figure 1B) included, the multispecies biofilm model was simulated for a biofilm thickness of 40 microns and the healthy nutrient levels (Table 1). The model was tuned such that the mature biofilm obtained after 300 h of simulation produced *B. thetaiotaomicron*:*F. prausnitzii*:*E. coli*:*C. difficile* abundances of 71%:21%:7%:1% when averaged across the biofilm (see Materials and Methods). These abundances were consistent with data from in vivo studies [57,58].

We analyzed species biomass concentrations (Figure 2A) and local growth rates (Figure 2B) with respect to location in the biofilm with nutrients supplied at the biofilm–stool interface ($z = 0$). *C. difficile* was predicted to have the highest growth rates in the nutrient-rich bottom half of the biofilm, but the lowest growth rates in the nutrient-lean top half. The local growth rates of the three commensal bacteria were comparable across the biofilm, with *B. thetaiotaomicron* having the highest growth rates in the bottom half and *F. prausnitzii* having a slight advantage in the top half. Due to its growth advantage in the nutrient-rich bottom half and slow cellular diffusion, *B. thetaiotaomicron* produced much higher biomass concentrations across the entire biofilm. *F. prausnitzii* and *E. coli* established lower biomass concentrations, while *C. difficile* was present at small concentrations due to its very small growth rate in the nutrient-lean top half. The spatial distributions of supplied nutrients, species biomass, and secreted byproducts were similar to those reported in our previous studies [50,51] and are omitted here. This simulation suggests that the commensal bacteria can sublimate *C. difficile* propagation through nutrient competition and may help explain how healthy individuals can be asymptomatically colonized.

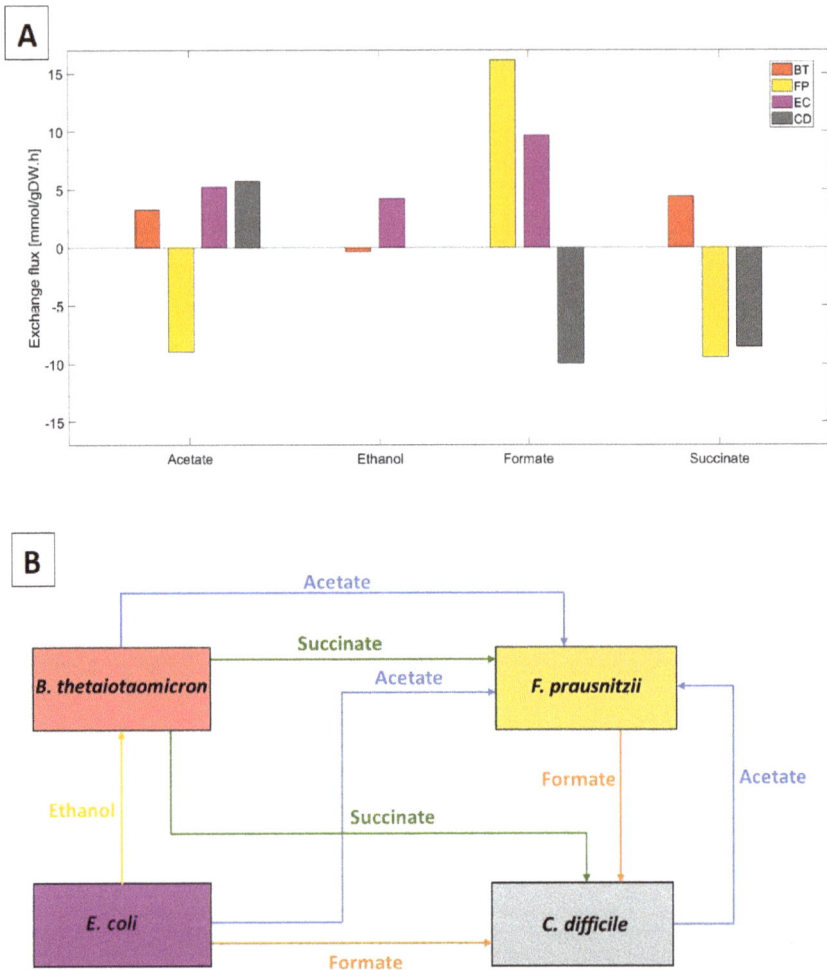

Figure 1. Predicted cross-feeding of byproducts between the four species. (**A**) Species exchange rates specified in mmol/gDW/h. Secretion rates are positive, and uptake rates are negative. (**B**) Byproduct cross-feeding patterns identified from the species uptake and secretion fluxes in (A).

The biofilm model also was tuned for healthy nutrient levels to produce acetate:propionate: butyrate fractions of 60%:20%:20% when averaged across the biofilm to be consistent with in vivo studies [5,59] (see Materials and Methods). The model predicted the total SCFA concentration to be 32.5 mmol/L (Figure 2C), which was in reasonable agreement with an in vivo study with a control diet that yielded 41.1 mmol/L of total SCFAs [60]. One possible explanation for the lower SCFA levels predicted by our model is the simplified diet (glucose, eight amino acids) compared to the control diet used experimentally.

Ethanol was present at a very low level (Figure 2D) due to limited synthesis by the small *E. coli* population and high consumption by the large *B. thetaiotaomicron* population. Of the two organic acids (OAs) produced, formate was predicted to be present at a high level because synthesis by *F. prausnitzii* and *E. coli* substantially exceeded consumption by *C. difficile*. Succinate was present at a moderate level since it was consumed by both *C. difficile* and *F. prausnitzii*. These predictions suggest that plentiful formate and succinate could be available to promote *C. difficile* propagation under in vivo perturbations.

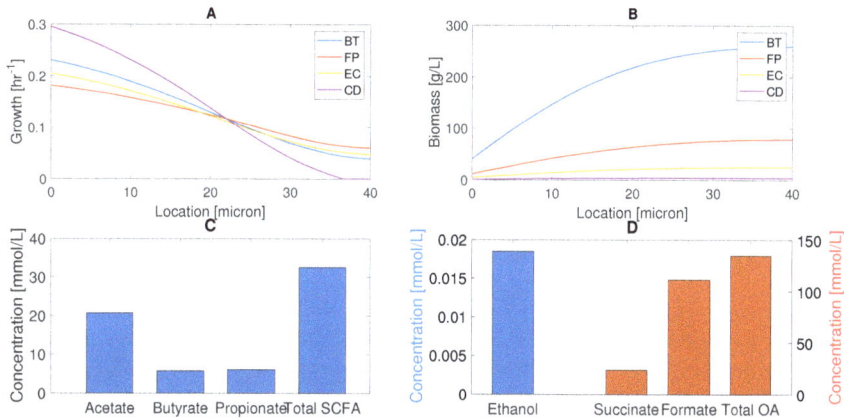

Figure 2. Predicted multispecies biofilm behavior in the absence of host-microbiota perturbations. (**A**) Species biomass concentrations across the thickness of the biofilm with nutrients supplied and biomass removed at z = 0 microns. (**B**) Local species growth rates across the thickness of the biofilm. (**C**) Acetate, butyrate, propionate, and total SCFA concentrations averaged across the biofilm. (**D**) Ethanol, succinate, formate, and total OA levels averaged across the biofilm.

2.3. Glucose and Amino Acid Perturbations

Various in vivo studies have shown that glucose concentration decreases and amino acid concentrations increase in the gut during *C. difficile* and other types of dysbiosis [12,61–64]. To investigate the effects of altered nutrient levels associated with host-microbiota perturbations, we performed simulations for a 40-micron biofilm with elevated amino acid and reduced glucose bulk concentrations (Table 1) under the assumption that *C. difficile* expansion is driven by these experimentally-observed nutrient changes. While in vivo nutrient levels are impacted by diet, host metabolism, and microbiota, this assumption was deemed reasonable given the simplified nature of our model. Given the uncertainty associated with the bulk nutrient concentrations, we performed a sensitivity analysis to explore their effects with respect to the species abundances (Figure S2). This analysis was consistent with the model predictions reported below as long as the glucose to amino acid ratio was sufficiently large. Compared to the healthy case, the local *C. difficile* growth rate decreased in the bottom half of the biofilm, but increased in the top half (Figure 3A). Similar trends were predicted for the three commensal species, which we attributed to reduced glucose, but increased amino acid penetration into the biofilm. *C. difficile* is known to grow efficiently on amino acids due to its ability to use amino acid pairs such as leucine and proline to generate ATP via Stickland metabolism [65–67].

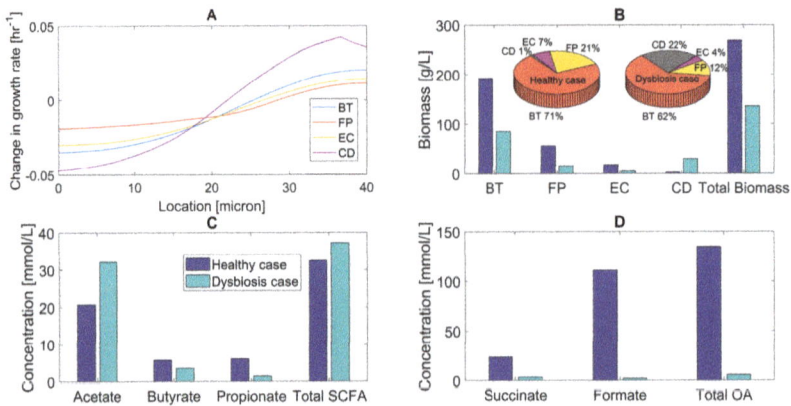

Figure 3. Predicted multispecies biofilm dysbiosis resulting from host–microbiota perturbations in glucose and amino acid concentrations. (**A**) Change in species growth rates across the biofilm plotted as the difference between the growth rates for the healthy and dysbiosis cases. (**B**) Biomass concentrations (bar graphs) and species abundances (pie chart) averaged across the biofilm for healthy and dysbiosis case. (**C**) Acetate, butyrate, propionate, and total SCFA concentrations averaged across the biofilm. (**D**) Succinate, formate, and total OA concentrations averaged across the biofilm.

As a result of its enhanced growth in the top half of the biofilm compared to the commensal species, *C. difficile* increased its average biomass concentration ten-fold and species abundance from 1%–22% compared to the healthy case (Figure 3A). The biomass concentration of each commensal species dropped due to reduced glucose availability. A substantial effect was predicted for *F. prausnitzii* with its species abundance decreasing from 21%–12%, partially due to increased competition for succinate with *C. difficile*. These predictions are in agreement with in vivo studies [29,68–70], with the exception that dysbiosis during CDI should be accompanied by an increase in *E. coli* abundance [13,15,71–73]. The model predicted reduced total biomass production due to reduced growth of the three commensal species.

Dysbiosis was predicted to result in increased acetate, decreased butyrate and propionate, and lower total SCFA levels compared to the healthy case (Figure 3C). We attributed reduced total SCFA synthesis to lower glucose availability and increased acetate and decreased butyrate levels to a change in the balance of acetate-producing *C. difficile* and acetate-to-butyrate converting *F. prausnitzii*. Experimental studies have shown that dysbiosis is associated with reduced butyrate concentrations in the gut [69,74]. The model predicted large changes in organic acid levels, with succinate, formate, and total OA concentrations dropping due to reduced glucose fermentation. These predictions suggest that the combination of decreased carbohydrate and increased amino acid levels could play a role in *C. difficile*-associated dysbiosis.

2.4. Primary Bile Acid Perturbations

Primary bile acids such as taurocholate are secreted by the liver and transported into the intestines where anaerobic bacteria degrade them into secondary bile acids [75–77]. Broad spectrum antibiotics are known to reduce gut microbiota diversity [30–33,78], including the possible loss of bacterial species from families *Lachnospiraceae* and *Ruminococcaceae* responsible for the conversion of primary bile acids. Various in vitro [77,79,80] and in vivo [16,81] studies have shown that *C. difficile* spores can use primary bile acids for germination. Sodium taurocholate is the typical reagent used to grow *C. difficile* in vitro [82,83]. We investigated the impact of such perturbations with the multispecies biofilm model by adding taurocholate as a representative primary bile acid (Table 1). While primary bile acids are known to promote *C. difficile* transition from spores to a vegetative state [79,84], we assumed that

C. difficile was already vegetative and investigated the effect of taurocholate on *C. difficile* growth. Preliminary FBA calculations with the *C. difficile* metabolic reconstruction showed that taurocholate uptake increased the growth rate, while taurocholate uptake was not possible with the three commensal species reconstructions.

Compared to the healthy case, the introduction of taurocholate was predicted to increase the local *C. difficile* growth rate across the biofilm (Figure 4A). *B. thetaiotaomicron* and *E. coli* growth were largely unaffected, while the *F. prausnitzii* growth rate decreased due to increased competition for succinate from *C. difficile*. As a result, the *C. difficile* abundance increased from 1%–18%, while the *F. prausnitzii* abundance decreased by 38% (Figure 4B). The *B. thetaiotaomicron* and *E. coli* abundances exhibited relatively small decreases, although experimental studies showed that *E. coli* abundance should increase during dysbiosis [71,73]. The total biomass concentration was predicted to remain almost constant, showing that taurocholate was responsible for changing the species distribution of the biomass.

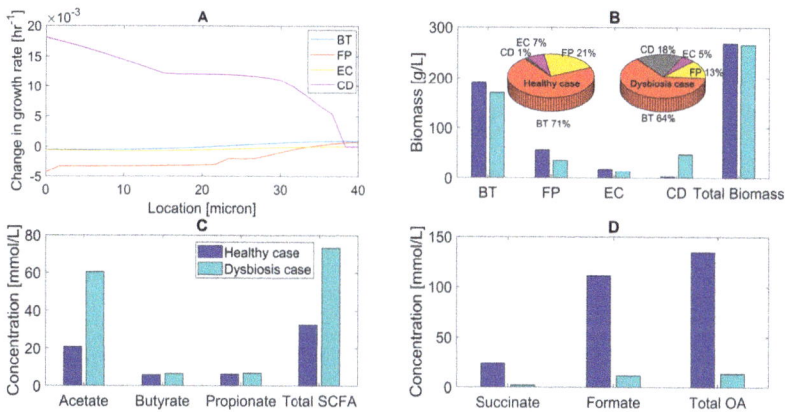

Figure 4. Predicted multispecies biofilm dysbiosis resulting from host-microbiota perturbations in the concentration of the primary bile acid taurocholate. (**A**) Change in species growth rates across the biofilm plotted as the difference between the growth rates for the healthy and dysbiosis case. (**B**) Biomass concentrations (bar graphs) and species abundances (pie charts) averaged across the biofilm for the healthy and dysbiosis case. (**C**) Acetate, butyrate, propionate, and total SCFA concentrations averaged across the biofilm. (**D**) Succinate, formate, and total OA concentrations averaged across the biofilm.

The predicted trends for SCFA and OA levels were similar to those observed for the combined glucose/amino acid perturbation. Acetate and total SCFA concentrations increased compared to the healthy case due to increased acetate synthesis by *C. difficile* and decreased acetate consumption by *F. prausnitzii* (Figure 4C). The formate concentration decreased because of the same mechanism, while we attributed the reduced succinate concentration to increased succinate consumption by *C. difficile* (Figure 4D). Butyrate (produced by *F. prausnitzii* and *C. difficile*) and propionate (produced by *B. thetaiotaomicron* and *C. difficile*) concentrations remained almost constant as *C. difficile* compensated for reduced SCFA synthesis by the two commensal species. We also simulated a host-microbiota perturbation with decreased glucose/increased amino acids and increased taurocholate to examine the combined effects of these nutrient changes. Compared to either perturbation alone, the model predicted a further increase in *C. difficile* abundance and a decrease in *F. prausnitzii* abundance (Figure S3). Overall, these results support the hypothesis that increased primary bile acid levels could contribute to *C. difficile* propagation in vivo.

2.5. Host-Derived Nitrate Perturbations

The human host is known to secrete nitrate in response to inflammation in the gut [17]. Preliminary FBA calculations showed that nitrate uptake increased the *E. coli* growth rate, while the other three community members were unable to use nitrate as an electron acceptor. Therefore, we hypothesized that host-derived nitrate would increase *E. coli* abundance during simulated *C. difficile*-associated dysbiosis and yield better agreement with experimental studies [71,73]. To quantify the effects of nitrate availability, biofilm simulations were performed with and without nitrate for a dysbiosis case with reduced glucose, increased amino acids, and available taurocholate (Table 1).

As hypothesized, the main impact of host-derived nitrate was to substantially increase *E. coli* abundance from 4% without nitrate to 20% with nitrate (Figure 5A). The *F. prausnitzii* abundance decreased from 7% to 2%, while the abundances of *B. thetaiotaomicron* and *C. difficile* decreased modestly to accommodate the increased *E. coli*. The species abundances predicted with nitrate are in good agreement with experimental studies for *C. difficile*-associated dysbiosis showing large increases in *C. difficile* and *E. coli*, large decreases in *F. prausnitzii*, and modest changes in *B. thetaiotaomicron* [85–87].

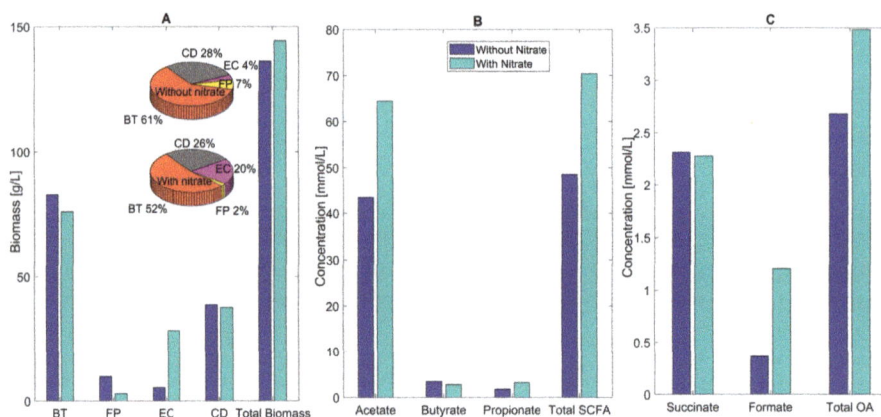

Figure 5. Predicted multispecies biofilm dysbiosis with and without host-derived nitrate. (**A**) Biomass concentrations (bar graphs) and species abundances (pie charts) averaged across the biofilm for the healthy and dysbiosis case. (**B**) Acetate, butyrate, propionate, and total SCFA concentrations (mmol/L) averaged across the biofilm. (**C**) Succinate, formate, and total OA concentrations averaged across the biofilm.

Nitrate availability was predicted to increase the acetate and total SCFA concentrations substantially due to large changes in *E. coli* and *F. prausnitzii* abundances (Figure 5B). Decreased succinate consumption by *F. prausnitzii* and increased formate synthesis by *E. coli* results in increased levels of individual and total OAs (Figure 5C). These predictions implicate a role for host-derived nitrate in *C. difficile*-associated dysbiosis.

We investigated the robustness of the four-species community during dysbiosis with available nitrate by removing selected cross-feeding relationships and varying the biofilm thickness from the nominal value of 40 microns. When *C. difficile* uptake of formate or succinate was eliminated, the *C. difficile* abundance dropped substantially (Figure S4), further suggesting that these cross-feeding relationships could be important for *C. difficile* propagation in vivo. Consistent with our previous study [50], cross-feeding of ethanol was important for *B. thetaiotaomicron* growth, and cross-feeding of both acetate and succinate was necessary for *F. prausnitzii* co-existence. For biofilm thicknesses of 30–60 microns, the species abundances were predicted to vary substantially with the most important trend being that thinner biofilms enhanced *C. difficile* growth (Figures S5 and S6). The growth rate profiles for 60 microns (Figure S7) suggested that *C. difficile* spores might be formed in the upper half of the biofilm where *C. difficile* was unable to sustain vegetative growth. Since such spores could be activated by favorable nutrient conditions, the incorporation of *C. difficile* spore formation and activation could be an interesting direction for future research. Overall, our results could help explain the role of broad spectrum antibiotics during CDI, as antibiotics could be expected to reduce the diversity and density of commensal bacteria that protect the gut from *C. difficile* expansion.

To gain insights into the internal pathway fluxes associated with the healthy and dysbiosis states, we determined for each species the eight internal fluxes that varied the most between the healthy (Figure 2) and *C. difficile* dysbiosis (Figure 5) states and identified the internal pathways associated with each of these fluxes. Using simulation data from the stool-biofilm interface at 300 h, the most variable fluxes were determined by computing for each individual flux the difference between the healthy and dysbiosis values and scaling the result by the healthy value (Figure S8). Pathways associated with amino acid metabolism were upregulated in *B. thetaiotaomicron* and *C. difficile*, demonstrating the ability of these two species to take advantage of increased amino acid availability. Similarly, the internal flux through the cysteine metabolism pathway was predicted to increase in *E. coli*. Most internal pathway fluxes in *F. prausnitzii* were predicted to decrease, suggesting that the dysbiosis environment was unfavorable for its growth, resulting in decreased abundance.

3. Discussion

The gut microbiota serve a broad array of important functions for the human host, including providing colonization resistance to opportunistic pathogens. Unhealthy changes in the microbiota composition, commonly termed dysbiosis, have been correlated to a wide variety of gut and metabolic diseases including inflammatory bowel disease, Crohn's disease, obesity, diabetes, and chronic gut infections. The opportunistic gut pathogen *Clostridium difficile* has been estimated to asymptomatically colonize 3%–15% of healthy adults [28]. A common cause of symptomatic *C. difficile* infection (CDI) is the use of broad spectrum antibiotics, which induce dysbiosis by reducing the diversity and density of gut commensal bacteria that provide resistance to *C. difficile* expansion [30–33,78]. Improved understanding of the complex interactions between commensal species, *C. difficile*, the gut environment, and the human host are needed to treat CDI more rationally.

To help unravel the metabolic determinants of *C. difficile*-associated dysbiosis, we developed a multispecies biofilm model by combining genome-scale metabolic reconstruction of *C. difficile* [88] and commensal species representing the three dominant phyla in the gut: *Bacteroides thetaiotaomicron* (Bacteroidetes) [89], *Faecalibacterium prausnitzii* (Firmicutes) [90], and *Escherichia coli* (Proteobacteria) [91]. The chosen species are well-studied representatives of the most dominant phyla in the human gut microbiome, and curated metabolic reconstructions of these species were available. While our four-species model represented a substantial reduction in complexity compared to the actual gut microbiota, the number of species and extracellular metabolites included were limited by computational considerations. Community models with substantially more species and cross-fed metabolites can be formulated and solved by neglecting spatial and temporal variations, as shown in our recent study of the gut microbiota [92]. However, these assumptions are not appropriate for biofilm simulations. Furthermore, our four-species model could be useful for designing in vitro systems for experimentally testing model predictions.

While specific spatial organization of gut microbes is currently unknown, the structure likely includes biofilm growth associated with host mucosa and epithelial tissue [93]. The literature provides significant evidence to support the hypothesis that some gut microbes develop spatially-structured multispecies biofilms [40,43]. We sought to understand how the commensal species could sublimate *C. difficile* expansion and under what gut conditions colonization resistance could become compromised. The biofilm model was tuned to represent a healthy state with species abundances and concentrations of short-chain fatty acids (SCFAs; acetate, butyrate, propionate) consistent with experimental studies for healthy individuals [5,57,59]. Because our model lacked an explicit description of the human host, we mimicked host-microbiota perturbations associated with CDI by varying nutrient levels guided by experimental observations. More specifically, dysbiosis states were modeled through changes in the concentrations of available glucose, amino acids [12,61–64], primary bile acids [16,77,81], and nitrate [17].

Our model predicted that cross-feeding of secreted byproducts plays an important role in *C. difficile* sublimation and expansion. *C. difficile* consumed formate synthesized by *F. prausnitzii* and *E. coli* and succinate synthesized by *B. thetaiotaomicron* and *F. prausnitzii*. The existence of both cross-feeding relationships is supported by the experimental literature [55,56]. In silico removal of either cross-feeding relationship was predicted to provide *C. difficile* colonization resistance, demonstrating the complexity and importance of cross-feeding networks even in this simplified four-species community. These predictions could be tested experimentally through the development of an in vitro model system of the four species. More importantly, these results suggest that therapeutic strategies that target species–species interactions could be promising alternatives to conventional antibiotics that target *C. difficile* directly.

Host–microbiota perturbations modeled as increases in glucose and decreases in amino acid concentrations reproduced several features of *C. difficile*-associated dysbiosis including substantially reduced *F. prausnitzii* and increased *C. difficile* abundances and an imbalance in SCFA synthesis characterized by increased acetate and reduced butyrate levels [94]. The predicted decrease in anti-inflammatory butyrate would be expected to exasperate dysbiosis and accelerate disease progression [69,74]. Similar results were obtained when glucose and amino acid changes were replaced by increases in the primary bile acid taurocholate, which was predicted to be used as an electron acceptor by *C. difficile* in vivo to provide a growth advantage in the absence of commensal bacteria that degrade primary bile acids to secondary bile acids [61,95,96]. Taurocholate availability was predicted to have less effect on butyrate and propionate synthesis, but the SCFA imbalance remained due to high acetate synthesis. Our model predicted that dysbiosis could be induced with moderate changes in nutrient concentrations, a prediction that could be tested in vitro and suggesting the possible promise of therapeutic strategies that aim to alter the gut nutritional environment.

Despite their many consistencies with experimental studies [12,97,98], our simulations with glucose, amino acids, and taurocholate changes were unable to reproduce the large increase in *E. coli* abundance observed during CDI [71,73]. The addition of host-derived nitrate [17,99] to the other nutrient changes rectified this inconsistency and reproduced the key microbiota signatures of *C. difficile*-associated dysbiosis during CDI: large increases in *C. difficile* and *E. coli* abundances, large decreases in health-promoting *F. prausnitzii* abundance, and moderate changes in *B. thetaiotaomicron* abundance. The model generated high acetate levels associated with dysbiosis states, a prediction that could be tested through in vitro experiments. We believe further development of our multispecies biofilm model could yield a general computational platform for in silico investigation of CDI, other gut infections, and chronic inflammation disorders such as inflammatory bowel and Crohn's diseases. Some possibilities include the modeling of *C. difficile* spore formation/germination, the inclusion of more commensal gut species (e.g., [100]) including those from other phyla [101–103], the addition of a broader array of gut nutrients including fibers, oligosaccharides, and fats resulting from realistic diets [12–15,104], and modeling of the human host through incorporation of available metabolic reconstructions such as Recon 2 or Recon 3D [105–107]. A possible drawback of our modeling approach is the lack of species-specific parameters for nutrient uptake kinetics and metabolite-dependent mass transfer coefficients.

4. Materials and Methods

4.1. Biofilm Model Formulation and Solution

The multispecies biofilm model was constructed by combining genome-scale metabolic reconstructions of *C. difficile* (Strain 630Δerm) [88] and three commensal gut species: *B. thetaiotaomicron* [89],

F. prausnitzii (Strain A2-165) [90], and *E. coli* (Strain K-12 MG1655) [91]. The biofilm was considered to be attached to the colon lining defined as the top of the biofilm (Figure 6A). A minimal defined media (MDM) containing glucose, cysteine, isoleucine, leucine, methionine, proline, serine, tryptophan, and valine along with essential vitamins and minerals was used for all simulations. The amino acids cysteine, isoleucine, leucine, proline, serine, and tryptophan are essential for in vivo *C. difficile* growth [66,67], while the amino acids methionine, tryptophan, and serine are essential for in vivo *F. prausnitzii* growth [108]. To simulate various host-microbiota perturbations, the primary bile acid taurocholate and/or the electron acceptor nitrate were added to the media. The diffusion of nutrients, byproducts, and species biomass was assumed to occur only in the axial direction *z*. Therefore, each variable was considered to be changing with respect to space *z* and time *t* over a fixed biofilm thickness *L*.

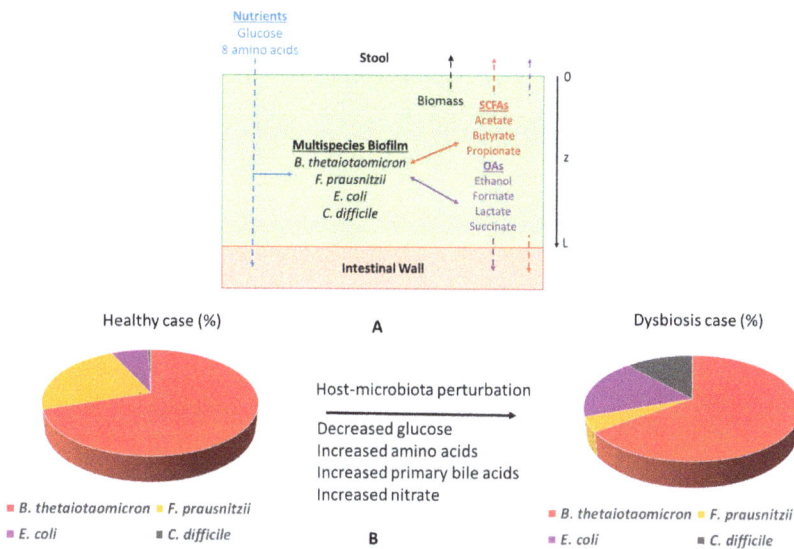

Figure 6. Schematic representation of the in silico gut community. (**A**) The model assumed biofilm attachment to the intestinal wall and described diffusion of glucose, amino acids, short-chain fatty acids, organic acids, ethanol, CO_2, and species biomass in and/or out of the biofilm along the axial direction *z*. (**B**) Host-microbiota perturbations were modeled through changes in the bulk concentrations of glucose, amino acids, primary bile acids, and nitrate at the biofilm–stool interface to predict species abundances in healthy and *C. difficile*-infected guts.

The nutrients were supplied at the top of the biofilm (Figure 6A). SCFAs, ethanol, organic acids, and CO_2 produced by the four species were allowed to diffuse and be removed from both ends of the biofilm. Biomass was assumed to move slowly through the biofilm by diffusion and be removed from the biofilm–stool interface according to a continuous erosion mechanism, as described in our previous publications [50,51,109]. This assumption provided a reasonable mechanism to ensure that biomass generation would be balanced by biomass loss such that a steady-state solution could be obtained. The multispecies biofilm model was tuned with nominal glucose and amino acid concentrations to reproduce species abundances and SCFA levels consistent with experimental studies on healthy individuals [57,58]. This tuned model was referred to as the "healthy case". Host-microbiota perturbations were simulated by altering glucose/amino acid concentrations and/or by introducing primary bile acids and nitrate as nutrients to predict the resulting species abundances (Figure 6B). These models were collectively referred to as the "dysbiosis case." In vivo concentrations of glucose and AA in the guts of healthy and *C. difficile*-infected patients are not commonly available. We have specified the glucose and AA concentrations for the healthy case based on limited experimental data [12,61–64] and have reduced the glucose concentration and increased AA concentrations for the dysbiosis case consistent with experimental observation [33,110]. We performed a sensitivity analysis of these concentrations to show that a similar behavior (i.e., healthy state) as that reported for the nominal values occurred if the glucose to AA ratio was sufficiently large (Figure S2). By contrast, a CDI dysbiosis-like state was obtained when the glucose to AA ratio was sufficiently small.

Uptake rates of nutrients and byproducts were assumed to follow Michaelis–Menten kinetics. Due to lack of available data, maximum uptake rates and Michaelis–Menten constants were assumed to be independent of species and metabolite. Calculated uptake rates were imposed as lower bounds of the exchange fluxes in the species metabolic reconstructions. The calculated growth rate, uptake fluxes, and secretion fluxes from each reconstruction served as inputs to reaction-diffusion-type equations for the biomass concentration of each species and the molar concentration of each nutrient and byproduct. This formulation yielded a set of 23 partial differential equations (PDEs) in the time and the axial direction z with embedded linear programs (LPs) for species metabolism (see Appendix S1). Following our previous methodology [50,51], lexicographic optimization with growth rate maximization as the primary objective was used to avoid alternative optima that would render the biofilm model non-smooth. This approach yielded a total of 71 LPs.

The biofilm model equations were solved by spatially discretizing the PDEs into a large set of ordinary differential equations (ODEs) [111,112]. We used 25 spatial node points to achieve a suitable compromise between solution accuracy and computational efficiency, which produced a discretized model with 575 ODEs and 1775 LPs that was solved with the MATLAB code DFBAlab [113]. We used Gurobi 6.5.2 for the LP solution, the stiff MATLAB solver ode15s for ODE integration, and DFBAlab running in MATLAB 9.0 (R2016a). Although not explored here, our biofilm modeling method can be extended to more species and extracellular metabolites. For N spatial discretization points, the addition of each new extracellular metabolite would generate N additional ODEs. For m total extracellular metabolites, the addition of each new species would generates N additional ODEs and $m+1$ LPs. Because the LP solution scales more favorably than the ODE solution, we anticipated that models with approximately 1000 ODEs and 7500 LPs would remain computationally viable on a typical desktop computer. These equation numbers translate into approximately 10 species and 30 extracellular metabolites.

4.2. Biofilm Model Parameterization and Tuning

Nominal parameter values used in the multispecies biofilm model are shown in Table 2. The parameters were obtained from the experimental literature to the extent possible and from our previous modeling studies [50,51] as necessary. The bulk glucose and amino acid concentrations at the biofilm–stool interface were specified to reflect healthy gut conditions. Due to the lack of species-specific uptake data, we used published kinetic parameters reported for *E. coli* [114]. Due to the lack of data, all eight byproducts were assumed to have the same uptake parameters as glucose. For simplicity, all eight amino acids were assumed to have the same uptake parameters obtained as the average of amino acid-dependent values reported for *E. coli* [114].

With all other parameter values fixed, the biofilm model was qualitatively tuned to achieve biomass and SCFA fractions within experimental ranges for a healthy patient. The species abundances were tuned by adjusting the non-growth-associated ATP maintenance (ATPM) values of the four metabolic reconstructions following our previous studies [50,51]. Our justification for tuning these values was the simple nature of the biofilm model, which neglected other phyla (e.g., Actinobacteria), other nutrients (e.g., oligosaccharides, fats), other species interactions (e.g., Actinobacteria cross-feeding of SCFAs and organic acids), as well as host metabolism present in the actual gut environment. These ATPM values listed in Table 2 produced *B. thetaiotaomicron*:*F. prausnitzii*:*E. coli*:*C. difficile* abundances of 71%:21%:7%:1%, which were deemed reasonable based on published data [57,58]. We found that the coexistence of the four species was achieved over a range of ATPM values (not shown here).

Table 2. Nominal parameter values for the multispecies biofilm model.

Symbol	Parameter	Value	Units	Source
L	Biofilm thickness	40	μm	[115]
X_b	Biomass bulk concentrations	0	g/L	[50]
P_b	Byproduct bulk concentrations	0	mmol/L	[50]
D_i	Diffusion coefficient			
D_X	Biomass	2×10^{-10}	cm^2/s	[50]
D_N	Glucose	2.01×10^{-6}	cm^2/s	[116]
	Cysteine	2.45×10^{-6}	cm^2/s	[116]
	Isoleucine	2.19×10^{-6}	cm^2/s	[116]
	Leucine	2.19×10^{-6}	cm^2/s	[116]
	Methionine	2.21×10^{-6}	cm^2/s	[116]
	Proline	2.51×10^{-6}	cm^2/s	[116]
	Serine	2.64×10^{-6}	cm^2/s	[116]
	Tryptophan	1.89×10^{-6}	cm^2/s	[116]
	Valine	2.49×10^{-6}	cm^2/s	[116]
D_P	Acetate	3.03×10^{-6}	cm^2/s	[116]
	Butyrate	1.74×10^{-6}	cm^2/s	[116]
	CO_2	1.15×10^{-5}	cm^2/s	[116]
	Ethanol	3.97×10^{-6}	cm^2/s	[116]
	Formate	4.23×10^{-6}	cm^2/s	[116]
	Lactate	3.1×10^{-6}	cm^2/s	[116]
	Propionate	4.03×10^{-6}	cm^2/s	[116]
	Succinate	2.82×10^{-6}	cm^2/s	[116]
	Nitrate	1.29×10^{-5}	cm^2/s	[116]
	Taurocholate	7.29×10^{-7}	cm^2/s	[116]
	Mass transfer coefficient			
k_X	Biomass	6×10^{-7}	cm/s	[50]
k_N	Glucose	2×10^{-4}	cm/s	[50]
	Amino acid	2×10^{-4}	cm/s	[50]
k_P	Byproduct	5×10^{-6}	cm/s	[50]
	Butyrate	8.5×10^{-5}	cm/s	Tuned
	Propionate	1.35×10^{-5}	cm/s	Tuned
	Nitrate	1.5×10^{-5}	cm/s	Tuned
	Taurocholate	2×10^{-3}	cm/s	Tuned
v_{max}	Maximum uptake rate			
	Glucose	10	mmol/gDW/h	[114]
	Amino acid	1	mmol/gDW/h	[114]
	Byproduct	10	mmol/gDW/h	[50]
K_m	Michaelis–Menten constant			
	Glucose	0.5	mmol/L	[114]
	Amino acids	0.1	mmol/L	[114]
	Byproduct	0.5	mmol/L	[50]
$ATPM$	ATP maintenance			
	B. thetaiotaomicron	4.25	mmol/gDW/h	Tuned
	F. prausnitzii	3.4	mmol/gDW/h	Tuned
	E. coli	2.75	mmol/gDW/h	Tuned
	C. difficile	8.43	mmol/gDW/h	Tuned

We adjusted the SCFA mass transfer coefficients controlling metabolite removal from the biofilm to tune the acetate, butyrate, and propionate concentrations for the healthy case. Starting with a value of 5×10^{-6} cm/s, the butyrate and propionate values were decreased until approximate fractions of 60%:20%:20% consistent with published data [5,59] were obtained. We justified the use of SCFA-dependent values by noting that our model neglected host-microbiota interactions, which would be expected to strongly affect SCFA levels in vivo. Biofilm simulations were performed for four combinations of bulk glucose, amino acid, nitrate, and taurocholate concentrations chosen to mimic a healthy gut environment and three unhealthy nutrient environments (high amino acids, high primary bile acids, high nitrate) experimentally correlated with *C. difficile*-associated dysbiosis (Table 1). We deemed the actual concentrations used to be less important than the concentration trends (e.g., decreasing glucose and increasing amino acids in the high amino acids case) since our goal was to assess qualitatively the effects of nutrient levels on community behavior.

5. Conclusions

Clostridium difficile infection (CDI) is a common problem in hospital settings, with almost 500,000 CDI cases diagnosed within the U.S. annually in acute care facilities alone. CDI involves dysbiosis of the commensal gut microbiota characterized by a significant reduction of butyrate-producing species, e.g., *Faecalibacterium prausnitzii*, and a large increase in Proteobacteria, e.g., *Escherichia coli*, along with uncontrolled propagation of *C. difficile*. Motivated by recent experimental studies demonstrating the ability of *C. difficile* and commensal gut bacteria to form biofilms, we developed a multispecies biofilm model with a minimal representation of the gut microbiota containing *C. difficile* and one species each from the three dominant phyla (*F. prausnitzii, E. coli, Bacteroides thetaiotaomicron*). The model was used to investigate possible metabolic determinants of CDI mediated through host–microbiota perturbations, modeled as decreased carbohydrate levels and increased amino acid, primary bile acid, and nitrate levels compared to the healthy gut. These nutrient perturbations were shown to mimic microbiota changes characteristic of CDI, namely marked increases in *C. difficile* and *E. coli* abundances and a sharp decrease in *F. prausnitzii* abundance. *C. difficile* propagation was strongly dependent on cross-feeding of formate and succinate secreted by the commensal species, a prediction in agreement with experimental studies and that provides possible targets for the development of novel therapeutic strategies. While our model is a simplified representation of a complex disease process, the results presented emphasized the importance of metabolic interactions between *C. difficile* and commensal species in CDI progression.

Supplementary Materials: The following are available online at http://www.mdpi.com/2227-9717/7/2/97/s1. Additional File 1. Model equations and description. Figure S1. Predicted cross-feeding of byproducts with *C. difficile* removed from the community. Figure S2. Predicted species abundances at various nutrient concentrations. Figure S3. Predicted multispecies biofilm dysbiosis resulting from host–microbiota perturbations in the concentrations of amino acids and the primary bile acid taurocholate. Figure S4. Effect of removing individual cross-feeding relationships on predicted species abundances. Figure S5. Effect of the biofilm length on predicted species abundances for the healthy case. Figure S6. Predicted multispecies biofilm behavior under healthy nutrient conditions for a 30 micron-thick biofilm. Figure S7. Predicted multispecies biofilm behavior under healthy nutrient conditions for a 60 micron-thick biofilm. Figure S8. Differences between healthy case and bile acid dysbiosis case internal pathway fluxes.

Author Contributions: P.P. and M.A.H. conceived of the study. P.P. and M.A.H. developed the model and model solution method. P.P. and M.A.H. performed the simulations and analyzed the results. P.P. and M.A.H. prepared the manuscript. All authors read and approved the final manuscript.

Funding: This research was partially funded by NIH (Award U01EB019416). This work was funded in part by a Fellowship from the University of Massachusetts to P.P. as part of an NIH-funded Training Program (National Research Service Award T32 GM108556). Funds from these grants were used to cover the costs to publish in open access.

Conflicts of Interest: The authors declare no conflict of interest.

Abbreviations

The following abbreviations are used in this manuscript:

ATPM	ATP maintenance
BT	*Bacteroides thetaiotaomicron*
CD	*Clostridium difficile*
CDI	*Clostridium difficile* infection
DFBAlab	Dynamic flux balance analysis laboratory
EC	*Escherichia coli*
FBA	Flux balance analysis
FP	*Faecalibacterium prausnitzii*
IBD	Inflammatory bowel diseases
LP	Linear program
MDM	Minimal defined media
OA	Organic acid
ODE	Ordinary differential equation
PDE	Partial differential equation
SCFA	Short chain fatty acid

References

1. Gill, S.R.; Pop, M.; DeBoy, R.T.; Eckburg, P.B.; Turnbaugh, P.J.; Samuel, B.S.; Gordon, J.I.; Relman, D.A.; Fraser-Liggett, C.M.; Nelson, K.E. Metagenomic analysis of the human distal gut microbiome. *Science* **2006**, *312*, 1355–1359. [CrossRef] [PubMed]

2. Bäckhed, F.; Ley, R.E.; Sonnenburg, J.L.; Peterson, D.A.; Gordon, J.I. Host-bacterial mutualism in the human intestine. *Science* **2005**, *307*, 1915–1920. [CrossRef] [PubMed]

3. Macpherson, A.J.; Harris, N.L. Interactions between commensal intestinal bacteria and the immune system. *Nat. Rev. Immunol.* **2004**, *4*, 478–485. [CrossRef] [PubMed]

4. Bäumler, A.J.; Sperandio, V. Interactions between the microbiota and pathogenic bacteria in the gut. *Nature* **2016**, *535*, 85–93. [CrossRef] [PubMed]

5. Den Besten, G.; van Eunen, K.; Groen, A.K.; Venema, K.; Reijngoud, D.J.; Bakker, B.M. The role of short-chain fatty acids in the interplay between diet, gut microbiota, and host energy metabolism. *J. Lipid Res.* **2013**, *54*, 2325–2340. [CrossRef] [PubMed]

6. Macfarlane, S.; Steed, H.; Macfarlane, G.T. Intestinal bacteria and inflammatory bowel disease. *Crit. Rev. Clin. Lab. Sci.* **2009**, *46*, 25–54. [CrossRef] [PubMed]

7. Morrison, D.J.; Preston, T. Formation of short chain fatty acids by the gut microbiota and their impact on human metabolism. *Gut Microbes* **2016**, *7*, 189–200. [CrossRef] [PubMed]

8. Rajilić-Stojanović, M.; Smidt, H.; De Vos, W.M. Diversity of the human gastrointestinal tract microbiota revisited. *Environ. Microbiol.* **2007**, *9*, 2125–2136. [CrossRef]

9. Zoetendal, E.; Rajilić-Stojanović, M.; De Vos, W. High-throughput diversity and functionality analysis of the gastrointestinal tract microbiota. *Gut* **2008**, *57*, 1605–1615. [CrossRef]

10. Gerritsen, J.; Smidt, H.; Rijkers, G.T.; Vos, W.M. Intestinal microbiota in human health and disease: The impact of probiotics. *Genes Nutr.* **2011**, *6*, 209. [CrossRef]

11. Young, V.B.; Schmidt, T.M. Antibiotic-associated diarrhea accompanied by large-scale alterations in the composition of the fecal microbiota. *J. Clin. Microbiol.* **2004**, *42*, 1203–1206. [CrossRef] [PubMed]

12. Brown, K.; DeCoffe, D.; Molcan, E.; Gibson, D.L. Diet-induced dysbiosis of the intestinal microbiota and the effects on immunity and disease. *Nutrients* **2012**, *4*, 1095–1119. [CrossRef] [PubMed]

13. Anitha, M.; Reichardt, F.; Tabatabavakili, S.; Nezami, B.G.; Chassaing, B.; Mwangi, S.; Vijay-Kumar, M.; Gewirtz, A.; Srinivasan, S. Intestinal dysbiosis contributes to the delayed gastrointestinal transit in high-fat diet fed mice. *CMGH Cell. Mol. Gastroenterol. Hepatol.* **2016**, *2*, 328–339. [CrossRef] [PubMed]

14. Alou, M.T.; Lagier, J.C.; Raoult, D. Diet influence on the gut microbiota and dysbiosis related to nutritional disorders. *Hum. Microb. J.* **2016**, *1*, 3–11. [CrossRef]

15. Agus, A.; Denizot, J.; Thévenot, J.; Martinez-Medina, M.; Massier, S.; Sauvanet, P.; Bernalier-Donadille, A.; Denis, S.; Hofman, P.; Bonnet, R.; et al. Western diet induces a shift in microbiota composition enhancing susceptibility to Adherent-Invasive *E. coli* infection and intestinal inflammation. *Sci. Rep.* **2016**, *6*, 19032. [CrossRef] [PubMed]

16. Theriot, C.M.; Bowman, A.A.; Young, V.B. Antibiotic-induced alterations of the gut microbiota alter secondary bile acid production and allow for *Clostridium difficile* spore germination and outgrowth in the large intestine. *MSphere* **2016**, *1*, e00045-15. [CrossRef]

17. Winter, S.E.; Winter, M.G.; Xavier, M.N.; Thiennimitr, P.; Poon, V.; Keestra, A.M.; Laughlin, R.C.; Gomez, G.; Wu, J.; Lawhon, S.D.; et al. Host-derived nitrate boosts growth of *E. coli* in the inflamed gut. *Science* **2013**, *339*, 708–711. [CrossRef]

18. Frank, D.N.; Amand, A.L.S.; Feldman, R.A.; Boedeker, E.C.; Harpaz, N.; Pace, N.R. Molecular-phylogenetic characterization of microbial community imbalances in human inflammatory bowel diseases. *Proc. Natl. Acad. Sci. USA* **2007**, *104*, 13780–13785. [CrossRef] [PubMed]

19. Ley, R.E.; Turnbaugh, P.J.; Klein, S.; Gordon, J.I. Microbial ecology: Human gut microbes associated with obesity. *Nature* **2006**, *444*, 1022–1023. [CrossRef] [PubMed]

20. Carding, S.; Verbeke, K.; Vipond, D.T.; Corfe, B.M.; Owen, L.J. Dysbiosis of the gut microbiota in disease. *Microb. Ecol. Health Dis.* **2015**, *26*, 26191. [CrossRef]

21. Surawicz, C.M.; Brandt, L.J.; Binion, D.G.; Ananthakrishnan, A.N.; Curry, S.R.; Gilligan, P.H.; McFarland, L.V.; Mellow, M.; Zuckerbraun, B.S. Guidelines for diagnosis, treatment, and prevention of *Clostridium difficile* infections. *Am. J. Gastroenterol.* **2013**, *108*, 478. [CrossRef] [PubMed]

22. Shim, J.K.; Johnson, S.; Samore, M.H.; Bliss, D.Z.; Gerding, D.N. Primary symptomless colonisation by *Clostridium difficile* and decreased risk of subsequent diarrhoea. *Lancet* **1998**, *351*, 633–636. [CrossRef]

23. Kyne, L.; Warny, M.; Qamar, A.; Kelly, C.P. Asymptomatic carriage of *Clostridium difficile* and serum levels of IgG antibody against toxin A. *N. Engl. J. Med.* **2000**, *342*, 390–397. [CrossRef] [PubMed]

24. Barbut, F.; Petit, J.C. Epidemiology of *Clostridium difficile*-associated infections. *Clin. Microbiol. Infect.* **2001**, *7*, 405–410. [CrossRef]

25. Kato, H.; Kita, H.; Karasawa, T.; Maegawa, T.; Koino, Y.; Takakuwa, H.; Saikai, T.; Kobayashi, K.; Yamagishi, T.; Nakamura, S. Colonisation and transmission of *Clostridium difficile* in healthy individuals examined by PCR ribotyping and pulsed-field gel electrophoresis. *J. Med. Microbiol.* **2001**, *50*, 720–727. [CrossRef] [PubMed]

26. Ozaki, E.; Kato, H.; Kita, H.; Karasawa, T.; Maegawa, T.; Koino, Y.; Matsumoto, K.; Takada, T.; Nomoto, K.; Tanaka, R.; et al. *Clostridium difficile* colonization in healthy adults: transient colonization and correlation with enterococcal colonization. *J. Med. Microbiol.* **2004**, *53*, 167–172. [CrossRef]

27. Poutanen, S.M.; Simor, A.E. *Clostridium difficile*-associated diarrhea in adults. *Can. Med. Assoc. J.* **2004**, *171*, 51–58. [CrossRef]

28. Furuya-Kanamori, L.; Marquess, J.; Yakob, L.; Riley, T.V.; Paterson, D.L.; Foster, N.F.; Huber, C.A.; Clements, A.C. Asymptomatic *Clostridium difficile* colonization: epidemiology and clinical implications. *BMC Infect. Dis.* **2015**, *15*, 516. [CrossRef] [PubMed]

29. Chang, J.Y.; Antonopoulos, D.A.; Kalra, A.; Tonelli, A.; Khalife, W.T.; Schmidt, T.M.; Young, V.B. Decreased diversity of the fecal microbiome in recurrent *Clostridium difficile*-associated diarrhea. *J. Infect. Dis.* **2008**, *197*, 435–438. [CrossRef]

30. Bartlett, J.G. *Clostridium difficile*: History of its role as an enteric pathogen and the current state of knowledge about the organism. *Clin. Infect. Dis.* **1994**, *18*, S265–S272. [CrossRef]

31. Pérez-Cobas, A.E.; Moya, A.; Gosalbes, M.J.; Latorre, A. Colonization resistance of the gut microbiota against *Clostridium difficile*. *Antibiotics* **2015**, *4*, 337–357. [CrossRef] [PubMed]

32. Pothoulakis, C. Pathogenesis of *Clostridium difficile*-associated diarrhoea. *Eur. J. Gastroenterol. Hepatol.* **1996**, *8*, 1041–1047. [CrossRef] [PubMed]

33. Theriot, C.M.; Young, V.B. Interactions between the gastrointestinal microbiome and *Clostridium difficile*. *Annu. Rev. Microbiol.* **2015**, *69*, 445–461. [CrossRef] [PubMed]

34. Mylonakis, E.; Ryan, E.T.; Calderwood, S.B. *Clostridium difficile*-associated diarrhea: A review. *Arch. Intern. Med.* **2001**, *161*, 525–533. [CrossRef] [PubMed]

35. Jarrad, A.M.; Karoli, T.; Blaskovich, M.A.; Lyras, D.; Cooper, M.A. *Clostridium difficile* drug pipeline: Challenges in discovery and development of new agents. *J. Med. Chem.* **2015**, *58*, 5164–5185. [CrossRef] [PubMed]
36. Lessa, F.C.; Mu, Y.; Bamberg, W.M.; Beldavs, Z.G.; Dumyati, G.K.; Dunn, J.R.; Farley, M.M.; Holzbauer, S.M.; Meek, J.I.; Phipps, E.C.; et al. Burden of *Clostridium difficile* infection in the United States. *N. Engl. J. Med.* **2015**, *372*, 825–834. [CrossRef]
37. Dubberke, E.R.; Olsen, M.A. Burden of *Clostridium difficile* on the healthcare system. *Clin. Infect. Dis.* **2012**, *55*, S88–S92. [CrossRef]
38. Dawson, L.F.; Valiente, E.; Faulds-Pain, A.; Donahue, E.H.; Wren, B.W. Characterisation of *Clostridium difficile* biofilm formation, a role for Spo0A. *PLoS ONE* **2012**, *7*, e50527. [CrossRef]
39. Dhapa, T.; Leuzzi, R.; Ng, Y.K.; Baban, S.T.; Adamo, R.; Kuehne, S.A.; Scarselli, M.; Minton, N.P.; Serruto, D.; Unnikrishnan, M. Multiple factors modulate biofilm formation by the anaerobic pathogen *Clostridium difficile*. *J. Bacteriol.* **2013**, *195*, 545–555.
40. Donelli, G.; Vuotto, C.; Cardines, R.; Mastrantonio, P. Biofilm-growing intestinal anaerobic bacteria. *FEMS Immunol. Med. Microbiol.* **2012**, *65*, 318–325. [CrossRef]
41. Semenyuk, E.G.; Laning, M.L.; Foley, J.; Johnston, P.F.; Knight, K.L.; Gerding, D.N.; Driks, A. Spore formation and toxin production in *Clostridium difficile* biofilms. *PLoS ONE* **2014**, *9*, e87757. [CrossRef] [PubMed]
42. Swidsinski, A.; Weber, J.; Loening-Baucke, V.; Hale, L.P.; Lochs, H. Spatial organization and composition of the mucosal flora in patients with inflammatory bowel disease. *J. Clin. Microbiol.* **2005**, *43*, 3380–3389. [CrossRef] [PubMed]
43. Macfarlane, S.; Dillon, J. Microbial biofilms in the human gastrointestinal tract. *J. Appl. Microbiol.* **2007**, *102*, 1187–1196. [CrossRef] [PubMed]
44. Costerton, J.W.; Stewart, P.S.; Greenberg, E.P. Bacterial biofilms: A common cause of persistent infections. *Science* **1999**, *284*, 1318–1322. [CrossRef] [PubMed]
45. Anderl, J.N.; Franklin, M.J.; Stewart, P.S. Role of antibiotic penetration limitation in Klebsiella pneumoniae biofilm resistance to ampicillin and ciprofloxacin. *Antimicrob. Agents Chemother.* **2000**, *44*, 1818–1824. [CrossRef] [PubMed]
46. Stewart, P.S.; Costerton, J.W. Antibiotic resistance of bacteria in biofilms. *Lancet* **2001**, *358*, 135–138. [CrossRef]
47. Stewart, P.S. Mechanisms of antibiotic resistance in bacterial biofilms. *Int. J. Med. Microbiol.* **2002**, *292*, 107–113. [CrossRef] [PubMed]
48. Zuroff, T.R.; Bernstein, H.; Lloyd-Randolfi, J.; Jimenez-Taracido, L.; Stewart, P.S.; Carlson, R.P. Robustness analysis of culturing perturbations on *Escherichia coli* colony biofilm beta-lactam and aminoglycoside antibiotic tolerance. *BMC Microbiol.* **2010**, *10*, 185. [CrossRef] [PubMed]
49. Shreiner, A.B.; Kao, J.Y.; Young, V.B. The gut microbiome in health and in disease. *Curr. Opin. Gastroenterol.* **2015**, *31*, 69. [CrossRef] [PubMed]
50. Henson, M.A.; Phalak, P. Byproduct Cross Feeding and Community Stability in an In Silico Biofilm Model of the Gut Microbiome. *Processes* **2017**, *5*, 13. [CrossRef]
51. Henson, M.A.; Phalak, P. Microbiota dysbiosis in inflammatory bowel diseases: *in silico* investigation of the oxygen hypothesis. *BMC Syst. Biol.* **2017**, *11*, 145. [CrossRef] [PubMed]
52. Rivière, A.; Selak, M.; Lantin, D.; Leroy, F.; De Vuyst, L. Bifidobacteria and butyrate-producing colon bacteria: Importance and strategies for their stimulation in the human gut. *Front. Microbiol.* **2016**, *7*, 979. [CrossRef] [PubMed]
53. Ríos-Covián, D.; Ruas-Madiedo, P.; Margolles, A.; Gueimonde, M.; de los Reyes-Gavilán, C.G.; Salazar, N. Intestinal short chain fatty acids and their link with diet and human health. *Front. Microbiol.* **2016**, *7*, 185. [CrossRef] [PubMed]
54. Jakobsdottir, G.; Xu, J.; Molin, G.; Ahrne, S.; Nyman, M. High-fat diet reduces the formation of butyrate, but increases succinate, inflammation, liver fat and cholesterol in rats, while dietary fibre counteracts these effects. *PLoS ONE* **2013**, *8*, e80476. [CrossRef] [PubMed]
55. Ferreyra, J.A.; Wu, K.J.; Hryckowian, A.J.; Bouley, D.M.; Weimer, B.C.; Sonnenburg, J.L. Gut microbiota-produced succinate promotes *C. difficile* infection after antibiotic treatment or motility disturbance. *Cell Host Microbe* **2014**, *16*, 770–777. [CrossRef] [PubMed]
56. Köpke, M.; Straub, M.; Dürre, P. *Clostridium difficile* is an autotrophic bacterial pathogen. *PLoS ONE* **2013**, *8*, e62157. [CrossRef] [PubMed]

57. Spor, A.; Koren, O.; Ley, R. Unravelling the effects of the environment and host genotype on the gut microbiome. *Nat. Rev. Microbiol.* **2011**, *9*, 279–290. [CrossRef]
58. De Filippo, C.; Cavalieri, D.; Di Paola, M.; Ramazzotti, M.; Poullet, J.B.; Massart, S.; Collini, S.; Pieraccini, G.; Lionetti, P. Impact of diet in shaping gut microbiota revealed by a comparative study in children from Europe and rural Africa. *Proc. Natl. Acad. Sci. USA* **2010**, *107*, 14691–14696. [CrossRef]
59. Byrne, C.; Chambers, E.; Morrison, D.; Frost, G. The role of short chain fatty acids in appetite regulation and energy homeostasis. *Int. J. Obes.* **2015**, *39*, 1331–1338. [CrossRef]
60. Campbell, J.M.; Fahey, G.C.; Wolf, B.W. Selected indigestible oligosaccharides affect large bowel mass, cecal and fecal short-chain fatty acids, pH and microflora in rats. *J. Nutr.* **1997**, *127*, 130–136. [CrossRef]
61. Theriot, C.M.; Koenigsknecht, M.J.; Carlson Jr, P.E.; Hatton, G.E.; Nelson, A.M.; Li, B.; Huffnagle, G.B.; Li, J.Z.; Young, V.B. Antibiotic-induced shifts in the mouse gut microbiome and metabolome increase susceptibility to *Clostridium difficile* infection. *Nat. Commun.* **2014**, *5*, 3114. [CrossRef] [PubMed]
62. Utzschneider, K.M.; Kratz, M.; Damman, C.J.; Hullarg, M. Mechanisms linking the gut microbiome and glucose metabolism. *J. Clin. Endocrinol. Metab.* **2016**, *101*, 1445–1454. [CrossRef] [PubMed]
63. Carmody, R.N.; Gerber, G.K.; Luevano, J.M.; Gatti, D.M.; Somes, L.; Svenson, K.L.; Turnbaugh, P.J. Diet dominates host genotype in shaping the murine gut microbiota. *Cell Host Microbe* **2015**, *17*, 72–84. [CrossRef] [PubMed]
64. Zhang, X.; Shen, D.; Fang, Z.; Jie, Z.; Qiu, X.; Zhang, C.; Chen, Y.; Ji, L. Human gut microbiota changes reveal the progression of glucose intolerance. *PLoS ONE* **2013**, *8*, e71108. [CrossRef] [PubMed]
65. Jackson, S.; Calos, M.; Myers, A.; Self, W.T. Analysis of proline reduction in the nosocomial pathogen *Clostridium difficile*. *J. Bacteriol.* **2006**, *188*, 8487–8495. [CrossRef] [PubMed]
66. Karlsson, S.; Burman, L.G.; Åkerlund, T. Suppression of toxin production in *Clostridium difficile* VPI 10463 by amino acids. *Microbiology* **1999**, *145*, 1683–1693. [CrossRef] [PubMed]
67. Neumann-Schaal, M.; Hofmann, J.D.; Will, S.E.; Schomburg, D. Time-resolved amino acid uptake of *Clostridium difficile* 630Δerm and concomitant fermentation product and toxin formation. *BMC Microbiol.* **2015**, *15*, 281. [CrossRef] [PubMed]
68. Antonopoulos, D.A.; Huse, S.M.; Morrison, H.G.; Schmidt, T.M.; Sogin, M.L.; Young, V.B. Reproducible community dynamics of the gastrointestinal microbiota following antibiotic perturbation. *Infect. Immunity* **2009**, *77*, 2367–2375. [CrossRef] [PubMed]
69. Antharam, V.C.; Li, E.C.; Ishmael, A.; Sharma, A.; Mai, V.; Rand, K.H.; Wang, G.P. Intestinal dysbiosis and depletion of butyrogenic bacteria in *Clostridium difficile* infection and nosocomial diarrhea. *J. Clin. Microbiol.* **2013**, *51*, 2884–2892. [CrossRef]
70. Vincent, C.; Manges, A. Antimicrobial use, human gut microbiota and *Clostridium difficile* colonization and infection. *Antibiotics* **2015**, *4*, 230–253. [CrossRef]
71. Schippa, S.; Conte, M.P. Dysbiotic events in gut microbiota: Impact on human health. *Nutrients* **2014**, *6*, 5786–5805. [CrossRef]
72. Tamboli, C.; Neut, C.; Desreumaux, P.; Colombel, J. Dysbiosis in inflammatory bowel disease. *Gut* **2004**, *53*, 1–4. [CrossRef] [PubMed]
73. Honda, K.; Littman, D.R. The microbiome in infectious disease and inflammation. *Annu. Rev. Immunol.* **2012**, *30*, 759–795. [CrossRef] [PubMed]
74. Ling, Z.; Liu, X.; Jia, X.; Cheng, Y.; Luo, Y.; Yuan, L.; Wang, Y.; Zhao, C.; Guo, S.; Li, L.; et al. Impacts of infection with different toxigenic *Clostridium difficile* strains on faecal microbiota in children. *Sci. Rep.* **2014**, *4*, 7485. [CrossRef] [PubMed]
75. Ridlon, J.M.; Kang, D.J.; Hylemon, P.B. Bile salt biotransformations by human intestinal bacteria. *J. Lipid Res.* **2006**, *47*, 241–259. [CrossRef]
76. Setchell, K.; Lawson, A.; Tanida, N.; Sjövall, J. General methods for the analysis of metabolic profiles of bile acids and related compounds in feces. *J. Lipid Res.* **1983**, *24*, 1085–1100. [PubMed]
77. Winston, J.A.; Theriot, C.M. Impact of microbial derived secondary bile acids on colonization resistance against *Clostridium difficile* in the gastrointestinal tract. *Anaerobe* **2016**, *41*, 44–50. [CrossRef] [PubMed]
78. Rea, M.C.; Dobson, A.; O'Sullivan, O.; Crispie, F.; Fouhy, F.; Cotter, P.D.; Shanahan, F.; Kiely, B.; Hill, C.; Ross, R.P. Effect of broad-and narrow-spectrum antimicrobials on *Clostridium difficile* and microbial diversity in a model of the distal colon. *Proc. Natl. Acad. Sci. USA* **2011**, *108*, 4639–4644. [CrossRef]

79. Sorg, J.A.; Sonenshein, A.L. Bile salts and glycine as cogerminants for *Clostridium difficile* spores. *J. Bacteriol.* **2008**, *190*, 2505–2512. [CrossRef] [PubMed]
80. Sorg, J.A.; Sonenshein, A.L. Inhibiting the initiation of *Clostridium difficile* spore germination using analogs of chenodeoxycholic acid, a bile acid. *J. Bacteriol.* **2010**, *192*, 4983–4990. [CrossRef] [PubMed]
81. Allegretti, J.R.; Kearney, S.; Li, N.; Bogart, E.; Bullock, K.; Gerber, G.K.; Bry, L.; Clish, C.B.; Alm, E.; Korzenik, J. Recurrent *Clostridium difficile* infection associates with distinct bile acid and microbiome profiles. *Aliment. Pharmacol. Ther.* **2016**, *43*, 1142–1153. [CrossRef] [PubMed]
82. Wilson, K.H.; Kennedy, M.J.; Fekety, F.R. Use of sodium taurocholate to enhance spore recovery on a medium selective for *Clostridium difficile*. *J. Clin. Microbiol.* **1982**, *15*, 443–446. [PubMed]
83. Buggy, B.; Hawkins, C.; Fekety, R. Effect of adding sodium taurocholate to selective media on the recovery of *Clostridium difficile* from environmental surfaces. *J. Clin. Microbiol.* **1985**, *21*, 636–637. [PubMed]
84. Wilson, K.H. Efficiency of various bile salt preparations for stimulation of *Clostridium difficile* spore germination. *J. Clin. Microbiol.* **1983**, *18*, 1017–1019. [PubMed]
85. Seekatz, A.M.; Young, V.B. *Clostridium difficile* and the microbiota. *J. Clin. Investig.* **2014**, *124*, 4182–4189. [CrossRef]
86. Thursby, E.; Juge, N. Introduction to the human gut microbiota. *Biochem. J.* **2017**, *474*, 1823–1836. [CrossRef]
87. Engevik, M.A.; Engevik, K.A.; Yacyshyn, M.B.; Wang, J.; Hassett, D.J.; Darien, B.; Yacyshyn, B.R.; Worrell, R.T. Human *Clostridium difficile* infection: Inhibition of NHE3 and microbiota profile. *Am. J. Physiol. Gastrointest. Liver Physiol.* **2015**, *308*, G497–G509. [CrossRef]
88. Dannheim, H.; Will, S.E.; Schomburg, D.; Neumann-Schaal, M. *Clostridioides difficile* 630Δerm in silico and in vivo–quantitative growth and extensive polysaccharide secretion. *FEBS Open Biol.* **2017**, *7*, 602–615. [CrossRef]
89. Heinken, A.; Sahoo, S.; Fleming, R.M.; Thiele, I. Systems-level characterization of a host-microbe metabolic symbiosis in the mammalian gut. *Gut Microbes* **2013**, *4*, 28–40. [CrossRef]
90. Heinken, A.; Khan, M.T.; Paglia, G.; Rodionov, D.A.; Harmsen, H.J.; Thiele, I. Functional metabolic map of *Faecalibacterium prausnitzii*, a beneficial human gut microbe. *J. Bacteriol.* **2014**, *196*, 3289–3302. [CrossRef]
91. Baumler, D.J.; Peplinski, R.G.; Reed, J.L.; Glasner, J.D.; Perna, N.T. The evolution of metabolic networks of *E. coli*. *BMC Syst. Biol.* **2011**, *5*, 182. [CrossRef] [PubMed]
92. Henson, M.A.; Phalak, P. Suboptimal community growth mediated through metabolite cross-feeding promotes species diversity in the gut microbiota. *PLoS Comput. Biol.* **2018**, *14*, e1006558. [CrossRef] [PubMed]
93. Ouwerkerk, J.P.; de Vos, W.M.; Belzer, C. Glycobiome: Bacteria and mucus at the epithelial interface. *Best Pract. Res. Clin. Gastroenterol.* **2013**, *27*, 25–38. [CrossRef] [PubMed]
94. Ross, C.L.; Spinler, J.K.; Savidge, T.C. Structural and functional changes within the gut microbiota and susceptibility to *Clostridium difficile* infection. *Anaerobe* **2016**, *41*, 37–43. [CrossRef] [PubMed]
95. Browne, H.P.; Forster, S.C.; Anonye, B.O.; Kumar, N.; Neville, B.A.; Stares, M.D.; Goulding, D.; Lawley, T.D. Culturing of 'unculturable' human microbiota reveals novel taxa and extensive sporulation. *Nature* **2016**, *533*, 543–546. [CrossRef] [PubMed]
96. Staley, C.; Weingarden, A.R.; Khoruts, A.; Sadowsky, M.J. Interaction of gut microbiota with bile acid metabolism and its influence on disease states. *Appl. Microbiol. Biotechnol.* **2017**, *101*, 47–64. [CrossRef] [PubMed]
97. Myers, S.P. The causes of intestinal dysbiosis: A review. *Altern. Med. Rev.* **2004**, *9*, 180–197.
98. Singh, R.K.; Chang, H.W.; Yan, D.; Lee, K.M.; Ucmak, D.; Wong, K.; Abrouk, M.; Farahnik, B.; Nakamura, M.; Zhu, T.H.; et al. Influence of diet on the gut microbiome and implications for human health. *J. Transl. Med.* **2017**, *15*, 73. [CrossRef]
99. Kamada, N.; Chen, G.Y.; Inohara, N.; Núñez, G. Control of pathogens and pathobionts by the gut microbiota. *Nat. Immunol.* **2013**, *14*, 685–690. [CrossRef]
100. Weingarden, A.R.; Chen, C.; Bobr, A.; Yao, D.; Lu, Y.; Nelson, V.M.; Sadowsky, M.J.; Khoruts, A. Microbiota transplantation restores normal fecal bile acid composition in recurrent *Clostridium difficile* infection. *Am. J. Physiol. Gastrointest. Liver Physiol.* **2013**, *306*, G310–G319. [CrossRef]
101. Heinken, A.; Thiele, I. Systematic prediction of health-relevant human-microbial co-metabolism through a computational framework. *Gut Microbes* **2015**, *6*, 120–130. [CrossRef] [PubMed]

102. Shoaie, S.; Nielsen, J. Elucidating the interactions between the human gut microbiota and its host through metabolic modeling. *Front. Genet.* **2014**, *5*, 86. [CrossRef]

103. Thiele, I.; Heinken, A.; Fleming, R.M. A systems biology approach to studying the role of microbes in human health. *Curr. Opin. Biotechnol.* **2013**, *24*, 4–12. [CrossRef] [PubMed]

104. Glick-Bauer, M.; Yeh, M.C. The health advantage of a vegan diet: Exploring the gut microbiota connection. *Nutrients* **2014**, *6*, 4822–4838. [CrossRef] [PubMed]

105. Swainston, N.; Mendes, P.; Kell, D.B. An analysis of a 'community-driven'reconstruction of the human metabolic network. *Metabolomics* **2013**, *9*, 757–764. [CrossRef] [PubMed]

106. Thiele, I.; Swainston, N.; Fleming, R.M.; Hoppe, A.; Sahoo, S.; Aurich, M.K.; Haraldsdottir, H.; Mo, M.L.; Rolfsson, O.; Stobbe, M.D.; et al. A community-driven global reconstruction of human metabolism. *Nat. Biotechnol.* **2013**, *31*, 419. [CrossRef] [PubMed]

107. Thiele, I.; Sahoo, S.; Heinken, A.; Heirendt, L.; Aurich, M.K.; Noronha, A.; Fleming, R.M. When metabolism meets physiology: Harvey and Harvetta. *bioRxiv* **2018**. . [CrossRef]

108. Dai, Z.L.; Wu, G.; Zhu, W.Y. Amino acid metabolism in intestinal bacteria: Links between gut ecology and host health. *Front. Biosci.* **2011**, *16*, 1768–1786. [CrossRef]

109. Horn, H.; Lackner, S. Modeling of biofilm systems: A review. In *Productive Biofilms*; Springer: Cham, Swizterland, 2014; pp. 53–76.

110. Lawley, T.D.; Clare, S.; Walker, A.W.; Stares, M.D.; Connor, T.R.; Raisen, C.; Goulding, D.; Rad, R.; Schreiber, F.; Brandt, C.; et al. Targeted restoration of the intestinal microbiota with a simple, defined bacteriotherapy resolves relapsing *Clostridium difficile* disease in mice. *PLoS Pathog.* **2012**, *8*, e1002995. [CrossRef]

111. Chen, J.; Gomez, J.A.; Höffner, K.; Phalak, P.; Barton, P.I.; Henson, M.A. Spatiotemporal modeling of microbial metabolism. *BMC Syst. Biol.* **2016**, *10*, 21. [CrossRef]

112. Phalak, P.; Chen, J.; Carlson, R.P.; Henson, M.A. Metabolic modeling of a chronic wound biofilm consortium predicts spatial partitioning of bacterial species. *BMC Syst. Biol.* **2016**, *10*, 90. [CrossRef] [PubMed]

113. Gomez, J.A.; Hoffner, K.; Barton, P.I. DFBAlab: A fast and reliable MATLAB code for dynamic flux balance analysis. *BMC Bioinform.* **2014**, *15*, 409. [CrossRef] [PubMed]

114. Meadows, A.L.; Karnik, R.; Lam, H.; Forestell, S.; Snedecor, B. Application of dynamic flux balance analysis to an industrial *Escherichia coli* fermentation. *Metab. Eng.* **2010**, *12*, 150–160. [CrossRef] [PubMed]

115. Stewart, P.S. A review of experimental measurements of effective diffusive permeabilities and effective diffusion coefficients in biofilms. *Biotechnol. Bioeng.* **1998**, *59*, 261–272. [CrossRef]

116. Stewart, P.S. Diffusion in biofilms. *J. Bacteriol.* **2003**, *185*, 1485–1491. [CrossRef] [PubMed]

processes

MDPI

Article

Multiscale Agent-Based and Hybrid Modeling of the Tumor Immune Microenvironment

Kerri-Ann Norton [1,2,*,†], Chang Gong [1,†], Samira Jamalian [1,†] and Aleksander S. Popel [1,3]

[1] Department of Biomedical Engineering, School of Medicine, Johns Hopkins University, Baltimore, MD 21205, USA; cgong5@jhu.edu (C.G.); samira.jamalian@jhu.edu (S.J.)

[2] Computer Science Program, Department of Science, Mathematics, and Computing, Bard College, Annandale-on-Hudson, NY 12504, USA

[3] Department of Oncology and the Sidney Kimmel Comprehensive Cancer Center, School of Medicine, Johns Hopkins University, Baltimore, MD 21205, USA; apopel@jhu.edu

* Correspondence: knorton@bard.edu; Tel.: +845-752-2307

† These authors contributed equally to this work.

Received: 11 December 2018; Accepted: 10 January 2019; Published: 13 January 2019

check for updates

Abstract: Multiscale systems biology and systems pharmacology are powerful methodologies that are playing increasingly important roles in understanding the fundamental mechanisms of biological phenomena and in clinical applications. In this review, we summarize the state of the art in the applications of agent-based models (ABM) and hybrid modeling to the tumor immune microenvironment and cancer immune response, including immunotherapy. Heterogeneity is a hallmark of cancer; tumor heterogeneity at the molecular, cellular, and tissue scales is a major determinant of metastasis, drug resistance, and low response rate to molecular targeted therapies and immunotherapies. Agent-based modeling is an effective methodology to obtain and understand quantitative characteristics of these processes and to propose clinical solutions aimed at overcoming the current obstacles in cancer treatment. We review models focusing on intra-tumor heterogeneity, particularly on interactions between cancer cells and stromal cells, including immune cells, the role of tumor-associated vasculature in the immune response, immune-related tumor mechanobiology, and cancer immunotherapy. We discuss the role of digital pathology in parameterizing and validating spatial computational models and potential applications to therapeutics.

Keywords: multiscale systems biology; computational biology; quantitative systems pharmacology (QSP); immuno-oncology; immunotherapy; immune checkpoint inhibitor; mathematical modeling

1. Introduction

In recent years it has become increasingly evident that studying the tumor microenvironment (TME), in addition to studying cancer cell transformation, is crucial to understanding tumor growth, progression and dissemination. TME is a complex and heterogeneous milieu where cancer cells and stromal cells (including immune cells and other cells resident in the tissue) interact with each other and with the extracellular matrix (ECM), Figure 1. One of the critical elements of the TME is the tumor's interaction with the host immune system. Hanahan and Weinberg described evasion of the immune system as one of the hallmarks of cancer [1]. The importance of the stromal microenvironment in tumor progression was also recognized in the classical paper by Paget [2]. It has become clear that the tumor stromal component, and specifically, the host immune system, contributes to tumor growth, and new therapeutics are now being aimed at altering the immune system as a cancer target (see reviews [3,4]).

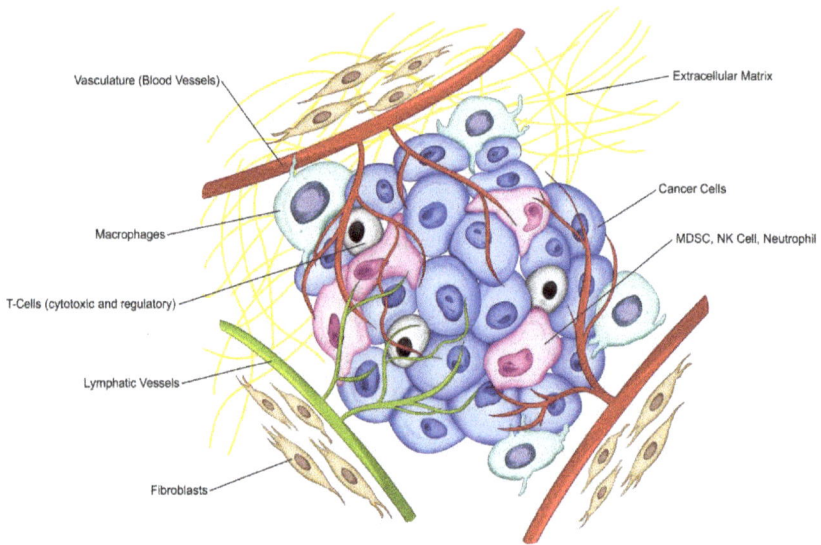

Figure 1. The Tumor Microenvironment (TME). The tumor microenvironment consists of different types of cells (cancer and stromal including immune cells), the extracellular matrix (ECM), and the myriad molecules such as chemokines, cytokines, microRNAs, and growth factors. Cancer cells (including cancer stem cells and progenitor cells), the tumor vessels (blood vessels and lymphatic vessels), immune cells (including tumor-associated macrophages (TAM) and T-cells (cytotoxic and regulatory), myeloid-derived suppressor cells (MDSC), natural killer cells (NK cell), neutrophils and other stromal components (including the extracellular matrix and cancer-associated fibroblasts (CAF)) are shown.

The cellular components of the TME can vary in different regions of the tumor [5], as well as between patients and even between tumors in a single patient [6]. Each of these cellular components has its own behavior in terms of migration, proliferation, differentiation, apoptosis, adhesion, and response to treatment. Cancer is so difficult to treat partly because of this degree of complexity that results in a highly unpredictable tumor behavior, partially due to the complex microenvironment [7], emphasizing the pressing need for personalized treatment for individual patients. Mathematical and computational modeling techniques provide a powerful tool in understanding the TME and predicting cancer progression [8]. Below, we provide a brief overview of the role of the immune system in cancer and introduce computational approaches to study the tumor immune microenvironment (TIME).

2. Immune System Biology and Cancer

The immune system consists of two major parts: the innate immune system, and the adaptive immune system. The innate immune system is the body's immediate defense against foreign antigens. The immunity via the innate immune system is nonspecific and short-lived, whereas the adaptive immune response is a late-stage immune response that is highly specific and can provide long-lasting defense [9]. The innate immune cells have pattern recognition receptors that recognize entities that are non-self and then induce an inflammatory response. The innate immune response is immediate, although some studies suggest these cells also have the capability for memory [10], and may be followed by an adaptive immune response [11]. The adaptive immune response is more specific than the innate immune response and can be antibody-mediated or cell-mediated, with T- and B-cells as the key cell types driving this response [12]. T-cells are a type of lymphocyte that matures in the thymus, with several different subtypes that play a distinct function in immune response [13]. Cytotoxic T-cells kill cancer cells [13], T-helper cells assist other cell types during the immune response, and

T-regulatory cells (Treg) play an important role in immunological tolerance. In the case of cancer, in addition to normal antigens (self), cancer cells express antigens unique to the tumor, which can result in an immune response. The computational models discussed in this review will largely focus on the adaptive immune system.

Although the immune system is well equipped to eliminate abnormal cells, cancer cells have several ways to evade the immune system. For example, cancer cells can become invisible to the immune system by downregulating major histocompatibility complex (MHC) class-I receptors on their cell surface, and in turn, not presenting the mutation associated antigen for detection by T-cell receptors. Cancer cells attract regulatory T-cells (Treg) [14] and myeloid-derived suppressor cells (MDSC) [15] to the tumors; an abundance of these cell types results in an immunosuppressive environment [16]. Furthermore, receptors such as CTLA-4 on Treg can bind to CD80 and CD86 on T-cells and antigen presenting cells (APC), which inhibits co-stimulation of these cells, and results in T-cell suppression. Treg can further inhibit the adaptive immune response by interfering with B-cell function and releasing immunosuppressive cytokines such as IL-10. In fact, under certain conditions, the immune response can contribute to tumor growth instead of inhibiting it [17]. Another way that tumors escape detection is through chronic inflammation [18]. During chronic inflammation, T-cells eventually lose their effectiveness over the course of the infection, called T-cell exhaustion [19]. T-cell exhaustion is also frequently found in the tumor microenvironment through the PD-L1/PD-1 pathway [20]. PD-1 blockage enhances T-cell and NK (Natural Killer) cell activity in tumors [21,22]. Several immune checkpoint inhibitors are currently being used in treatments of patients with cancer [23,24].

Innate immune cells, such as macrophages, neutrophils, and eosinophils, have also been shown to decrease or enhance tumor growth depending on their polarization state. Macrophages have a great deal of plasticity but are usually classified in one of two types: an M1-type that is immuno-enhancing, and an M2-type which is immuno-suppressing; though there is a spectrum of states between M1 and M2 [25]. Studies have shown that high numbers of tumor-associated macrophages (TAM) can lead to a worse clinical outcome [26]. TAM have been shown to promote tumor growth by increasing vascularization, cancer cell migration, cancer cell survival, and immuno-suppression [27]. Macrophages are recruited to hypoxic areas of the tumor [28], and aid in tumor progression [26,29]. Cancer cells recruit macrophages through Colony Stimulating Factor 1 (CSF1), and high CSF1 concentrations are correlated with poor prognoses [30]. Macrophages can be converted to TAM within the tumor by secreted factors, such as c-c chemokine receptor type-2 (CCR2) [31], which causes them to exhibit an M2-like phenotype [32]; this conversion of macrophages leads to a distinct subpopulation [33]. TAM secretion of c-c chemokine ligand type-18 (CCL18) can enable the epithelial-to-mesenchymal transition of breast cancer cells [34,35]. These TAM have also been shown to be associated with invasion, extravasation and metastasis [36–38]. Thus, M2-type macrophages are currently being targeted with therapeutics for tumor treatment [39]. Neutrophils are less abundant in tumors, but they are becoming more recognized for their duel role in the immune response to cancer [40]. Eosinophils are also commonly found within tumors [41], and have been found to enhance T-cell infiltration [42]. These cells also have a duel role in the immune response, and can promote or suppress tumor growth [43]. There is a complex interaction between cancer cells and the immune system. Thus, it is important to understand the conditions in which tumors are eliminated or enhanced by the immune system.

As is evident from the complex mechanisms of immune response and immune evasion described above, modeling the immune system is a challenging task [44]. For the specific case of cancer, immune cells can be found within the TME, the lymphatic system and the lymph nodes, resulting in spatial complexity. Molecular and cellular components themselves are complex and have patient specific features such as unique lymphocyte antigen receptors. In addition, different functions of the immune system occur at different time scales, ranging from minutes to years. For example, intracellular signaling occurs in minutes, whereas memory cells exist on the order of years. Revealing this complexity across different scales as discussed above is very challenging or impossible to achieve in an experimental setting. Thus, computational modeling platforms provide a powerful tool to complement experimental measurements for better understanding of the immune system in cancer. In the following section, we provide an overview of the current computational modeling approaches for the study of cancer.

3. Overview of Computational Modeling Methodologies including Agent-Based Modeling

Computational modeling has provided great insight into studying intra-tumor heterogeneity [45] and the interplay between the tumor and the microenvironment [46]. Modeling has the benefit of providing a quantitative time- and cost-effective means to study the physical and chemical interactions in tumor initiation and growth. Modeling efforts complement experimental platforms by providing an understanding of clonal dynamics and microenvironmental cues over time. There are several ways to classify mathematical/computational models in general, and cancer models in particular. One is deterministic vs. stochastic; another is continuum vs. discrete models. Deterministic models have an end state that does not change as long as the initial conditions remain the same, whereas stochastic models have randomness included, resulting in differences in end states, even with the same initial conditions. Continuum models treat cells as concentrations of cell types, whereas discrete models (such as agent-based or particle models) consider discrete cells; the cell behaviors, including interactions between cells, can be described as deterministic or stochastic. A multiscale setting (called a hybrid model, illustrated in Figure 2) can include both approaches, i.e., the discrete modeling of cells and continuous modeling of molecular species, such as oxygen, growth factors, chemokines, microRNAs, and drugs, but appropriate linking and calibration of such hybrid models should be performed. For continuum-based models, temporal ordinary differential equations (ODE) and spatio-temporal partial differential equations (PDE) have been used to model the immune response in cancer, e.g., [47–53]; here, we focus on discrete agent-based models (ABM) and hybrid models.

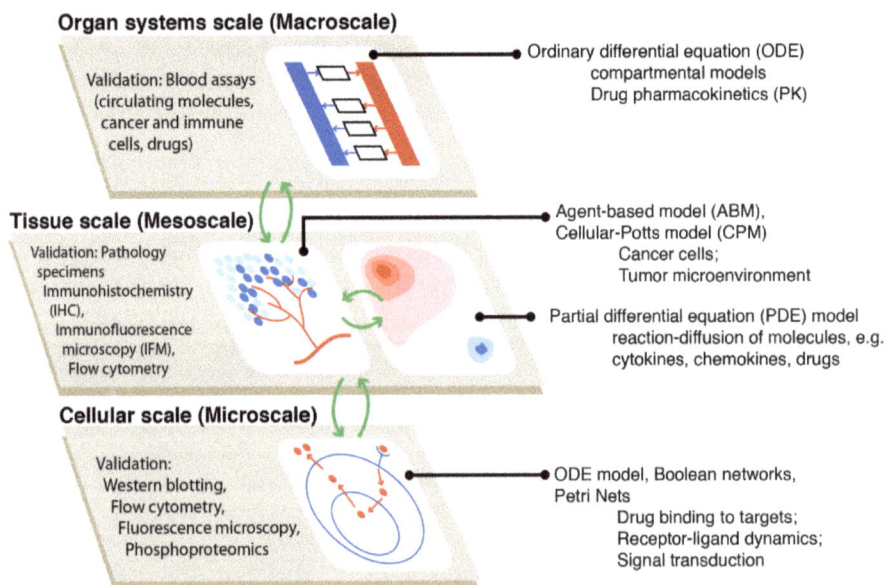

Figure 2. Using hybrid models to study immuno-oncology. While agent-based models are ideal tools to recapitulate the spatio-temporal dynamics of cancer cells and the tumor microenvironment at the tissue scale, the mechanisms at other biological scales can be efficiently embodied using other types of mathematical representations; however, agent-based models (ABM) can also be used at any scale. Such multi-scale hybrid models increase the flexibility in model construction, improve computational performance, and enhance model credibility by allowing comparison between model output and a wide range of experimental and clinical observations.

From another standpoint, the models that describe the immune system can be broadly categorized into top-down and bottom-up, and previous reviews have focused on computational modeling of the immune system [44,54]. The top-down approach models populations of cells, not single entities, and uses the mean behavior at the macroscopic level. ODE and PDE models are examples of this type of modeling where individual interactions are not simulated. Stochastic differential equation models are also a part of this class. On the other hand, the bottom-up approach focuses on the microscopic level. The model tracks each agent (e.g., a cell) and its interactions with the surrounding environment, and emergent behavior arises from all the entities and their local behavior. Features such as stochastic behavior, spatial distribution, and heterogeneity of entities are inherent to bottom up models, and thus, easier to capture with this approach. Drawbacks of these models are that they require more computational power because they track individual agents and their interactions over time and space; also, there are computational limitations on the number of agents that can be considered; it is thus impossible to consider an entire organ or patient. Therefore, both approaches will need to be combined to achieve both spatial cellular and sub-cellular resolutions and whole patient pharmacokinetics and pharmacodynamics.

Agent-based models are an example of a bottom-up approach with applications in immunology and immune related diseases such as cancer [55]. An agent-based model is a discrete mathematical and computational framework that is capable of capturing emergent behavior of its interacting agents, defined as large-scale spatio-temperal patterns resulting from local spatial interactions between agents. The behavior and function of these agents are driven by the information they sense in their local environment and the rules of the agent-based model. Some of the characteristics that separate agent-based models apart from other rule-based modeling systems (in which outcomes are based on a set of rules that govern decisions) are that (1) they are spatial, (2) they incorporate agents that interact with other agents and their environment, (3) they may incorporate stochasticity, (4) they are modular, and (5) they produce emergent behavior [56]. These models allow the individual agents to adapt to their local environment (i.e., agents are adaptive instead of reactive), and take part in local interactions with other agents [57]. These result in complex aggregate behavior stemming from simple rules and emergent properties from agent interactions. Agent-based models can be lattice-based or lattice-free, depending on whether agents reside and move on a (regular or irregular) spatially discretized lattice, or have their locations and velocities represented by continuous variables, usually governed by forces in the environment. For example, in lattice-based ABM, agents are placed on a lattice structure that defines the locations of cells and their neighbors for cellular interactions. There are several model types that, although they are not explicitly characterized as agent-based, are reviewed here for completeness; those include cellular automata, Potts models, and Petri net models.

Agent-based models are particularly suitable for capturing spatially-varying events and heterogeneities [58], and for understanding the immune system's function. With this aim in mind, several investigators have developed agent-based models of diseases with involvement of the immune system. Several ABM have simulated the immune system's involvement in maintaining homeostasis and disease conditions, such as bacterial infections [59], fungal infections [60], abnormal systemic inflammatory response [61], ulceration [62], allergens [63], ischemia [64], tuberculosis [65], sepsis [66], and wound healing [67]. For cancers, such models include tumor growth and invasion [68], as well as specific cancer types such as hepatocellular carcinoma [69], breast cancer [70], melanoma [71], colorectal [72], lung cancer [73,74], and metastasis [75]. Software packages have been developed based on the ABM framework to study the immune system; these include ImmSim [76–79], Immunogrid [80,81], Simmune [82], Cycell [83], and PhysiCell [84].

Now we discuss the latest agent-based and hybrid models that investigate the effects of the immune system on cancer progression and immunotherapy, see Table 1. In order to be included in the review, the work needs to have an immune component, a tumor component, and include ABM. We limited our focus to papers that were published within the last ten years. However, we also included some studies on diseases other than cancer where we feel that the methodology is relevant and could be applied to cancer; we also refer to a few general software tools that can be readily adapted to cancer.

We break the review into the following sections, although there may be significant overlap:

(1)   Models focusing on immune-related tumor mechanobiology
(2)   Models focusing on tumor-associated vasculature in the immune response
(3)   Models focusing on tumor-associated lymphatics and lymph nodes
(4)   Models focusing on tumor immunotherapy
(5)   Models focusing on tumor-enhancing immune cells
(6)   Models focusing on intra-tumor heterogeneity

3.1. Models Focusing on Tumor Mechanobiology

Changes in the tumor extracellular matrix (ECM) have been known to contribute to tumor progression and metastasis [85], with several computational models focusing on investigating glioma invasion [86–88], but less is known about its contribution to immune response. Computational modeling has been used to shed light on the interactions between the ECM and the immune system in cancer dynamics. A hybrid agent-based model was used to investigate the role of cellular adhesion to the ECM in tumor and immune system dynamics [89]. Frascoli et al. found that the greater the motility of the cancer cells, the more likely they will escape from immunotherapy. They also found that intermediate levels of adhesion in general led to less successful outcomes, but these results were variable.

Kather et al. used ABM to investigate the combination of adoptive cell transfer and therapy that permeabilized the fibrotic stromal component in colorectal cancer [72]. Adoptive cell transfer is a therapeutic strategy that aims to increase the number of immune cells to strengthen immuno-surveillance and counter tumor development. Kather et al. simulated various conditions of immune surveillance. In their model, T-cell killing of tumor cells occurred in a purely stochastic manner, with killing probability representing the effect of tumor specificity, immunogenicity, stimulatory and inhibitory effects all in one parameter. An immune rich environment promoted immune escape, but tumor growth slowed in a lymphocyte deprived environment. Tumor control was observed in a subgroup of tumors with less stroma and a high numbers of immune cells. They found that high levels of fibrosis and low numbers of lymphocytes reduced overall survival. Their findings were validated with data from colorectal cancer patients, where low density stroma and high lymphocyte level correlated with better overall survival. In this study, Kather et al. simulated the effect of immunotherapy by boosting the number of immune cells by 2–8 fold. Therapy was intended to enhance fibrotic stromal permeability; this was implemented by modifying the corresponding parameter by a factor of 4% to 16%. The model predicted that optimal tumor eradication requires a combination of therapeutics aiming at both activating adaptive immune system and stromal depletion.

3.2. Models Focusing on Tumor-Associated Vasculature in the Immune Response

Tumor-associated vasculature is an important aspect of the tumor-immune complex because it not only provides oxygen and nutrients for the tumor to grow, but it is also the source of tumor dissemination via circulating tumor cells (CTC), and recruitment for many immune cells, such as monocytes/macrophages and T-cells. Studies have aimed to provide a better understanding of these processes [90,91]. An ABM of Early Metastasis (ABMEM) framework was used to model the interactions between tumor cells, platelets, neutrophils, and endothelial cells [92]. Receptor binding to Mac-1 (macrophage antigen-1) by endothelial cells, platelets, or tumor cells leads to reactive oxygen species (ROS) production by neutrophils. Uppal et al. examined two types of platelet inhibition: inhibition of thromboxane, inhibition of adenosine diphosphate (ADP) receptors and inhibition of both [92]. They found that thromboxane inhibition alone resulted in the best outcome.

Alfonso et al. developed an agent-based model of immune cell-epithelial cell interactions in breast lobular epithelium [93]. The model investigated the effect of menstrual cycle length and hormone status on inflammatory response to cell turnover in breast tissue. Blood vessels were homogeneously distributed in the intra- and interlobular stroma. The model accounts for myoepithelial and luminal cells. Cellular processes (i.e., epithelial cell proliferation, cell death via effector cells, programmed cell death, removal of dead cells, immune cell motility, and inhibition of effector cells by regulatory cells) are modeled as stochastic events. Effector CD8+ cells are the only cells responsible for killing of damaged epithelial cells. Regulatory CD4+ and CD8+ cells act by inducing inactivation of effector dependent response. Chemokines from damaged epithelial cells activate the immune cells. Immune cells become ineffective when such chemokines are absent, or due to the suppression via regulatory cells. The outcome of the model identified novel prognostic information for breast cancer, such as the number of immune clusters being associated with the degree of epithelial damage.

3.3. Models Focusing on Tumor-Associated Lymphatics and Lymph Nodes

Reddy developed the first mathematical model of the lymphatic system in 1977 [94]. In recent years, various computational modeling approaches have been used to study the lymphatic vessels [95–99] and lymph nodes [100–102] in general, and in the application to infectious disease [103,104], and simulating fluid and chemokine transport in the lymphatic system as it relates to health and disease conditions [105]. Agent-based models have more recently been used to simulate various processes that occur during an adaptive immune response in a lymph node. Meyer-Hermann developed an ABM of germinal centers of the lymph node [106,107]. The authors studied B-cell germinal center reactions and how they contribute to germinal center deregulation [106]. They expanded this model to study B-cell affinity maturation in the lymph node germinal centers [108]. They found that competition for T-cell rescue and increased refractory time leads to a more robust affinity maturation.

A series of studies on modeling of T-cell behavior in the lymph node have been conducted by Bogle and Dunbar [44,109–112]. They modeled T-cell trafficking, activation, and proliferation in the lymph node paracortex using an agent-based approach. The model included chemokine and cytokine gradients. Using this lattice-based approach, they were able to model the movement and behavior of T-cells in the lymph node paracortex [112]. In the next step, they expanded the agent-based model of the lymph node paracortex in three dimensions to include T-cells and dendritic cells (DC). The model allows simulation of a large number of T-cells at physiologic densities. The virtual lymph node can shrink or swell, depending on the dynamics of cell trafficking. The model was able to simulate T-cell activation in agreement with in-vivo observations, and provide new understanding on T-cell-DC interactions. Not all the parameters of the model were experimentally measured; thus, the model can be refined by more accurate measurement of those parameters [110]. Next, the authors built on their previous models to simulate T-cell ingress and egress, as well as chemotaxis in the lymph node, by incorporating new numerical methods. The new model allows simulation of expansion and contraction of T-cells in the lymph node paracortex during an immune response. The ability to model chemotaxis could be useful in studying other biological processes involving chemotaxis [111].

Moreau et al. constructed a virtual lymph node using agent-based modeling to study T-cell activation by synapses (long-lasting contacts) and kinapses (transient interactions) [113]. The model incorporated T-cell migration and T-cell-DC interactions. Additionally, virtual fluorescence-activated cell sorting (FACs) profiles were obtained from modeling by visualizing T-cell proliferation. This virtual lymph node model provides new opportunities for understanding the mechanisms of T-cell regulation in infection or vaccine application [113].

The ABM developed for studies of the lymphatic system thus far mainly focus on the lymph node, and in particular, T-cell processes. Folcik et al. developed the basic immune simulator (BIS), which is an agent-based platform that includes parenchymal tissue, secondary lymphoid tissue and the lymphatic/humoral circulation [114]. Using agent-based and hybrid models, lymph node dynamics are studied in the context of infectious diseases and cancer. Kim and Lee used a hybrid model to study the efficacy of preventative cancer vaccines. The model comprised two compartments for interactions of tumor and immune cells at the tissue site and in the draining lymph nodes [115]. Jacob et al. developed a three-compartment ABM that includes lymph nodes, blood vessels, and organ/tissue. The model was used to study immune response against viruses in these compartments [116]. Marino et al. developed a hybrid model where the lymph node and blood compartment were simulated using ordinary differential equations and the lung compartment was simulated using agents. They focused on the formation of granulomas in the lung, which are organized structures of immune cells in the lung, and are a hallmark of infection. The model focused on the recruitment of APCs in the lymph node from the lung for *Mycobacterium tuberculosis* (Mtb, the causative bacterium of TB) infection [117]. In another study, they investigated the role of DC in Mtb infection [118]. The growth and dissemination of bacteria were highly affected by CD8+ and CD4+ T-cell proliferation rates and DC migration. Such multiscale models allow the study of tissue level dynamics during adaptive immune response [118], and although they focus on infectious disease, many of the components and processes involved in anti-cancer immunity and adaptive immunity against infection are shared. For example, T-cells specific to tumor antigens are primed and expanded in a similar fashion to that in which T-cells specific to foreign antigens are during their response to infection; the immune suppressive mechanisms that cancer cells hijack to evade immune surveillance are also deployed during an immune response against infection to prevent excessive tissue damage. Since the body reacts similarly in response to an infection as it does in response to cancer (e.g., activation of similar signaling pathways), cancer models can heavily borrow from this literature.

3.4. Models Focusing on Tumor Immunotherapy

A variety of cancer immunotherapy strategies exist that range from boosting the overall immune response to specifically targeting cancer immunity. Some examples of immunotherapies are treatment vaccines, adoptive cell transfer, and immune checkpoint inhibitor treatments. Agent-based and hybrid models are developed to help understand these therapies when applied separately or in combination with other cancer treatments. One type of therapy that has been explored is cancer vaccines. Therapeutic cancer vaccines treat existing cancers by delivering immunogenic and tumor specific antigens to the patient to induce cellular and/or humoral anti-tumor immunity. Pennisi et al. have developed several hybrid models investigating the immune system effects on tumors. They developed a hybrid model to study the development of lung metastases from mammary carcinoma [75]. Pennisi et al. also developed a hybrid model MetastaSim to simulate the protection against lung metastases in mouse using Triplex cell vaccine [73]. In this simulation, macrophages could capture tumor-associated antigen and immunocomplexes, breaking them down and eliminating them from the system. This vaccine elicited a combination of three stimuli: the p185neu antigen expressed by the HER2/neu gene, allogeneic major histocompatibility complex (MHC) molecules, and IL-12 which enhances antigen presentation. Using this model, after calibration and validation, the authors were able to evaluate different protocols of vaccine administration. The simulation results suggested that in order to maximize protection while reducing the number of administrations, the vaccination strategy should include a significant dosage early on and a few recalls afterwards.

Dreau et al. developed an ABM model of solid tumor progression to understand the interplay between solid tumor growth, tumor vascular growth, and the host's immune system [119]. The model includes tumor and immune cells, vasculature, tumor cell proliferation, and immune system response. Their model supported immunotherapy as an effective cancer treatment in individuals with functioning immune systems. They concluded that a strong immune response limits tumor growth in a way that cannot be achieved under a weaker immune response. Another study focused on the role of T-cells in the effectiveness of response to immunotherapy in B-16 melanoma [120]. The model includes macrophages, DC, tumor vasculature, and interactions between these components. It was found that early entry of T-cells effectively eliminated the tumor and was dependent on CD137 (a co-stimulatory protein that helps in tumor rejection [121]) expression in tumor vasculature.

Oncolytic virus therapy is a strategy that utilizes viral infection to kill cancer cells, but not normal cells, with the potential of enhancing T-cell recruitment to the tumor and increasing their access to cancer cells. Several computational models have examined the conditions of success for this type of therapeutic in silico [122]. Walker et al. developed an agent-based model of pancreatic tumors to study the synergy between chimeric antigen receptor (CAR) T-cell therapy and oncolytic virus therapy [123]. CAR T-cell therapy is one type of adoptive cell transfer treatment involving genetically engineered T-cells specifically targeting cancer cells, and has been the subject of several computational models [124]. The agent-based model recapitulates treatment mechanisms including cancer specific CAR T-cell recruitment to the tumor site via vasculature and the injection and spread of oncolytic virus. Rohrs et al. demonstrated the ability of the model to track the dynamics of cancer cells and stromal cells in space in the presence of the treatment combinations; optimization of the combination therapy requires more accurate calibration [124].

Immune checkpoint inhibitors are used in cancer immunotherapy that enhances anti-tumor immune response by targeting cancer immune evasion mechanisms. In many cancer types, tumor neoantigens are sufficiently immunogenic to promote the expansion of antitumor immune cells [125]; however, these immune cells are not functional due to the inhibitory signals from molecules adaptively induced during cancer development [24,126]. Among them, one of the most prominent mechanisms is PD-1/PD-L1 interaction, where T-cells are suppressed through PD-1 signaling upon contact with induced PD-L1 in the tumor microenvironment. Gong et al. developed an ABM of tumor-immune interaction in 3D to study the spatio-temporal dynamics of cancer cells and cytotoxic T-cells [127]. In this study, the inhibitor to the checkpoint molecules were modeled as a factor which modulates the parameter governing the suppression of tumor specific T-cells by PD-L1+ cancer cells. They found that patient responsiveness to such therapy could be associated with the level of mutational burden of the cancer and antigen strength among patients. They also found that tumor growth is insensitive to the vascular density of the tumor core. From these results, a scoring method was proposed to predict anti-PDL1 treatment efficacy in patients.

3.5. Models Focusing on Tumor-Enhancing Immune Cells

While the immune system has evolved to kill off tumor cells, there are many ways in which cancer cells can avoid immune detection. In addition, there is mounting evidence that immune cells can stimulate tumor growth under certain conditions. Several agent-based models have focused on understanding the tumor-enhancing contributions of the immune system. Enderling and colleagues explored the interactions between tumor cell death and the immune system using a cellular automata model focused on the interplay between cancer stem cells and the immune system [128]. They showed that immune system-induced tumor cell death led to stem cell selection, and thus, more aggressive tumors [128,129]. In this model, even though immune cells effectively killed off tumor cells, they also affected progenitor cells. This resulted in the creation of a space for cancer stem cells to proliferate and produce more cancer stem cells. This ultimately resulted in a larger stem cell population and a more aggressive tumor.

Several studies have specifically focused on the tumor-enhancing contribution of immune cells, such as tumor-associated macrophages (TAM). Macrophages are one of the most abundant immune cells found in tumors, but their population is heterogeneous [130]. M1-type macrophages have been shown to be tumor inhibiting, whereas M2-type macrophages have been shown to be tumor enhancing [26]. One model looked at the transition from the M1 to M2 macrophage phenotype on tumor growth and then predicted targeted therapies [131]. Knútsdóttir et al. used a hybrid model to investigate epidermal growth factor (EGF) and macrophage colony-stimulating factor 1 (CSF-1) signaling between macrophages and cancer cells during macrophage aggregation [132]. They found that CSF-1/CSFR1 autocrine signaling affects the ratio of tumor cells to macrophages during tumor growth. In a further study, they found that the macrophage/tumor cell ratio was most sensitive to the strength of EGF signaling, but usually maintained a 1:3 ratio [133].

Another ABM of triple-negative breast cancer examined the tumor enhancing effects of macrophages [134]. Norton et al. investigated the interplay between tumor growth, blood vessel recruitment and macrophage recruitment through tumor vasculature. They observed that while macrophages increase tumor growth, excessive macrophage recruitment conversely leads to a decrease in tumor growth due to the inhibition of proliferation resulting from overcrowding.

3.6. Models Focusing on Intra-Tumor Heterogeneity

Intra-tumor heterogeneity and the characteristics of the tumor microenvironment are found to have important implications in the outcome of disease progression [135]. Patients often have varied responses to treatment because each patient is unique in their genome, microbiome, disease history, lifestyle, and environment. The case of tumors is especially complex, because this heterogeneity is observed not only between tumors, but also between subpopulations of cells from the same tumor, resulting in different response to drugs [136]. While capturing this degree of heterogeneity may be difficult in experiments and clinical trials, especially the temporal dynamics of spatial heterogeneity, computational models are especially suited to tackling this challenge. This section focuses on the models that have aimed to capture intra-tumor heterogeneity.

A 2D agent-based model was used to study the interactions between an avascular tumor and immune cells (NK cells and cytotoxic T-cells) [137]. They examined the effects of cancer cell proliferation on overall tumor growth under two conditions: the first, where cancer cells do not consider the microenvironment when deciding when to proliferate, and the second, where they proliferate based on the number of healthy cells surrounding them. Tumor-immune cell interaction can have three outcomes: tumor cell death, immune cell death, or no cell death based on the state of the tumor. The predicted growth of the tumor was then compared to a xenograft tumor growth. Spatial heterogeneity was also examined in a different model where cancer cells use glycolysis instead of oxidative phosphorylation to increase their energy production. In order to study how this increased energy production affects the surrounding stroma, a combination of computational modeling and in vitro/in vivo experiments was used [138]. They used agent-based modeling to understand tumor growth in a vascularized area of the tumor. They found that tumors develop spatial patterns where macrophages and tumor cells coexisted in areas with high levels of oxygen, but that only tumor cells survived in ischemic regions. They then used an in vitro tissue-mimetic system to create the directional gradients for oxygen and lactate, which also allowed for the co-culture of tumor cells and macrophages.

Figueredo et al. created a series of hybrid models to study the interplay between the immune system (including macrophages) and tumor cells [139,140]. An agent-based on-lattice model for tumors was created using Chaste (Cancer, Heart and Soft Tissue Environment), part of the Virtual Physiological Human (VPH) Toolkit; the model consists of three layers: a diffusible layer, a cellular layer, and a subcellular layer [141]. The diffusible layer consists of diffusible species such as oxygen, the cellular layer consists of normal cells, tumor cells, and macrophages, and the subcellular layer governs apoptosis and cell-cycle in each cell. In this model, macrophages were M1-like, they supported the immune system, and aided immunotherapies. They investigated the growth of the tumor under oxygen-dependent proliferation. They found that the emergent behavior of agent-based models allowed for the generation of additional tumor architectures over other modeling methodologies [142].

Table 1. Summary of Section 3: ABM and hybrid models discussed in each section.

3.1. Models Focusing on Immune-Related Tumor Mechanobiology		3.2 Models Focusing on Tumor-Associated Vasculature in the Immune Response		3.3 Models Focusing on Tumor-Associated Lymphatics		3.4 Models Focusing on Tumor Immunotherapy		3.5 Models Focusing on Tumor-Enhancing Immune Cells		3.6 Models Focusing on Intra-Tumor Heterogeneity	
Study	Ref	Study	Ref	Study	Ref	Study	Ref	Study	Ref	Study	Ref
Cellular adhesion to ECM	(Frascoli et al. 2016) [89]	Early metastasis	(Uppal et al. 2017) [92]	Germinal centers of LN	(Meyer-Hermann et al. 2002, 2005) [106–108]	Lung met in mammary carcinoma	(Pennisi et al. 2009) [73]	Cancer stem cell-immune cell interaction	(Hillen 2013) (Enderling 2012) [128,129]	Tumor, NK cell, cytotoxic T-cell interactions	(Pourhasanzade et al. 2017) [137]
Adoptive cell transfer in colorectal cancer	(Kather et al. 2017) [72]	Immune-epithelial cell interactions in breast epithelium	(Alfonso et al. 2016) [93]	T-cell behavior in LN	(Bogle et al. 2010, 2012, 2008) [108–113]	Effect of vaccine on lung metastasis	(Pennisi et al. 2010) [75]	Effect of M1 and M2 macrophages on tumor growth	(Wells 2015) [131]	Effect of stroma on tumor spatial patterns	(Carmona-Fontaine et al. 2013) [138]
				T-cell activation in virtual LN	(Moreau, 2016) [113]	Immunotherapy in solid tumors	(Dréau et al. 2009) [119]	Signaling between macrophages and cancer cells	(Knútsdóttir et al. 2014, 2016) [132,133]	Immune cell, macrophage, tumor cell interactions	(Figueredo 2011, 2013) [139,140]
				Model of LN to study cancer vaccines	(Kim et al. 2009) [114]	Role of T-cells in response to immunotherapy	(Pappalardo et al. 2011) [120]	Effect of macrophages on TNBC tumor growth	(Norton et al. 2018) [134]	Tumors under oxygen-dependent proliferation	(Figueredo 2013, 2014) [141,142]
				Immune response against viruses	(Jacob et al. 2011) [116]	Effect of different therapies on pancreatic tumors	(Walker et al. 2016) [123]				
				Recruitment of APCs in the LN from lung	(Marino et al. 2011) [117]	Spatio-temporal dynamics of tumor-immune cell interactions	(Gong et al. 2017) [127]				
				T-cell trafficking and proliferation	(Marino et al. 2016) [118]						

4. Discussion and Emerging Applications

The immune system is made up of many interacting components that together drive a complex spatio-temporal behavior during immune response. Thus, agent-based modeling is particularly suitable for understanding the immune systems function in health and in disease conditions such as cancer. Here, we reviewed the latest agent-based and hybrid models that investigate the contributions of the immune system to cancer growth and the effect of immunotherapy. In this context, we focused on models of immune-related tumor mechanobiology, tumor-associated vasculature, tumor-associated lymphatics, tumor immunotherapies, tumor-enhancing immune cells, and finally, models focusing on intra-tumor heterogeneity. Overall, ABM can generate novel hypotheses to be validated and refined by future experiments. Development and refinement of multiscale agent-based models along with experiments through an iterative process can improve our understanding of biological processes in cancer and lead to the identification of novel prognostic and predictive biomarkers that can improve therapies and help design and interpret the results of clinical trials [143].

Models investigating the tumor-enhancing effects of the immune system can provide useful insights into managing tumor-immune interactions. Since the tumor microenvironment can be very heterogeneous, care must be taken to appropriately model cell-cell interactions between cancer, stromal, and immune cells, the extracellular matrix, and the secreted factors. Accurate data from in vitro and in vivo experiments must be used to understand the transition from tumor-inhibiting to tumor-enhancing immune cell types. In addition, since immune cells such as macrophages and T-cells are usually recruited to the tumor by secreted factors, an evolving tumor vasculature is necessary to accurately model these processes. Agent-based models of the tumor enhancing effects of the immune system can help us better understand how to prevent or revert the immune system back into a tumor-inhibiting phenotype. Thus, these models will help improve immunotherapies for cancer treatment.

One limitation of the immunotherapy studies mentioned above is that although the models are quantitative in many aspects, the modules governing drug delivery and response is relatively qualitative or semi-quantitative in nature. This could potentially be resolved by combining spatio-temporal agent-based models with traditional model types, such as Physiologically Based Pharmacokinetic (PBPK) models to track drug distribution in different physiological compartments, and pharmacodynamic (PD) models for individual cellular agents to represent effects of drugs on target cells [144]. Such hybrid quantitative systems pharmacology (QSP) models can be utilized not only as a platform for basic science research, but also as a potential complement to the clinical research and drug discovery pipeline. A schematic of such a hybrid model based on the research in our own group is presented in Figure 3. Compared with continuous models, the discretely represented agents allow the flexibility to track intra-tumor heterogeneity, such as tumor neoantigen profile, T-cell clonality and local expression of immune checkpoint molecules with preferred levels of granularity. By running multiple simulations in parallel using different parameter values and initial conditions accounting for genetic background and environmental exposure, the models can represent cohorts of patients with desired population scale heterogeneity. These properties render agent-based and hybrid models a powerful platform for conducting virtual clinical trials.

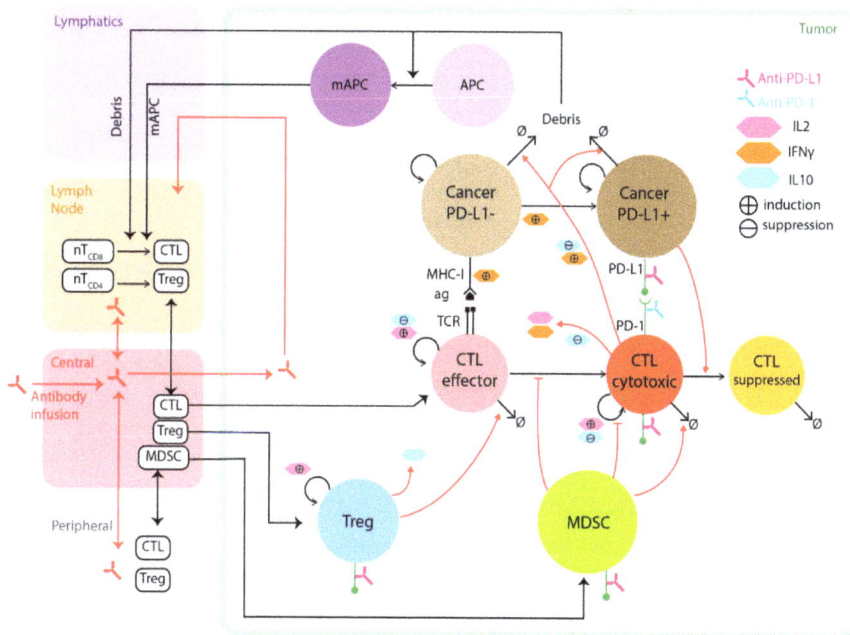

Figure 3. Diagram of a multi-compartment hybrid model capturing tumor development and anti-tumor immune response. Dynamics of cells and pharmacokinetics of drug (e.g., antibody) in the lymphatics, tumor-draining lymph node, central (blood) and peripheral compartments are modeled using ordinary differential equation systems. Spatial dynamics of cells and molecules in the tumor compartment are captured using agent-based model and partial differential equations. Death of cancer cells produces antigens which drive maturation of APC and their migration to the tumor draining LN, where CD8+ and CD4+ T-cells go through priming and proliferation before they enter blood circulation and extravasate to the tumor microenvironment. Effector CD8+ T-cells can be further activated and expanded when they encounter tumor antigens. These cytotoxic cells kill cancer cells and also release various cytokines, including IL2 which drives further proliferation of T-cells, and IFNγ which is proinflammatory and induces PD-L1 expression on cancer cells. PD-L1 can then bind to PD-1 molecules on cytotoxic T-cells, resulting in T-cell exhaustion. Both PD-L1 and PD-1 molecules are potential targets for immune checkpoint blockade antibodies. Regulatory cell types in the ABM include Treg and MDSC, which can inhibit cytotoxic T lymphocytes (CTL) through different mechanisms.

Computer models provide large-scale predictive power by allowing us to simulate clinical trials with sufficient details to study response to various conditions. Using these models, it is possible to test and predict drug failures in simulations rather than in patients, which could result in improved drug design, reduced risks and side effects, and can dramatically decrease costs of drug development. Importantly, models can predict how the immune-tumor system evolves during the course of the treatment [136]. The challenge is that available data for individual patients is limited. To address this problem, machine learning approaches can be used to build statistical models based on available patient data, and these models can be employed to simulate virtual populations to predict the effects of therapies [145]. These approaches have already been expanded to identify biomarkers and find important mutations that affect response to treatment with drugs in cancer cell lines [146–148].

Mechanistic models are another suitable approach that provides large-scale predictive capabilities based on the available information on the interactions between various components of a biological system. These models can be discrete or continuous. A certain type of model is chosen based on the application and availability of the data [149,150]. Depending on the type of the model, predictions can be made for the behavior of the signaling system in a qualitative, semi-quantitative or quantitative manner. For example, for quantitative predictions of signaling and regulatory gene networks, continuous variables need to be modeled on continuous time scales using ODEs [151].

Such detailed modeling requires knowledge of important biological reactions at every step. This can only be achieved in several iterative steps that include the implementation of various components such as signaling events and defining values for related parameters and appropriate initial conditions. In recent years, numerous models have been developed that simulate individual signaling pathways [152–154]. The challenge is that these models often do not fully capture crosstalk mechanisms that are crucial in predicting patient response to treatment, as each drug perturbs multiple biological processes. ODE-based models can be combined with agent-based models to capture the dynamics of the system being modeled in a more complex fashion. Stochasticity is one of the main advantages of agent-based models, as it applies to biological processes [155]. In comparison to ODE models that make predictions of concentrations and other events over time, ABM allows the study of each agent, as it interacts with other agents in their proximity and the ways in which that affects the large-scale behavior. These models, however, are computationally more expensive. There are also challenges in validating results from ABM due to insufficient spatio-temporal data on tumor development [156].

One aspect that is typically missing from mechanistic knowledge-based models including QSP and ABM is an input from high-throughput data, genomic or proteomic; such data can inform the models and can supplement the data obtained at the cellular and tissue levels [157]. Examples include immune landscape information from different sources including patients' databases such as TCGA (The Cancer Genome Atlas) [33,158,159]. Another source of "Big Data" for parameterization and validation of the models, including ABM, will be the emerging methodologies of digital pathology, such as using multiplex immunofluorescence (mIF) of patients' biopsies and resected tumors, with subsequent analysis of cellular and molecular spatial patterns. Steps in this direction are already underway [160]. Another source of data is image-based, using microCT, confocal, multiphoton, and super resolution microscopy, both ex vivo and in vivo; examples include imaging entire tumor vasculature with subsequent computer simulation of blood flow and molecular transport [161,162].

In addition to modeling approaches used to simulate response to treatment, virtual patients are a key component of virtual clinical trials. A virtual population has the characteristics of the original patient population but also includes individual diversity, usually comprising parameter sets weighted by a clinical or response distribution [163]. This diversity allows testing a broad range of responses that can be missed in a clinical trial. In contrast to traditional clinical trials that can only be performed after costly and lengthy development, in silico trials could be performed at every stage of the drug development. In silico design of treatments can be conducted with data-driven or mechanistic-based (knowledge-based) approaches [164]. It should be noted that in silico clinical trials require integration of data at different scales via a multi-model approach using virtual patients. Similar to a traditional clinical trial, rigorous statistical approaches are needed at various steps of virtual clinical trials. The ability to test treatment via in silico clinical trials can significantly reduce the cost and increase the efficacy of drug development.

The main challenge in the way of predictive models for virtual clinical trials is the availability of input data for the model for each patient. Detailed knowledge about the situation at the start of the simulation can significantly affect the predictive power of that model. Such input information is being generated at a growing rate and a lower cost. Furthermore, proteomic data enable the modeling of interactions of different subgroups of cells from the same tumor with each other as well as immune cells and other stromal cells, allowing modeling tumors for individual patients. Computational models can make predictions for the optimal treatments, making it safer, faster, and cheaper to complement current clinical trials. These models will improve by continuous comparison of predicted and actual response to therapy. Additionally, as more detailed information on biological parameters and disease mechanisms become available, the accuracy of the models will increase.

Currently, only 1 in 10s of clinical trials results in drugs that make it to the market [165]. The process takes 10–12 years, costing billions of dollars, sometimes with low effectiveness when used by real patients [136]. Although virtual clinical trials with virtual patients and virtual cohorts cannot replace clinical trials, they can inform design of such trials to improve success rates and increase the efficacy of the process of drug development. Virtual trials can also better address the need for personalized therapy [166]. Finally, combining agent-based models and data-driven artificial intelligence (AI) methods (e.g., machine learning, including deep learning), we can better understand the gap between preclinical findings and clinical outcomes.

In summary, in silico modeling and specifically agent-based modeling are powerful tools of cancer systems biology and cancer immune systems biology. Combined with novel measurement methodologies and increasing amounts and sophistication of data available from clinical trials, they should bring a better mechanistic understanding and predictive capabilities of therapeutic interventions in cancer, including immunotherapies. The field is ripe for conducting predictive virtual clinical trials as a prerequisite to clinical trials in patients.

Author Contributions: All authors conceptualized the work, participated in draft preparation, and editing and finalizing the manuscript. K.A.N., C.G., and S.J. contributed equally to the paper.

Acknowledgments: This work was supported by the National Institutes of Health grants R01CA138264, U01CA212007 and R01CA196701 and American Cancer Society postdoctoral fellowship PF-13-174-01-CSM (KAN). The authors thank Amanda Figueroa for her expert drawing of Figure 1. The authors thank Mohammad Jafarnejad and Richard Sove for critical comments on the manuscript.

Conflicts of Interest: The authors declare no conflict of interest.

References

1. Hanahan, D.; Weinberg, R.A. Hallmarks of cancer: The next generation. *Cell* **2011**, *144*, 646–674. [CrossRef] [PubMed]
2. Paget, S. The distribution of secondary growths in cancer of the breast. *Lancet* **1889**, *133*, 571–573. [CrossRef]
3. Wei, S.C.; Duffy, C.R.; Allison, J.P. Fundamental mechanisms of immune checkpoint blockade therapy. *Cancer Discov.* **2018**, *8*, 1069–1086. [CrossRef] [PubMed]
4. Chen, D.S.; Mellman, I. Elements of cancer immunity and the cancer-Immune set point. *Nature* **2017**, *541*, 321–330. [CrossRef] [PubMed]
5. Quail, D.F.; Joyce, J.A. Microenvironmental regulation of tumor progression and metastasis. *Nat. Med.* **2013**, *19*, 1423–1437. [CrossRef] [PubMed]
6. Ansell, S.M.; Vonderheide, R.H. Cellular composition of the tumor microenvironment. *Am. Soc. Clin. Oncol. Educ. B* **2013**, *33*, e91–e97. [CrossRef]
7. Pitt, J.M.; Marabelle, A.; Eggermont, A.; Soria, J.C.; Kroemer, G.; Zitvogel, L. Targeting the tumor microenvironment: Removing obstruction to anticancer immune responses and immunotherapy. *Ann. Oncol.* **2016**, *27*, 1482–1492. [CrossRef]
8. Crespo, I.; Coukos, G.; Doucey, M.; Xenarios, I. Modelling approaches to discovery in the tumor microenvironment. *J. Cancer Immunol. Ther.* **2018**, *1*, 23–37.
9. Netea, M.G.; Joosten, L.A.B.; Latz, E.; Mills, K.H.G.; Natoli, G.; Stunnenberg, H.G.; O'Neill, L.A.J.; Xavier, R.J. Trained immunity: A program of innate immune memory in health and disease. *Science* **2016**, *352*, aaf1098. [CrossRef]
10. Netea, M.G.; Quintin, J.; Van Der Meer, J.W.M. Trained immunity: A memory for innate host defense. *Cell Host Microbe* **2011**, *9*, 355–361. [CrossRef]
11. Kumar, H.; Kawai, T.; Akira, S. Pathogen recognition by the innate immune system. *Int. Rev. Immunol.* **2011**, *30*, 16–34. [CrossRef]
12. Gajewski, T.F.; Schreiber, H.; Fu, Y.-X. Innate and adaptive immune cells in the tumor microenvironment. *Nat. Immunol.* **2013**, *14*, 1014–1022. [CrossRef]
13. Dykes, S.S.; Hughes, V.S.; Wiggins, J.M.; Fasanya, H.O.; Tanaka, M.; Siemann, D. Stromal cells in breast cancer as a potential therapeutic target. *Oncotarget* **2018**, *9*, 23761–23779. [CrossRef] [PubMed]

14. Chaudhary, B.; Elkord, E. Regulatory T cells in the tumor microenvironment and cancer progression: Role and therapeutic targeting. *Vaccines* **2016**, *4*, 28. [CrossRef]
15. Fleming, V.; Hu, X.; Weber, R.; Nagibin, V.; Groth, C.; Altevogt, P.; Utikal, J.; Umansky, V. Targeting myeloid-derived suppressor cells to bypass tumor-induced immunosuppression. *Front. Immunol.* **2018**, *9*, 398. [CrossRef]
16. Beyer, M.; Schultze, J.L. Regulatory T cells in cancer. *Blood* **2006**, *108*, 804–811. [CrossRef] [PubMed]
17. De Visser, K.E.; Eichten, A.; Coussens, L.M. Paradoxical roles of the immune system during cancer development. *Nat. Rev. Cancer* **2006**, *6*, 24–37. [CrossRef] [PubMed]
18. Palucka, A.K.; Coussens, L.M. The basis of oncoimmunology. *Cell* **2016**, *164*, 1233–1247. [CrossRef]
19. Blank, C.; Mackensen, A. Contribution of the PD-L1/PD-1 pathway to T-cell exhaustion: An update on implications for chronic infections and tumor evasion. *Cancer Immunol. Immunother.* **2007**, *56*, 739–745. [CrossRef]
20. Jiang, Y.; Li, Y.; Zhu, B. T-cell exhaustion in the tumor microenvironment. *Cell Death Dis.* **2015**, *6*, e1792. [CrossRef]
21. Seifert, A.M.; Zeng, S.; Zhang, J.Q.; Kim, T.S.; Cohen, N.A.; Beckman, M.J.; Medina, B.D.; Maltbaek, J.H.; Loo, J.K.; Crawley, M.H.; et al. PD-1/PD-L1 blockade enhances T-cell activity and antitumor efficacy of Imatinib in gastrointestinal stromal tumors. *Clin. Cancer Res.* **2017**, *23*, 454–465. [CrossRef] [PubMed]
22. Oyer, J.L.; Gitto, S.B.; Altomare, D.A.; Copik, A.J. PD-L1 blockade enhances anti-tumor efficacy of NK cells. *Oncoimmunology* **2018**, *7*, e1509819. [PubMed]
23. Abdel-Wahab, N.; Shah, M.; Lopez-Olivo, M.A.; Suarez-Almazor, M.E. Use of immune checkpoint inhibitors in the treatment of patients with cancer and preexisting autoimmune disease: A systematic review. *Ann. Intern. Med.* **2018**, *168*, 121–130. [CrossRef] [PubMed]
24. Sharma, P.; Allison, J.P. Immune checkpoint targeting in cancer therapy: Toward combination strategies with curative potential. *Cell* **2015**, *161*, 205–214. [CrossRef]
25. Raggi, F.; Pelassa, S.; Pierobon, D.; Penco, F.; Gattorno, M.; Novelli, F.; Eva, A.; Varesio, L.; Giovarelli, M.; Bosco, M.C. Regulation of human macrophage M1-M2 polarization balance by hypoxia and the triggering receptor expressed on myeloid cells-1. *Front. Immunol.* **2017**, *8*, 1097. [CrossRef]
26. Choi, J.; Gyamfi, J.; Jang, H.; Koo, J.S. The role of tumor-associated macrophage in breast cancer biology. *Histol. Histopathol.* **2018**, *33*, 133–145. [PubMed]
27. Seager, R.J.; Hajal, C.; Spill, F.; Kamm, R.D.; Zaman, M.H. Dynamic interplay between tumour, stroma and immune system can drive or prevent tumour progression. *Converg. Sci. Phys. Oncol.* **2017**, *3*, 034002. [CrossRef]
28. Tripathi, C.; Tewari, B.N.; Kanchan, R.K.; Baghel, K.S.; Nautiyal, N.; Shrivastava, R.; Kaur, H.; Bhatt, M.L.B.; Bhadauria, S. Macrophages are recruited to hypoxic tumor areas and acquire a pro-angiogenic M2-polarized phenotype via hypoxic cancer cell derived cytokines Oncostatin M and Eotaxin. *Oncotarget* **2014**, *5*, 5350–5368. [CrossRef]
29. Williams, C.B.; Yeh, E.S.; Soloff, A.C. Tumor-associated macrophages: Unwitting accomplices in breast cancer malignancy. *NPJ Breast Cancer* **2016**, *2*, 15025. [CrossRef]
30. Luo, H.; Tu, G.; Liu, Z.; Liu, M.; Noy, R.; Pollard, J.W.W. Tumor-associated macrophages: From mechanisms to therapy. *Immunity* **2014**, *41*, 49–61.
31. Hollmén, M.; Roudnicky, F.; Karaman, S.; Detmar, M. Characterization of macrophage—Cancer cell crosstalk in estrogen receptor positive and triple-negative breast cancer. *Sci. Rep.* **2015**, *5*, 9188. [CrossRef] [PubMed]
32. Sousa, S.; Brion, R.; Lintunen, M.; Kronqvist, P.; Sandholm, J.; Mönkkönen, J.; Kellokumpu-Lehtinen, P.-L.; Lauttia, S.; Tynninen, O.; Joensuu, H.; et al. Human breast cancer cells educate macrophages toward the M2 activation status. *Breast Cancer Res.* **2015**, *17*, 101. [CrossRef] [PubMed]
33. Gubin, M.M.; Esaulova, E.; Ward, J.P.; Malkova, O.N.; Runci, D.; Wong, P.; Noguchi, T.; Arthur, C.D.; Meng, W.; Alspach, E.; et al. High-dimensional analysis delineates myeloid and lymphoid compartment remodeling during successful immune-checkpoint cancer therapy. *Cell* **2018**, *175*, 1014–1030. [CrossRef] [PubMed]
34. Su, S.; Liu, Q.; Chen, J.; Chen, J.; Chen, F.; He, C.; Huang, D.; Wu, W.; Lin, L.; Huang, W.; et al. A Positive feedback loop between mesenchymal-like cancer cells and macrophages is essential to breast cancer metastasis. *Cancer Cell* **2014**, *25*, 605–620. [CrossRef]

35. Burger, G.A.; Danen, E.H.J.; Beltman, J.B. Deciphering epithelial—Mesenchymal transition regulatory networks in cancer through computational approaches. *Front. Oncol.* **2017**, *7*, 162. [CrossRef] [PubMed]
36. Mitchem, J.B.; Brennan, D.J.; Knolhoff, B.L.; Belt, B.A.; Zhu, Y.; Sanford, D.E.; Belaygorod, L.; Carpenter, D.; Collins, L.; Piwnica-Worms, D.; et al. Targeting tumor-infiltrating macrophages decreases tumor-initiating cells, relieves immunosuppression, and improves chemotherapeutic responses. *Cancer Res.* **2013**, *73*, 1128–1141. [CrossRef] [PubMed]
37. Qian, B.; Deng, Y.; Im, J.H.; Muschel, R.J.; Zou, Y.; Li, J.; Lang, R.A.; Pollard, J.W. A distinct macrophage population mediates metastatic breast cancer cell extravasation, establishment and growth. *PLoS ONE* **2009**, *4*, e6562. [CrossRef]
38. Lohela, M.; Casbon, A.-J.; Olow, A.; Bonham, L.; Branstetter, D.; Weng, N.; Smith, J.; Werb, Z. Intravital imaging reveals distinct responses of depleting dynamic tumor-associated macrophage and dendritic cell subpopulations. *Proc. Natl. Acad. Sci. USA* **2014**, *111*, E5086–E5095. [CrossRef]
39. Ngambenjawong, C.; Cieslewicz, M.; Schellinger, J.G.; Pun, S.H. Synthesis and evaluation of multivalent M2pep peptides for targeting alternatively activated M2 macrophages. *J. Control. Release* **2016**, *224*, 103–111. [CrossRef]
40. Mishalian, I.; Granot, Z.; Fridlender, Z.G. The diversity of circulating neutrophils in cancer. *Immunobiology* **2017**, *222*, 82–88. [CrossRef]
41. Davis, B.P.; Rothenberg, M.E. Eosinophils and cancer. *Cancer Immunol. Res.* **2014**, *2*, 1–8. [CrossRef] [PubMed]
42. Hämmerling, G.J.; Carretero, R.; Beckhove, P.; Salgado, O.C.; Sektioglu, I.M.; Garbi, N. Eosinophils orchestrate cancer rejection by normalizing tumor vessels and enhancing infiltration of CD8+ T cells. *Nat. Immunol.* **2015**, *16*, 609–617.
43. Sakkal, S.; Miller, S.; Apostolopoulos, V.; Nurgali, K. Eosinophils in cancer: Favourable or unfavourable? *Curr. Med. Chem.* **2016**, *23*, 650–666. [CrossRef] [PubMed]
44. Shinde, S.B.; Kurhekar, M.P. Review of the systems biology of the immune system using agent-based models. *IET Syst. Biol.* **2018**, *12*, 83–92. [CrossRef] [PubMed]
45. Altrock, P.M.; Liu, L.L.; Michor, F. The mathematics of cancer: Integrating quantitative models. *Nat. Rev. Cancer* **2015**, *15*, 730–745. [CrossRef] [PubMed]
46. Rejniak, K.A.; McCawley, L.J. Current trends in mathematical modeling of tumor-microenvironment interactions: A survey of tools and applications. *Exp. Biol. Med.* **2010**, *235*, 411–423. [CrossRef] [PubMed]
47. Eftimie, R.; Bramson, J.L.; Earn, D.J.D. Interactions between the immune system and cancer: A brief review of non-spatial mathematical models. *Bull. Math. Biol.* **2011**, *73*, 2–32. [CrossRef]
48. Alemani, D.; Pappalardo, F.; Pennisi, M.; Motta, S.; Brusic, V. Combining cellular automata and lattice Boltzmann method to model multiscale avascular tumor growth coupled with nutrient diffusion and immune competition. *J. Immunol. Methods* **2012**, *376*, 55–68. [CrossRef]
49. Bellomo, N.; Delitala, M. From the mathematical kinetic, and stochastic game theory to modelling mutations, onset, progression and immune competition of cancer cells. *Phys. Life Rev.* **2008**, *5*, 183–206. [CrossRef]
50. Eladdadi, A.; de Pillis, L.; Kim, P. Modelling tumour-immune dynamics, disease progression and treatment. *Lett. Biomath.* **2018**, *5*, S1–S5. [CrossRef]
51. Dritschel, H.; Waters, S.L.; Roller, A.; Byrne, H.M. A mathematical model of cytotoxic and helper T cell interactions in a tumour microenvironment. *Lett. Biomath.* **2018**, *5*, S36–S68. [CrossRef]
52. Nikolopoulou, E.; Johnson, L.R.; Harris, D.; Nagy, J.D.; Stites, E.C.; Kuang, Y. Tumour-immune dynamics with an immune checkpoint inhibitor. *Lett. Biomath.* **2018**, *5*, S137–S159. [CrossRef]
53. Salgia, R.; Mambetsariev, I.; Hewelt, B.; Achuthan, S.; Li, H.; Poroyko, V.; Wang, Y.; Sattler, M. Modeling small cell lung cancer (SCLC) biology through deterministic and stochastic mathematical models. *Oncotarget* **2018**, *9*, 26226–26242. [CrossRef] [PubMed]
54. Konstorum, A.; Vella, A.T.; Adler, A.J.; Laubenbacher, R.C. Addressing current challenges in cancer immunotherapy with mathematical and computational modelling. *J. R. Soc. Interface* **2017**, *14*, 20170150. [CrossRef] [PubMed]
55. Chiacchio, F.; Pennisi, M.; Russo, G.; Motta, S.; Pappalardo, F. Agent-based modeling of the immune system: NetLogo, a promising framework. *BioMed Res. Int.* **2014**, *907171*. [CrossRef] [PubMed]
56. An, G.; Mi, Q.; Dutta-Moscato, J.; Vodovotz, Y. Agent-based models in translational systems biology. *Wiley Interdiscip. Rev. Syst. Biol. Med.* **2009**, *1*, 159–171. [CrossRef]

57. Chavali, A.K.; Gianchandani, E.P.; Tung, K.S.; Lawrence, M.B.; Peirce, S.M.; Papin, J.A. Characterizing emergent properties of immunological systems with multi-cellular rule-based computational modeling. *Trends Immunol.* **2008**, *29*, 589–599. [CrossRef]
58. Shi, Z.Z.; Wu, C.-H.; Ben-Arieh, D. Agent-based model: A surging tool to simulate infectious diseases in the immune system. *Open J. Model. Simul.* **2014**, *02*, 12–22. [CrossRef]
59. Segovia-Juarez, J.L.; Ganguli, S.; Kirschner, D. Identifying control mechanisms of granuloma formation during M. tuberculosis infection using an agent-based model. *J. Theor. Biol.* **2004**, *231*, 357–376. [CrossRef]
60. Tokarski, C.; Hummert, S.; Mech, F.; Figge, M.T.; Germerodt, S.; Schroeter, A.; Schuster, S.; Linde, J.; Hu, G. Agent-based modeling approach of immune defense against spores of opportunistic human pathogenic fungi. *Front. Microbiol.* **2012**, *3*, 129. [CrossRef]
61. Dong, X.; Foteinou, P.T.; Calvano, S.E.; Lowry, S.F.; Androulakis, I.P. Agent-based modeling of endotoxin-induced acute inflammatory response in human blood leukocytes. *PLoS ONE* **2010**, *5*, e9249. [CrossRef] [PubMed]
62. Solovyev, A.; Mi, Q.; Tzen, Y.T.; Brienza, D.; Vodovotz, Y. Hybrid equation/agent-based model of ischemia-induced hyperemia and pressure ulcer formation predicts greater propensity to ulcerate in subjects with spinal cord injury. *PLoS Comput. Biol.* **2013**, *9*, e1003070. [CrossRef] [PubMed]
63. Santoni, D.; Pedicini, M.; Castiglione, F. Implementation of a regulatory gene network to simulate the TH1/2 differentiation in an agent-based model of hypersensitivity reactions. *Bioinformatics* **2008**, *24*, 1374–1380. [CrossRef] [PubMed]
64. Bailey, A.M.; Lawrence, M.B.; Shang, H.; Katz, A.J.; Peirce, S.M. Agent-based model of therapeutic adipose-derived stromal cell trafficking during ischemia predicts ability to roll on p-selectin. *PLoS Comput. Biol.* **2009**, *5*, e1000294. [CrossRef] [PubMed]
65. D'Souza, R.M.; Lysenko, M.; Marino, S.; Kirschner, D.; Souza, R.M.D.; Arbor, A. Data-parallel algorithms for agent-based model simulation of tuberculosis on graphics processing units. In Proceedings of the 2009 Spring Simulation Multiconference, San Diego, CA, USA, 22–27 March 2009.
66. Song, S.O.; Hogg, J.; Peng, Z.Y.; Parker, R.; Kellum, J.A.; Clermont, G. Ensemble models of neutrophil trafficking in severe sepsis. *PLoS Comput. Biol.* **2012**, *8*, e1002422. [CrossRef] [PubMed]
67. Mi, Q.; Rivière, B.; Clermont, G.; Steed, D.L.; Vodovotz, Y. Agent-based model of inflammation and wound healing: Insights into diabetic foot ulcer pathology and the role of transforming growth factor-β1. *Wound Repair Regen.* **2007**, *15*, 671–682. [CrossRef] [PubMed]
68. Alarcón, T.; Byrne, H.M.; Maini, P.K. A mathematical model of the effects of hypoxia on the cell-cycle of normal and cancer cells. *J. Theor. Biol.* **2004**, *229*, 395–411. [CrossRef]
69. Hoehme, S.; Bertaux, F.; Weens, W.; Grasl-Kraupp, B.; Hengstler, J.G.; Drasdo, D. Model prediction and validation of an order mechanism controlling the spatiotemporal phenotype of early hepatocellular carcinoma. *Bull. Math. Biol.* **2018**, *80*, 1134–1171. [CrossRef]
70. Bianca, C.; Pennisi, M. The triplex vaccine effects in mammary carcinoma: A nonlinear model in tune with SimTriplex. *Nonlinear Anal. Real World Appl.* **2012**, *13*, 1913–1940. [CrossRef]
71. Wang, J.; Zhang, L.; Jing, C.; Ye, G.; Wu, H.; Miao, H.; Wu, Y.; Zhou, X. Multi-scale agent-based modeling on melanoma and its related angiogenesis analysis. *Theor. Biol. Med. Model.* **2013**, *10*, 41. [CrossRef]
72. Kather, J.N.; Poleszczuk, J.; Suarez-Carmona, M.; Krisam, J.; Charoentong, P.; Valous, N.A.; Weis, C.A.; Tavernar, L.; Leiss, F.; Herpel, E.; et al. In silico modeling of immunotherapy and stroma-targeting therapies in human colorectal cancer. *Cancer Res.* **2017**, *77*, 6442–6452. [CrossRef] [PubMed]
73. Pennisi, M.; Pappalardo, F.; Motta, S. Agent based modeling of lung metastasis-immune system competition. In *Lecture Notes in Computer Science (Including Subseries Lecture Notes in Artificial Intelligence and Lecture Notes in Bioinformatics)*; Springer: Berlin/Heidelberg, Germany, 2009; Volume 5666, pp. 1–3.
74. Jagiella, N.; Müller, B.; Müller, M.; Vignon-Clementel, I.E.; Drasdo, D. Inferring growth control mechanisms in growing multi-cellular spheroids of NSCLC cells from spatial-temporal image data. *PLoS Comput. Biol.* **2016**, *12*, e1004412. [CrossRef] [PubMed]
75. Pennisi, M.; Pappalardo, F.; Palladini, A.; Nicoletti, G.; Nanni, P.; Lollini, P.-L.; Motta, S. Modeling the competition between lung metastases and the immune system using agents. *BMC Bioinform.* **2010**, *11*, S13. [CrossRef] [PubMed]
76. Bezzi, M.; Celada, F.; Ruffo, S.; Seiden, P.E. The transition between immune and disease states in a cellular automaton model of clonal immune response. *Phys. A Stat. Mech. Its Appl.* **1997**, *245*, 145–163. [CrossRef]

77. Baldazzi, V.; Castiglione, F.; Bernaschi, M. An enhanced agent based model of the immune system response. *Cell Immunol.* **2006**, *244*, 77–79. [CrossRef] [PubMed]
78. Bernaschi, M.; Castiglione, F. Design and implementation of an immune system simulator. *Comput. Biol. Med.* **2001**, *31*, 303–331. [CrossRef]
79. Celada, F.; Seiden, P.E. A computer model of cellular interactions in the immune system. *Immunol. Today* **1992**, *13*, 56–62. [CrossRef]
80. Emerson, A.; Rossi, E. ImmunoGrid - the virtual human immune system project. *Stud. Heal. Technol. Inf.* **2007**, *126*, 87–92.
81. Halling-Brown, M.; Pappalardo, F.; Rapin, N.; Zhang, P.; Alemani, D.; Emerson, A.; Castiglione, F.; Duroux, P.; Pennisi, M.; Miotto, O.; et al. ImmunoGrid: Towards agent-based simulations of the human immune system at a natural scale. *Philos. Trans. R. Soc. A Math. Phys. Eng. Sci.* **2010**, *368*, 2799. [CrossRef]
82. Perelson, A.S.; Ribeiro, R.M. Modeling the within-host dynamics of HIV infection. *BMC Biol.* **2013**, *11*, 96. [CrossRef]
83. Warrender, C.; Forrest, S.; Koster, F. Modeling intercellular interactions in early Mycobacterium infection. *Bull. Math. Biol.* **2006**, *68*, 2233–2261. [CrossRef] [PubMed]
84. Ghaffarizadeh, A.; Heiland, R.; Friedman, S.H.; Mumenthaler, S.M.; Macklin, P. PhysiCell: An open source physics-based cell simulator for 3-D multicellular systems. *PLOS Comput. Biol.* **2018**, *14*, e1005991. [CrossRef] [PubMed]
85. Gilkes, D.M.; Semenza, G.L.; Wirtz, D. Hypoxia and the extracellular matrix: Drivers of tumour metastasis. *Nat. Rev. Cancer* **2014**, *14*, 430–439. [CrossRef] [PubMed]
86. Alfonso, J.C.L.; Talkenberger, K.; Seifert, M.; Klink, B.; Hawkins-Daarud, A.; Swanson, K.R.; Hatzikirou, H.; Deutsch, A. The biology and mathematical modelling of glioma invasion: A review. *J. R. Soc. Interface* **2017**, *14*, 20170490. [CrossRef] [PubMed]
87. Massey, S.C.; Rockne, R.C.; Hawkins-Daarud, A.; Gallaher, J.; Anderson, A.R.A.; Canoll, P.; Swanson, K.R. Simulating PDGF-driven glioma growth and invasion in an anatomically accurate brain domain. *Bull. Math. Biol.* **2018**, *80*, 1292–1309. [CrossRef] [PubMed]
88. Juliano, J.; Gil, O.; Hawkins-Daarud, A.; Noticewala, S.; Rockne, R.C.; Gallaher, J.; Massey, S.C.; Sims, P.A.; Anderson, A.R.A.; Swanson, K.R.; et al. Comparative dynamics of microglial and glioma cell motility at the infiltrative margin of brain tumours. *J. R. Soc. Interface* **2018**, *15*, 20170582. [CrossRef] [PubMed]
89. Frascoli, F.; Flood, E.; Kim, P.S. A model of the effects of cancer cell motility and cellular adhesion properties on tumour-immune dynamics. *Math. Med. Biol.* **2016**, *34*, dqw004. [CrossRef]
90. Noonan, D.M.; De Lerma Barbaro, A.; Vannini, N.; Mortara, L.; Albini, A. Inflammation, inflammatory cells and angiogenesis: Decisions and indecisions. *Cancer Metastasis Rev.* **2008**, *27*, 31–40. [CrossRef]
91. Tian, L.; Goldstein, A.; Wang, H.; Ching Lo, H.; Sun Kim, I.; Welte, T.; Sheng, K.; Dobrolecki, L.E.; Zhang, X.; Putluri, N.; et al. Mutual regulation of tumour vessel normalization and immunostimulatory reprogramming. *Nature* **2017**, *544*, 250–254. [CrossRef]
92. Uppal, A.; Wightman, S.C.; Ganai, S.; Weichselbaum, R.R.; An, G. Investigation of the essential role of platelet-tumor cell interactions in metastasis progression using an agent-based model. *Theor. Biol. Med. Model.* **2014**, *11*, 17. [CrossRef]
93. Alfonso, J.C.L.; Schaadt, N.S.; Schönmeyer, R.; Brieu, N.; Forestier, G.; Wemmert, C.; Feuerhake, F.; Hatzikirou, H. In-silico insights on the prognostic potential of immune cell infiltration patterns in the breast lobular epithelium. *Sci. Rep.* **2016**, *6*, 33322. [CrossRef] [PubMed]
94. Reddy, N.P.; Krouskop, T.A.; Newell, P.H. A computer model of the lymphatic system. *Comput. Biol. Med.* **1977**, *7*, 181–197. [CrossRef]
95. Jamalian, S.; Jafarnejad, M.; Zawieja, S.D.; Bertram, C.D.; Gashev, A.A.; Zawieja, D.C.; Davis, M.J.; Moore, J.E. Demonstration and analysis of the suction effect for pumping lymph from tissue beds at subatmospheric pressure. *Sci. Rep.* **2017**, *7*, 12080. [CrossRef] [PubMed]
96. Jamalian, S.; Davis, M.J.; Zawieja, D.C.; Moore, J.E. Network scale modeling of lymph transport and its effective pumping parameters. *PLoS ONE* **2016**, *11*, e0148384. [CrossRef] [PubMed]
97. Jamalian, S.; Bertram, C.D.; Richardson, W.J.; Moore, J.E. Parameter sensitivity analysis of a lumped-parameter model of a chain of lymphangions in series. *Am. J. Physiol. Circ. Physiol.* **2013**, *305*, H1709–H1717. [CrossRef] [PubMed]

98. Wilson, J.T.; van Loon, R.; Wang, W.; Zawieja, D.C.; Moore, J.E. Determining the combined effect of the lymphatic valve leaflets and sinus on resistance to forward flow. *J. Biomech.* **2015**, *48*, 3584–3590. [CrossRef] [PubMed]

99. Roose, T.; Swartz, M.A. Multiscale modeling of lymphatic drainage from tissues using homogenization theory. *J. Biomech.* **2012**, *45*, 107–115. [CrossRef] [PubMed]

100. Jafarnejad, M.; Zawieja, D.C.; Brook, B.S.; Nibbs, R.J.B.; Moore, J.E. A novel computational model predicts key regulators of chemokine gradient formation in lymph nodes and site-specific roles for CCL19 and ACKR4. *J. Immunol.* **2017**, *199*, ji1700377. [CrossRef]

101. Jafarnejad, M.; Woodruff, M.C.; Zawieja, D.C.; Carroll, M.C.; Moore, J.E. Modeling lymph flow and fluid exchange with blood vessels in lymph nodes. *Lymphat. Res. Biol.* **2015**, *13*, 234–247. [CrossRef]

102. Cooper, L.J.; Heppell, J.P.; Clough, G.F.; Ganapathisubramani, B.; Roose, T. An image-based model of fluid flow through lymph nodes. *Bull. Math. Biol.* **2016**, *78*, 52–71. [CrossRef]

103. Marino, S.; Gideon, H.P.; Gong, C.; Mankad, S.; McCrone, J.T.; Lin, P.L.; Linderman, J.J.; Flynn, J.A.L.; Kirschner, D.E. Computational and empirical studies predict mycobacterium tuberculosis-specific T cells as a biomarker for infection outcome. *PLoS Comput. Biol.* **2016**, *12*, e1004804. [CrossRef] [PubMed]

104. Gong, C.; Linderman, J.J.; Kirschner, D. Harnessing the heterogeneity of T cell differentiation fate to fine-tune generation of effector and memory T cells. *Front. Immunol.* **2014**, *5*, 57. [CrossRef] [PubMed]

105. Margaris, K.N.; Black, R.A. Modelling the lymphatic system: Challenges and opportunities. *J. R. Soc. Interface* **2012**, *9*, 601–612. [CrossRef] [PubMed]

106. Meyer-Hermann, M. A mathematical model for the germinal center morphology and affinity maturation. *J. Theor. Biol.* **2002**, *216*, 273–300. [CrossRef] [PubMed]

107. Meyer-Hermann, M.E.; Maini, P.K. Cutting edge: Back to "one-way" germinal centers. *J. Immunol.* **2005**, *174*, 2489–2493. [CrossRef] [PubMed]

108. Meyer-Hermann, M.E.; Maini, P.K.; Iber, D. An analysis of B cell selection mechanisms in germinal centers. *Math. Med. Biol. A J. IMA* **2006**, *23*, 255–277. [CrossRef] [PubMed]

109. Bogle, G.; Dunbar, P.R. T cell responses in lymph nodes. *Wiley Interdiscip. Rev. Syst. Biol. Med.* **2010**, *2*, 107–116. [CrossRef]

110. Bogle, G.; Dunbar, P.R. Agent-based simulation of T-cell activation and proliferation within a lymph node. *Immunol. Cell Biol.* **2010**, *88*, 172–179. [CrossRef]

111. Bogle, G.; Dunbar, P.R. On-lattice simulation of T cell motility, chemotaxis, and trafficking in the lymph node paracortex. *PLoS ONE* **2012**, *7*, e45258. [CrossRef]

112. Bogle, G.; Dunbar, P.R. Simulating T-cell motility in the lymph node paracortex with a packed lattice geometry. *Immunol. Cell Biol.* **2008**, *86*, 676–687. [CrossRef]

113. Moreau, H.D.; Bogle, G.; Bousso, P. A virtual lymph node model to dissect the requirements for T-cell activation by synapses and kinapses. *Immunol. Cell Biol.* **2016**, *94*, 680–688. [CrossRef] [PubMed]

114. Folcik, V.A.; An, G.C.; Orosz, C.G. The Basic Immune Simulator: An agent-based model to study the interactions between innate and adaptive immunity. *Theor. Biol. Med. Model.* **2007**, *4*, 39. [CrossRef] [PubMed]

115. Kim, P.S.; Levy, D.; Lee, P.P. Modeling and simulation of the immune system as a self-regulating network. *Methods Enzymol.* **2009**, *467*, 79–109. [PubMed]

116. Jacob, C.; Sarpe, V.; Gingras, C.; Feyt, R.P. Swarm-based simulations for immunobiology: What can agent-based models teach us about the immune system? In *Intelligent Systems Reference Library*; Springer: Berlin/Heidelberg, Germany, 2011; Volume 11, pp. 29–64.

117. Marino, S.; El-Kebir, M.; Kirschner, D. A hybrid multi-compartment model of granuloma formation and T cell priming in Tuberculosis. *J. Theor. Biol.* **2011**, *280*, 50–62. [CrossRef] [PubMed]

118. Marino, S.; Kirschner, D. A multi-compartment hybrid computational model predicts key roles for dendritic cells in Tuberculosis infection. *Computation* **2016**, *4*, 39. [CrossRef] [PubMed]

119. Dréau, D.; Stanimirov, D.; Carmichael, T.; Hadzikadic, M. An agent-based model of solid tumor progression. In *Lecture Notes in Computer Science (Including Subseries Lecture Notes in Artificial Intelligence and Lecture Notes in Bioinformatics)*; Springer: Berlin/Heidelberg, Germany, 2009; Volume 5462, pp. 187–198.

120. Pappalardo, F.; Forero, I.M.; Pennisi, M.; Palazon, A.; Melero, I.; Motta, S. Simb16: Modeling induced immune system response against B16-melanoma. *PLoS ONE* **2011**, *6*, e26523. [CrossRef] [PubMed]

121. Sanchez-Paulete, A.R.; Labiano, S.; Rodriguez-Ruiz, M.E.; Azpilikueta, A.; Etxeberria, I.; Bolaños, E.; Lang, V.; Rodriguez, M.; Aznar, M.A.; Jure-Kunkel, M.; et al. Deciphering CD137 (4-1BB) signaling in T-cell costimulation for translation into successful cancer immunotherapy. *Eur. J. Immunol.* **2016**, *46*, 513–522. [CrossRef] [PubMed]

122. Santiago, D.N.; Heidbuechel, J.P.W.; Kandell, W.M.; Walker, R.; Djeu, J.; Engeland, C.E.; Abate-Daga, D.; Enderling, H. Fighting cancer with mathematics and viruses. *Viruses* **2017**, *9*, 239. [CrossRef]

123. Walker, R.; Navas, P.E.; Friedman, S.H.; Galliani, S.; Karolak, A.; MacFarlane, F.; Noble, R.; Poleszczuk, J.; Russell, S.; Rejniak, K.A.; et al. Enhancing synergy of CAR T cell therapy and oncolytic virus therapy for pancreatic cancer. *bioRxiv* **2016**, 055988. [CrossRef]

124. Rohrs, J.A.; Zheng, D.; Graham, N.A.; Wang, P.; Finley, S.D. Computational model of chimeric antigen receptors explains site-specific phosphorylation kinetics. *Biophys. J.* **2018**, *115*, 1116–1129. [CrossRef]

125. Schumacher, T.N.; Schreiber, R.D. Neoantigens in cancer immunotherapy. *Science* **2015**, *348*, 69–74. [CrossRef] [PubMed]

126. Sharma, P.; Allison, J.P. The future of immune checkpoint therapy. *Science* **2015**, *348*, 56–61. [CrossRef] [PubMed]

127. Gong, C.; Milberg, O.; Wang, B.; Vicini, P.; Narwal, R.; Roskos, L.; Popel, A.S. A computational multiscale agent-based model for simulating spatio-temporal tumour immune response to PD1 and PDL1 inhibition. *J. R. Soc. Interface* **2017**, *14*, 20170320. [CrossRef] [PubMed]

128. Hillen, T.; Enderling, H.; Hahnfeldt, P. The tumor growth paradox and immune system-mediated selection for cancer stem cells. *Bull. Math. Biol.* **2013**, *75*, 161–184. [CrossRef] [PubMed]

129. Enderling, H.; Hlatky, L.; Hahnfeldt, P. Immunoediting: Evidence of the multifaceted role of the immune system in self-metastatic tumor growth. *Theor. Biol. Med. Model.* **2012**, *9*, 31. [CrossRef] [PubMed]

130. Dehne, N.; Mora, J.; Namgaladze, D.; Weigert, A.; Brüne, B. Cancer cell and macrophage cross-talk in the tumor microenvironment. *Curr. Opin. Pharmacol.* **2017**, *35*, 12–19. [CrossRef] [PubMed]

131. Wells, D.K.; Chuang, Y.; Knapp, L.M.; Brockmann, D.; Kath, W.L.; Leonard, J.N. Spatial and functional heterogeneities shape collective behavior of tumor-immune networks. *PLoS Comput. Biol.* **2015**, *11*, e1004181. [CrossRef]

132. Knútsdóttir, H.; Pálsson, E.; Edelstein-Keshet, L. Mathematical model of macrophage-facilitated breast cancer cells invasion. *J. Theor. Biol.* **2014**, *357*, 184–199. [CrossRef]

133. Knutsdottir, H.; Condeelis, J.S.; Palsson, E. 3-D individual cell based computational modeling of tumor cell-macrophage paracrine signaling mediated by EGF and CSF-1 gradients. *Integr. Biol.* **2016**, *8*, 104–119. [CrossRef]

134. Norton, K.A.; Jin, K.; Popel, A.S. Modeling triple-negative breast cancer heterogeneity: Effects of stromal macrophages, fibroblasts and tumor vasculature. *J. Theor. Biol.* **2018**, *452*, 56–68. [CrossRef]

135. Marusyk, A.; Almendro, V.; Polyak, K. Intra-tumour heterogeneity: A looking glass for cancer? *Nat. Rev. Cancer* **2012**, *12*, 323–334. [CrossRef] [PubMed]

136. Lehrach, H. Virtual clinical trials, an essential step in increasing the effectiveness of the drug development process. *Public Health Genom.* **2015**, *18*, 366–371. [CrossRef] [PubMed]

137. Pourhasanzade, F.; Sabzpoushan, S.; Alizadeh, A.M.; Esmati, E. An agent-based model of avascular tumor growth: Immune response tendency to prevent cancer development. *Simulation* **2017**, *93*, 641–657. [CrossRef]

138. Carmona-Fontaine, C.; Bucci, V.; Akkari, L.; Deforet, M.; Joyce, J.A.; Xavier, J.B. Emergence of spatial structure in the tumor microenvironment due to the Warburg effect. *Proc. Natl. Acad. Sci. USA* **2013**, *110*, 19402–19407. [CrossRef] [PubMed]

139. Figueredo, G.P.; Aickelin, U.; Siebers, P.O. Systems dynamics or agent-based modelling for immune simulation? In *Proceedings of the Lecture Notes in Computer Science (Including Subseries Lecture Notes in Artificial Intelligence and Lecture Notes in Bioinformatics)*; Springer: Berlin/Heidelberg, Germany, 2011; Volume 6825, pp. 81–94.

140. Figueredo, G.P.; Siebers, P.-O.; Aickelin, U. Investigating mathematical models of immuno-interactions with early-stage cancer under an agent-based modelling perspective. *BMC Bioinform.* **2013**, *14*, S6. [CrossRef]

141. Figueredo, G.P.; Joshi, T.V.; Osborne, J.M.; Byrne, H.M.; Owen, M.R. On-lattice agent-based simulation of populations of cells within the open-source Chaste framework. *Interface Focus* **2013**, *3*, 20120081. [CrossRef] [PubMed]

142. Figueredo, G.P.; Siebers, P.O.; Owen, M.R.; Reps, J.; Aickelin, U. Comparing stochastic differential equations and agent-based modelling and simulation for early-stage cancer. *PLoS ONE* **2014**, *9*, e95150. [CrossRef] [PubMed]
143. Yankeelov, T.E.; An, G.; Saut, O.; Luebeck, E.G.; Popel, A.S.; Ribba, B.; Vicini, P.; Zhou, X.; Weis, J.A.; Ye, K.; et al. Multi-scale modeling in clinical oncology: Opportunities and barriers to success. *Ann. Biomed. Eng.* **2016**, *44*, 2626–2641. [CrossRef]
144. Cosgrove, J.; Butler, J.; Alden, K.; Read, M.; Kumar, V.; Cucurull-Sanchez, L.; Timmis, J.; Coles, M. Agent-based modeling in systems pharmacology. *CPT Pharmacometrics Syst. Pharmacol.* **2015**, *4*, 615–629. [CrossRef]
145. Rieger, T.R.; Allen, R.J.; Bystricky, L.; Chen, Y.; Colopy, G.W.; Cui, Y.; Gonzalez, A.; Liu, Y.; White, R.D.; Everett, R.A.; et al. Improving the generation and selection of virtual populations in quantitative systems pharmacology models. *Prog. Biophys. Mol. Biol.* **2018**, *139*, 15–22. [CrossRef]
146. Barretina, J.; Caponigro, G.; Stransky, N.; Venkatesan, K.; Margolin, A.A.; Kim, S.; Wilson, C.J.; Lehár, J.; Kryukov, G.V.; Sonkin, D.; et al. The Cancer Cell Line Encyclopedia enables predictive modelling of anticancer drug sensitivity. *Nature* **2012**, *483*, 603–607. [CrossRef] [PubMed]
147. Garnett, M.J.; Edelman, E.J.; Heidorn, S.J.; Greenman, C.D.; Dastur, A.; Lau, K.W.; Greninger, P.; Thompson, I.R.; Luo, X.; Soares, J.; et al. Systematic identification of genomic markers of drug sensitivity in cancer cells. *Nature* **2012**, *483*, 570–575. [CrossRef] [PubMed]
148. Rubio-Perez, C.; Tamborero, D.; Schroeder, M.P.; Antolín, A.A.; Deu-Pons, J.; Perez-Llamas, C.; Mestres, J.; Gonzalez-Perez, A.; Lopez-Bigas, N. In silico prescription of anticancer drugs to cohorts of 28 tumor types reveals targeting opportunities. *Cancer Cell* **2015**, *27*, 382–396. [CrossRef] [PubMed]
149. Anderson, A.R.A.; Maini, P.K. Mathematical oncology. *Bull. Math. Biol.* **2018**, *80*, 945–953. [CrossRef] [PubMed]
150. Sun, X.; Hu, B. Mathematical modeling and computational prediction of cancer drug resistance. *Brief. Bioinform.* **2017**, *19*, 1382–1399. [CrossRef]
151. Arney, K. Improving brain-cancer therapies through mathematical modelling. *Nature* **2018**, *561*, S52–S53. [CrossRef] [PubMed]
152. Barua, D.; Hlavacek, W.S. Modeling the effect of APC truncation on destruction complex function in colorectal cancer cells. *PLoS Comput. Biol.* **2013**, *9*, e1003217. [CrossRef]
153. Klinger, B.; Sieber, A.; Fritsche-Guenther, R.; Witzel, F.; Berry, L.; Schumacher, D.; Yan, Y.; Durek, P.; Merchant, M.; Schafer, R.; et al. Network quantification of EGFR signaling unveils potential for targeted combination therapy. *Mol. Syst. Biol.* **2014**, *9*, 673. [CrossRef]
154. Kirouac, D.C.; Du, J.Y.; Lahdenranta, J.; Overland, R.; Yarar, D.; Paragas, V.; Pace, E.; McDonagh, C.F.; Nielsen, U.B.; Onsum, M.D. Computational modeling of ERBB2-amplified breast cancer identifies combined ErbB2/3 blockade as superior to the combination of MEK and AKT inhibitors (Science Signaling 6:288 (ra68)). *Sci. Signal.* **2014**, *7*, er5.
155. Carbo, A.; Bassaganya-Riera, J.; Pedragosa, M.; Viladomiu, M.; Marathe, M.; Eubank, S.; Wendelsdorf, K.; Bisset, K.; Hoops, S.; Deng, X.; et al. Predictive computational modeling of the mucosal immune responses during Helicobacter pylori infection. *PLoS ONE* **2013**, *8*, e73365. [CrossRef]
156. Zand, R.; Abedi, V.; Hontecillas, R.; Lu, P.; Noorbakhsh-Sabet, N.; Verma, M.; Leber, A.; Tubau-Juni, N.; Bassaganya-Riera, J. Development of synthetic patient populations and in silico clinical trials. In *Accelerated Path to Cures*; Springer International Publishing AG: Basel, Switzerland, 2018; pp. 57–77.
157. Edelman, L.B.; Eddy, J.A.; Price, N.D. In silico models of cancer. *Wiley Interdiscip. Rev. Syst. Biol. Med.* **2010**, *2*, 438–459. [CrossRef] [PubMed]
158. Jerby-Arnon, L.; Shah, P.; Cuoco, M.S.; Rodman, C.; Su, M.-J.; Melms, J.C.; Leeson, R.; Kanodia, A.; Mei, S.; Lin, J.-R.; et al. A cancer cell program promotes T cell exclusion and resistance to checkpoint blockade. *Cell* **2018**, *175*, 984–997.e24. [CrossRef] [PubMed]
159. Thorsson, V.; Gibbs, D.L.; Brown, S.D.; Wolf, D.; Bortone, D.S.; Ou Yang, T.-H.; Porta-Pardo, E.; Gao, G.F.; Plaisier, C.L.; Eddy, J.A.; et al. The immune landscape of cancer. *Immunity* **2018**, *48*, 812–830.e14. [CrossRef] [PubMed]
160. Gong, C.; Anders, R.A.; Zhu, Q.; Taube, J.M.; Green, B.; Cheng, W.; Bartelink, I.H.; Vicini, P.; Wang, B.; Popel, A.S. Quantitative Characterization of CD8+ T Cell Clustering and Spatial Heterogeneity in Solid Tumors. *Front Oncol.* **2018**, *8*, 649. [CrossRef]

161. D'Esposito, A.; Sweeney, P.W.; Ali, M.; Saleh, M.; Ramasawmy, R.; Roberts, T.A.; Agliardi, G.; Desjardins, A.; Lythgoe, M.F.; Pedley, R.B.; et al. Computational fluid dynamics with imaging of cleared tissue and of in vivo perfusion predicts drug uptake and treatment responses in tumours. *Nat. Biomed. Eng.* **2018**, *2*, 773–787. [CrossRef]

162. Stamatelos, S.K.; Kim, E.; Pathak, A.P.; Popel, A.S. A bioimage informatics based reconstruction of breast tumor microvasculature with computational blood flow predictions. *Microvasc. Res.* **2014**, *91*, 8–21. [CrossRef] [PubMed]

163. Kirouac, D.C. How do we "validate" a QSP model? *CPT Pharmacometrics Syst. Pharmacol.* **2018**, *7*, 547–548. [CrossRef]

164. Gadkar, K.; Kirouac, D.; Parrott, N.; Ramanujan, S. Quantitative systems pharmacology: A promising approach for translational pharmacology. *Drug Discov. Today Technol.* **2016**, *21–22*, 57–65. [CrossRef]

165. Workman, P.; Draetta, G.F.; Schellens, J.H.M.; Bernards, R. How much longer will we put up with 100,000 cancer drugs? *Cell* **2017**, *168*, 579–583. [CrossRef]

166. Zhang, X.; Li, Y.; Pan, X.; Xiaoqiang, L.; Mohan, R.; Komaki, R.; Cox, J.D.; Chang, J.Y. Intensity-modulated proton therapy reduces the dose to normal tissue compared with intensity-modulated radiation therapy or passive scattering proton therapy and enables individualized radical radiotherapy for extensive stage IIIB non-small-cell lung canc. *Int. J. Radiat. Oncol. Biol. Phys.* **2010**, *77*, 357–366. [CrossRef]

processes

MDPI

Article

Early Afterdepolarisations Induced by an Enhancement in the Calcium Current

André H. Erhardt

Department of Mathematics, University of Oslo, P.O. Box 1053 Blindern, 0316 Oslo, Norway; andreerh@math.uio.no

Received: 20 November 2018; Accepted: 29 December 2018; Published: 4 January 2019

check for
updates

Abstract: Excitable biological cells, such as cardiac muscle cells, can exhibit complex patterns of oscillations such as spiking and bursting. Moreover, it is well known that an enhancement in calcium currents may yield certain kind of cardiac arrhythmia, so-called early afterdepolarisations (EADs). The presence of EADs strongly correlates with the onset of dangerous cardiac arrhythmia. In this paper we study mathematically and numerically the dynamics of a cardiac muscle cell with respect to the calcium current by investigating a simplistic system of differential equations. For the study of this phenomena, we use bifurcation theory, numerical bifurcation analysis, geometric singular perturbation theory and computational methods to investigate a nonlinear multiple time scales system. It will turn out that EADs related to an enhanced calcium current are canard–induced and that we have to combine these theories to derive a better understanding of the dynamics behind EADs. Moreover, a suitable time scale separation argument determines the important and sensitive system parameters which are related to the occurrence of EADs.

Keywords: nonlinear dynamics; multiple time scales; geometric singular perturbation theory; bifurcation analysis; canard-induced EADs; calcium current

MSC: 37G15; 37N25; 65P30; 92B05

1. Introduction

The aim of this manuscript is the mathematical and numerical investigation of a four dimensional version of the model introduced in [1] with respect to an enhancement in the calcium current, which is already used to study early afterdepolarisations (EADs)—a special type of cardiac arrhythmia—induced by a reduced potassium current. We will show reasons for the occurrence of EADs via an enhancement in the calcium current, using numerical bifurcation analysis and geometric singular perturbation theory (GSPT). One main advantage of the GSPT, which is an analytic technique for multi-scale problems that combines asymptotic theory with dynamical techniques, is the study of a reduced model, i.e., a subsystem. This approach is very useful and shows some mechanisms yielding EADs. Moreover, this ansatz is very valuable to identify the sensitive parameters of the system. Nevertheless, it turns out that not all details can be explained using GSPT. Thus, a combination of both theories—bifurcation theory and geometric singular perturbation theory—is needed. We will explain our approach for this simplified model, but of course we can use this ansatz also for more complex models, cf. [2,3].

In general, EADs are additional small amplitude spikes during the plateau or the repolarisation phase of the action potential (AP), i.e., pathological voltage oscillations during one of these phases. They are caused by ion channel diseases, oxidative stress or drugs and are often associated with deficiencies in potassium currents or enhancements in calcium currents [4]. Furthermore, the presence of EADs strongly correlates with the onset of dangerous cardiac arrhythmias, including torsades de pointes (TdP), which is a specific type of abnormal heart rhythm that can lead to sudden cardiac death, see [5–10]. Furthermore, EADs are so-called mixed-mode oscillations (MMOs) [11], i.e., complex oscillatory waveforms that naturally occur in physiologically relevant dynamical processes. MMOs correspond to the switching between small amplitude oscillations and relaxation oscillations.

In this paper, we will use the geometric singular perturbation theory [11,12] and bifurcation analysis [13] to investigate reasons for the appearing of EADs. Here, we are focused on EADs related to an enhancement in the calcium currents, see [14,15]. The main novelty is the combination of these theories to study EADs and mainly the use of the needed time scale separation argument to derive the parameter sensitivity of the considered system. Moreover, we will show that the mathematical approach which is used for instance in [1] is limited to the study of EADs related to an inhibited potassium current. We will see that the considered system exhibits up to four different time scales depending on the different system parameters.

The paper is organised as follows. We start with a brief introduction into the topic of cardiac APs and arrhythmia, i.e., afterdepolarisations, see Section 1.1. Then, in Section 1.2 we will go on with the mathematical modelling of cardiac APs using a Hodgkin-Huxley type formalism. For our mathematical and numerical analysis of the dynamics of our model, we will use the GSPT and bifurcation analysis. Therefore, in Section 2.1 we will give a brief introduction into the topic of GSPT. This theory we will utilise in Section 2.2 and it turns out that EADs related to an enhanced calcium current are canard–induced MMOs. Nevertheless, in Section 2.3 we will show that the study of the reduced system does not show all details of the occurrence of EADs. Therefore, we are also using numerical bifurcation analysis. The desired bifurcation diagram we will derive utilising the MATLAB toolboxes MATCONT and CL_MATCONT [16–18], which are numerical continuation packages for the interactive bifurcation analysis of dynamical systems. Finally, in Section 3 we will discuss our results.

1.1. Biological and Mathematical Background

An AP is a temporary, characteristic variance in the membrane potential of an excitable biological cell, e.g., neuron or cardiac muscle cell, from its resting potential. The molecular mechanism of an AP is based on the interaction of voltage-sensitive ion channels. The reason for the formation and the special properties of the AP is established in the properties of different groups of ion channels in the plasma membrane. An initial stimulus activates the ion channels as soon as a certain threshold potential is reached. Then, these ion channels break open and/or up such that this interaction allows an ion current flow, which changes the membrane potential. A normal AP is always uniform and the cardiac muscle cell AP is typically divided into four phases, i.e., the resting phase, the upstroke phase, the (long) plateau phase and the repolarisation phase, see for more details [15]. The resting phase is designated by high potassium (K^+) currents. After the initial stimulus the sodium (Na^+) conductance increases rapidly and the Na^+ current flux into the cardiac muscle cell until a spike potential is achieved. Then, the Na^+ current inactivates rapidly followed by the activation of L-type calcium (Ca^{2+}) current. The Ca^{2+} current is more slowly than the Na^+ current and plays a key role in maintaining the long plateau phase, which is characteristic for the cardiac muscle cell. While the Ca^{2+} conductance increases the K^+ conductance decreases. The plateau phase is followed by a repolarisation phase, where the intrinsic K^+ ion channels are activated and this is connected with the reduction of the Ca^{2+} conductance. Finally, the K^+ current increases until the resting phase is reached. If there are depolarising variations of the membrane voltage, then we are speaking about afterdepolarisations. These afterdepolarisations are divided into EADs and delayed afterdepolarisations (DADs). This division depends on the timing obtaining of the AP. EADs occur either in the plateau or in the repolarisation phase of the AP and are benefited by an elongation of the AP, while DADs occur after the repolarisation phase is completed. EADs are resulting for example from a reduction of the repolarising K^+ currents. Triggers for this are congenital disorders of the ion channels or the ingestion of some medicament. The elongation of the AP can generate afterpolarisations by reactivation L-type

Ca^{2+} influx. Also chronic cardiac insufficiency may appear with an elongation of the AP by a reduction of the repolarising K^+ currents.

1.2. The Mathematical Model

The history of the modelling of APs of excitable biological cells as neurons and cardiac muscle cells starts with the famous and pioneering Hodgkin-Huxley model in 1952 [19]. In this paper, the authors established a mathematical approach that can be used to model an AP of excitable biological cells, i.e., one uses a Hodgkin-Huxley (type) formalism for the description of APs as systems of ordinary differential equations. The first model of a cardiac cell is the Noble model [20] of a generic Purkinje cell. In 1991, Luo and Rudy published an ionic model for cardiac action potential in guinea pig ventricular cells. Moreover, the Ten Tusscher-Noble-Noble-Panfilov model [21] from 2004 describes a model for human ventricular tissue, cf. also [2]. Such conductance-based models are based on an equivalent circuit representation of a cell membrane. These models represent a minimal biophysical interpretation for an excitable biological cell in which current flow across the membrane is due to charging of the membrane capacitance and movement of ions across ion channels. Ion channels are selective for particular ionic species, such as calcium (Ca^{2+}) or potassium (K^+), giving rise to currents $I_{Ca^{2+}}$ or I_{K^+}, respectively. Our simplistic model reads as follows:

$$\frac{dV}{dt} = -\frac{I_{K^+} + I_{Ca^{2+}}}{C_m}, \tag{1}$$

with the membrane capacity $C_m = 1\frac{\mu F}{m^2}$ and ion currents

$$I_{K^+} := G_{K^+} \cdot x \cdot (V - E_{K^+}) \quad \text{and} \quad I_{Ca^{2+}} := G_{Ca^{2+}} \cdot d \cdot f \cdot (V - E_{Ca^{2+}}), \tag{2}$$

where the different gating variables d, f and x are satisfying the differential equation

$$\frac{dy}{dt} = \frac{y_\infty(V) - y}{\tau_y} \tag{3}$$

and y represents the gating variables d, f and x, while

$$y_\infty := y_\infty(V) = \frac{1}{1 + \exp\left(\frac{V - V_{T_y}}{k_y}\right)} \tag{4}$$

with $V_{T_y} \in \mathbb{R}$, $k_y \in \mathbb{R} \setminus \{0\}$ denotes the equilibrium of the corresponding gating variable and τ_y is the corresponding relaxation time constant for each of d, f and x. The gating variables d, f and $x \in [0, 1]$ are important for the activation (opening) and inactivation (closing) of the ion channels and therefore for the ion current interaction, see [15]. Moreover, the Nernst potentials of these ion currents are denoted by $E_{Ca^{2+}}$ and E_{K^+}, while the corresponding conductance are represented by $G_{Ca^{2+}} = 0.025\frac{mS}{cm^2}$ and $G_{K^+} = 0.05\frac{mS}{cm^2}$, respectively. Furthermore, the relaxation time constants are given by $\tau_f = 80$ ms and $\tau_x = 300$ ms. We have to remark that in [1] it is assumed that the gating variable d is equal to its steady state. Please note that if τ_d tends to zero, we have the situation as in [1], since

$$\tau_d \frac{dd}{dt} = (d_\infty - d) \quad \Rightarrow \quad 0 = (d_\infty - d),$$

as $\tau_d \to 0$. In this paper, we will use the relaxation time constant of d, i.e., τ_d, as further non-zero parameter. Moreover, the choice $\tau_d = 0.1$ ms yields the same trajectory as in [1], but also smaller values of τ_d are conceivable. In Figure 1 some examples of EADs are presented with $\tau_d = 0.1$ ms and $G_{Ca^{2+}} \in \left\{0.029\frac{mS}{cm^2}; 0.03\frac{mS}{cm^2}; 0.031\frac{mS}{cm^2}; 0.035\frac{mS}{cm^2}\right\}$ (from left to right). Please compare Figure 6a in [15] with the second trajectory in Figure 1. Here, we see that the four dimensional system behaves very similar to the three dimensional system, provided τ_d is small enough and the other system parameters are the same. Moreover, we want to highlight that in [1] the authors basically studied the influence of $\tau_x \to \infty$, while in [15] the influence of mainly $G_{Ca^{2+}}$ and G_{K^+} is investigated. In this paper, we are focused on the influence of more system parameter and the identification of their importance. Therefore, we will consider in the following $\tau_d = 20$ ms. This will help to understand the complex dynamics of the considered system.

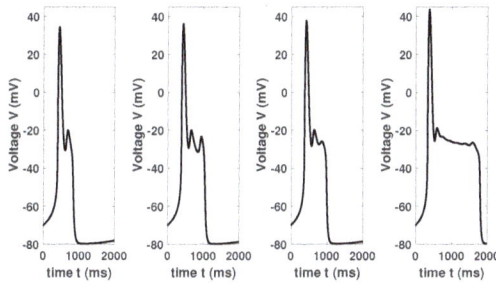

Figure 1. Trajectories (1) with different $G_{Ca^{2+}}$ values.

2. Investigation of EADs Using GSPT and Bifurcation Analysis

In this section, will study and analyse system (1). To this aim we will use the geometric singular perturbation theory, numerical bifurcation analysis and computational mathematics.

2.1. Brief Introduction into the GSPT

Here, we give a brief overview on the topic of GSPT. In general, a slow-fast system is of the form

$$\begin{cases} \varepsilon\dfrac{dx}{d\tau} &= F(x,y,p,\varepsilon), \\ \dfrac{dy}{d\tau} &= G(x,y,p,\varepsilon), \end{cases} \tag{5}$$

where $0 \le \varepsilon \ll 1$, $x \in \mathbb{R}^m$, $y \in \mathbb{R}^n$, $p \in \mathbb{R}^r$ with $m, n \ge 1$ and $r \ge 0$. We denote by x and y the state space variables and by p the system parameters, while the small parameter ε represents the ratio of time scales. Moreover, the functions $F : \mathbb{R}^m \times \mathbb{R}^n \times \mathbb{R}^r \times \mathbb{R} \to \mathbb{R}^m$ and $G : \mathbb{R}^m \times \mathbb{R}^n \times \mathbb{R}^r \times \mathbb{R} \to \mathbb{R}^n$ are assumed to be sufficiently smooth, typically C^∞. The space variables x are called fast variables, while the space variables y are called slow variables. Moreover, τ denotes the slow time scale and the fast time scale t is given by $t = \tau/\varepsilon$. If we rescale the system (5) in time—switching from the slow time scale to the fast one—we arrive at

$$\begin{cases} \dfrac{dx}{dt} &= F(x,y,p,\varepsilon), \\ \dfrac{dy}{dt} &= \varepsilon G(x,y,p,\varepsilon). \end{cases} \tag{6}$$

In general, solutions of slow-fast systems frequently exhibit slow and fast epochs characterised by the speed at which the solution advances. If ε tends to zero, the trajectories of (5) converge during the slow epochs to the solution of the slow flow/slow subsystem or reduced system

$$\begin{cases} 0 &= F(x,y,p,0), \\ \dfrac{dy}{d\tau} &= G(x,y,p,0), \end{cases} \tag{7}$$

while during fast epochs the trajectories of (6) converge to the fast subsystem or layer problem

$$
\begin{cases}
\dfrac{dx}{dt} = F(x,y,p,0), \\[2mm]
\dfrac{dy}{dt} = 0.
\end{cases}
\tag{8}
$$

The fast subsystem describes the evolution of the fast variables $x \in \mathbb{R}^m$ for fixed $y \in \mathbb{R}^n$, while the slow subsystem describes the evolution of the slow variables $y \in \mathbb{R}^n$. The phase space of the slow flow (reduced problem) is the critical manifold C_0, which is defined by $C_0 := \{(x,y) \in \mathbb{R}^m \times \mathbb{R}^n : F(x,y,p,0) = 0\}$. A subset $S \subset C_0$ is called normally hyperbolic if the $m \times m$ matrix (D_xF) of the first partial derivatives with respect to the fast variables x, i.e., the Jacobian of F with respect to x, has no eigenvalues with zero real part for all $(x,y) \in S$. Moreover, we call a normally hyperbolic subset $S^a \subset C_0$ attracting if all eigenvalues of (D_xF) have negative real parts, while we call a normally hyperbolic subset $S^r \subset C_0$ repelling if all eigenvalues of (D_xF) have positive real parts. If $S \subset C_0$ is normally hyperbolic and neither attracting nor repelling, it is of saddle type. Usually, the interesting dynamics are localised around these non-hyperbolic regions. There may be isolated points in C_0, i.e., folded singularities, satisfying $(D_yF)G(x,y,p,0) = 0 \in \mathbb{R}^m$ and $\mathrm{rk}(D_xF)(x,y,p,0) = m-1$, where the trajectories of the slow flow switch from incoming to outgoing. Away from fold points the implicit function theorem implies that C_0 is locally the graph of a function $h(y) = x$. Then, the reduced system (7) can be expressed as $\dot{y} = G(h(y),y,p,0)$, where $\dot{y} = dy/d\tau$. However, it is more convenient to write the slow flow in terms of the fast variables x and we can keep the differential-algebraic equations structure of (7). To this aim we determine the total (time) derivative of $F(x,y,p,0) = 0$. This yields $(D_xF)\dot{x} + (D_yF)\dot{y} = 0$ and we can write the slow flow (7) as the restriction to C_0 of the vector field

$$
\begin{cases}
\dot{x} = -(D_xF)^{-1}(D_yF)G(x,y,p,0), \\[1mm]
\dot{y} = G(x,y,p,0).
\end{cases}
\tag{9}
$$

This vector field blows up if F is singular and the slow flow is not defined on $\mathbb{F}$, i.e., the set of folded singularities, before desingularisation. Therefore, we consider the desingularised reduced system, which is given by

$$
\begin{cases}
\dfrac{dx}{d\tau_1} = (D_yF)G(x,y,p,0), \\[2mm]
\dfrac{dy}{d\tau_1} = -(D_xF)G(x,y,p,0)
\end{cases}
\tag{10}
$$

restricted to C_0, where we rescaled the time by $\tau = -(D_xF) \cdot \tau_1$. Moreover, ordinary singularities satisfy $G(x,y,p,0) = 0 \in \mathbb{R}^n$ are equilibria of the desingularised reduced system (10), the reduced system (9) and can be equilibria of the original system (5). Against it folded singularities are in general no equilibria of the reduced system (9) and of the original system (5). Notice that in the reduced system (9) folded singularities are special points, since both sides of the first equation vanish simultaneously. This means that there is potentially a cancellation of a simple zero, i.e., $\dot{x}$ is finite and non-zero at a folded singularity. This allows trajectories to cross the fold in finite time. Such solutions are called singular canards and their persistence under small perturbations gives rise to complex dynamics. If $n \geq 2$, the Jacobian of (10) evaluated at the folded singularities has $(n-2)$ zero eigenvalues and two remaining eigenvalues $\lambda_{1,2}$. Moreover, the folded singularities are classified as follows

$$\begin{cases} \text{folded saddle,} & \text{if } \lambda_{1,2} \in \mathbb{R} \text{ and } \lambda_1\lambda_2 < 0, \\ \text{folded saddle-node,} & \text{if } \lambda_{1,2} \in \mathbb{R} \text{ and } \lambda_1\lambda_2 = 0, \\ \text{folded node,} & \text{if } \lambda_{1,2} \in \mathbb{R} \text{ and } \lambda_1\lambda_2 > 0, \\ \text{folded focus,} & \text{if } \lambda_{1,2} \in \mathbb{C}. \end{cases} \tag{11}$$

Here, we have to highlight that folded saddles, folded nodes and folded foci are also known as canard points, see [22]. Even more, for sufficiently small values of the perturbation parameter ε it is possible to calculate the maximal number of small oscillations of a MMO pattern, see [23,24]. For instance, if λ_1 and λ_2 are the eigenvalues of the linearisation of the desingularised system at a folded node and $\mu = \lambda_1/\lambda_2$ with $|\lambda_1| < |\lambda_2|$, then the maximal number of small oscillations in the MMO (in a neighbourhood of the folded node) is given by

$$s_{max} := \left\lfloor \frac{\mu + 1}{2\mu} \right\rfloor, \tag{12}$$

i.e., the greatest integer less than or equal to $(\mu + 1)/2\mu$, provided $\sqrt{\varepsilon} \ll \mu$.

2.2. The Study of EADs as MMOs

After this short introduction into the topic of GSPT, we will go on with the investigation of the dynamics of our multiple time scale problem. To this aim we first have to derive a suitable model to be able to apply this theory. To determine the different time scales we use a certain type of time scale separation argument, cf. [25]. Thus, we introduce a new (dimensionless) time variable τ satisfying $t := k_t \cdot \tau$, where k_t is a reference time. Choosing $k_t = \tau_f$ and rewriting (1)–(3), we get:

$$\begin{cases} \varepsilon \cdot \dot{V} = -\bar{G}_{K^+} x(V - E_{K^+}) - \bar{G}_{Ca^{2+}} d \cdot f(V - E_{Ca^{2+}}) := F_1(V, d, f, x), \\ \varepsilon \cdot \dot{d} = (d_\infty - d) := F_2(V, d, f, x) \\ \dot{f} = (f_\infty(V) - f) := G_1(V, d, f, x), \\ \dot{x} = \delta(x_\infty(V) - x) := \delta G_2(V, d, f, x), \end{cases} \tag{13}$$

where we divided the first equation by $G := \max\{G_{K^+}, G_{Ca^{2+}}\}$ and defined $\bar{G}_{K^+} := G_{K^+}/G$ and $\bar{G}_{Ca^{2+}} := G_{Ca^{2+}}/G$ to derive the dimensionless singular perturbation parameters $\varepsilon_V := C_m/(\tau_f \cdot G)$, $\varepsilon_d := \tau_d/\tau_f$ and $\delta := \tau_f/\tau_x$. Using the setting from above we have that $\varepsilon \equiv \varepsilon_d \equiv \varepsilon_V$ with $0 \leq \varepsilon < \delta \ll 1$, which implies that the system exhibits three different time scales, where d and V are the fastest variables and x the slowest one. First of all, we have to notice that there are several system parameters, which have a huge influence on the time scale separation and the time scales, i.e., τ_d, τ_f, τ_x, $G_{Ca^{2+}}$, G_{K^+} and C_m, cf. (13). Our next step is to derive the critical manifold C_0. This yields

$$C_0 := \left\{ (V, f) : d = d_\infty(V), \ x = -\frac{\bar{G}_{Ca^{2+}}}{\bar{G}_{K^+}} \cdot d \cdot f \cdot \frac{(V - E_{Ca^{2+}})}{(V - E_{K^+})} \right\}. \tag{14}$$

We want to highlight that the critical manifold C_0 is the same in both cases (V and d are of the same time scale, or V is the fast variable and $d \equiv d_\infty$ in the 3D system). Since the critical manifold C_0 is the same in both cases and $d = d_\infty$ implying that $\dfrac{dd_\infty}{d\tau} = \dfrac{\partial d_\infty}{\partial V}\dfrac{dV}{d\tau}$ one can show that the desingularised slow flow of (13) restricted to C_0 is also the same as the one of the three dimensional system with $d \equiv d_\infty$. Moreover, for $\varepsilon \to 0$ we have the following slow subsystem:

$$0 = F_1(V, d_\infty, f, x), \quad \frac{df}{d\tau} = G_1(V, d_\infty, f, x), \quad \frac{dx}{d\tau} = \delta G_1(V, d_\infty, f, x), \tag{15}$$

and similarly the fast subsystem

$$\frac{dV}{d\tau_{\text{fast}}} = F_1(V, d, f, x), \quad \frac{dd}{d\tau_{\text{fast}}} = F_2(V, d, f, x), \quad \frac{df}{d\tau_{\text{fast}}} = 0, \quad \frac{dx}{d\tau_{\text{fast}}} = 0. \tag{16}$$

From (10), using $\dot{d}_\infty = \frac{\partial d_\infty}{\partial V}\dot{V}$ we can derive immediately the desingularised slow flow:

$$\begin{cases} \dfrac{dV}{d\tau_1} = \left(\dfrac{\partial F_1}{\partial f}\right) G_1(V, d_\infty, f, x) + \delta\left(\dfrac{\partial F_1}{\partial x}\right) G_2(V, d_\infty, f, x), \\ \dfrac{df}{d\tau_1} = -\left(\dfrac{\partial F_1}{\partial V}\right) G_1(V, d_\infty, f, x), \end{cases} \tag{17}$$

restricted to C_0, where $\tau = -\left(\dfrac{\partial F}{\partial V}\right)\tau_1$. Remember that system (17) is the desingularised version of

$$-\left(\frac{\partial F_1}{\partial V}\right)\frac{dV}{d\tau_1} = \left(\frac{\partial F_1}{\partial f}\right) G_1(V, d_\infty, f, x) + \delta\left(\frac{\partial F_1}{\partial x}\right) G_2(V, d_\infty, f, x),$$
$$\frac{df}{d\tau_1} = G_1(V, d_\infty, f, x), \tag{18}$$

restricted to C_0. An ordinary singularity of (17) is given if $G_1 = G_2 = 0$, while a fold point $z_\bullet = (V_\bullet, d_\infty(V_\bullet), f_\bullet, x_\bullet)^T \in \mathbb{F}$ is a folded singularity of (17) if

$$\frac{\partial F_1}{\partial f}(z_\bullet) \cdot G_1(z_\bullet) + \delta\frac{\partial F_1}{\partial x}(z_\bullet) \cdot G_2(z_\bullet) = 0 \quad \text{and} \quad \frac{\partial F_1}{\partial V}(z_\bullet) = 0.$$

This yields explicit expressions for $f_\bullet$ and $x_\bullet$ depending on $V_\bullet$, i.e.,

$$x_\bullet = -\frac{\bar{G}_{Ca^{2+}}}{\bar{G}_{K^+}} d_\infty(V_\bullet) f_\bullet \frac{(V_\bullet - E_{Ca^{2+}})}{(V_\bullet - E_{K^+})}$$

and

$$f_\bullet = \frac{1}{1-\delta} f_\infty(V_\bullet) + \frac{\delta}{1-\delta}\frac{\bar{G}_{K^+}(V_\bullet - E_{K^+})}{\bar{G}_{Ca^{2+}}(V_\bullet - E_{Ca^{2+}})}\frac{x_\infty(V_\bullet)}{d_\infty(V_\bullet)}.$$

At this stage we see that the shape of the critical manifold is not depending on δ or the choice of τ_f and τ_x, but the location of the folded singularities and their stability. Moreover, notice that the Jacobian of (17) has at least one zero eigenvalue. Furthermore, varying the ratio τ_f/τ_x changes the desingularised slow flow (17). Notice that for $\delta \to 0$ we have an one dimensional slow flow, where f is determined by the critical manifold and x is constant. Therefore, it does not make sense to consider both limits $\varepsilon \to 0$ and $\delta \to 0$ simultaneously. However, varying δ may compensate the effect of an enhanced calcium current, cf. [15]. Moreover, the critical manifold C_0 as well as the desingularised slow flow are depending on $\bar{G}_{K^+}$ and $\bar{G}_{Ca^{2+}}$, cf. (14) and (17). Hence, varying $\bar{G}_{K^+}$ and/or $\bar{G}_{Ca^{2+}}$ has an influence on (14) and (17). Furthermore, τ_d and C_m have only an influence on the time scale separation argument and after passing to the singular limit $\varepsilon \to 0$ our discussion is independent on τ_d and C_m.

In the following, we consider $G_{Ca^{2+}} = 0.032\frac{mS}{cm^2}$. Computing the critical manifold C_0 (14) together with two fold lines $L^\pm = \{(V, d, f, x) \in C_0 : F_{1V}(V, d, f, x) = 0, F_{1VV}(V, d, f, x) \neq 0\}$, the folded node $(V, f, x) \approx (-24.7923, 0.5804, 0.7027)$ with eigenvalues $\lambda_1 \approx -0.1974$ and $\lambda_2 \approx -1.7305$, an ordinary singularity $(V, f, x) \approx (-30.2250, 0.7666, 0.8760)$ and the singular orbit, we gain Figure 2. Notice that for τ_f and τ_x satisfying the ratio $\delta = \tau_f/\tau_x \equiv 4/15$ the folded node will be the same—similarly if $G_{Ca^{2+}}/G_{K^+} = 16/25$. The critical manifold is divided into two attracting sheets $S_a^\pm$ and one repelling sheet S^r, where S^r lies between the two fold lines $L^\pm$. The fold lines are nondegenerate since $\partial F_1/\partial f \neq 0$ or $\partial F_1/\partial x \neq 0$ or both is satisfied. Moreover, we have an ordinary singularity on S^r. Notice that spiking, bursting and plateauing are only possible provided that the ordinary singularity is unstable, i.e. the ordinary singularity lies on the repelling manifold S^r, cf. [26]. The singular or relaxation orbit consists of four distinct segments, i.e., two slow orbit Γ_S and two fast orbit Γ_F segments. Notice that in general, singular periodic orbits which are filtered into the folded node on L^+ are singular representations of MMOs. The aim of GSPT is now to combine information from the reduced and layer problems in order to understand the dynamics of the cell model (1), particularly the oscillatory behaviour. Thus, we use the reduced and the layer flows to construct singular periodic orbits, which—according to GSPT [26,27]—will perturb to nearby periodic orbits of the full system (1) for sufficiently small perturbations.

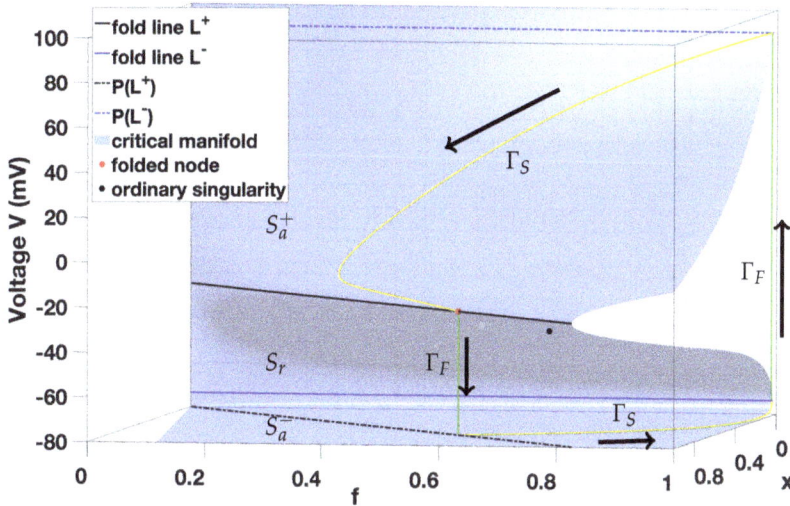

Figure 2. The critical manifold C_0, which is cubic shaped, i.e., $C_0 = S_a^- \cup L^- \cup S_r \cup L^+ \cup S_a^+$, including the singular orbit, which consists of four distinct segments, i.e., two slow orbit Γ_S (yellow line) and two fast orbit Γ_F (green line) segments, the fold lines $L^\pm$, the folded node and the ordinary singularity. In general, singular periodic orbits which are filtered into the folded node on L^+ are singular representations of mixed-mode oscillations (MMOs).

The singular orbit is constructed as follows. From lower fold line L^- there is a rapid evolution Γ_F described by (16) towards the upper attracting manifold S_a^+. Once the trajectory reaches S_a^+ the reduced flow Γ_S takes over until the trajectory reaches the upper fold line L^+. Then, at the fold line the reduced flow is singular and there is a finite time blow-up of the solution. The layer problem (16) becomes the appropriate descriptor and there is a fast down-jump to the lower attracting manifold. Here, the reduced system (15) describes the slow motions along the critical manifold until the trajectory once again hits the fold line. The GSPT guarantees that this singular orbit will persist as a nearby periodic relaxation oscillation corresponding to a spiking solution of (1). A folded node occurs in generic slow-fast systems with two (or more) slow variables [22,24,27]. Moreover, a folded node allows for an entire sector of trajectories to pass from the upper attracting branch S_a^+ of the critical manifold to the repelling branch S_r and to follow that repelling branch for an $\mathcal{O}(1)$ time on the slow time scale. Notice that solutions of the reduced problem (18) passing through a canard point from an attracting manifold S_a^+ to a repelling manifold S_r are called singular canards. The sector of canard solutions (the singular funnel, cf. Figure 3) is bounded by the fold line L^+ and by the strong canard γ_S, which is the unique trajectory tangential to the strong eigendirection of the folded node, cf. [26]. Two singular canards are related to the eigendirections of the folded node, i.e., the weak and strong canards. They correspond to the smallest and largest (in absolute value) eigenvalues respectively.

There are two major requirements that the singularly perturbed system has to satisfy to guarantee the existence of canard-induced MMOs, see [28] and cf. also [23], i.e.,

(a)   The reduced flow (desingularised slow flow) has to possess a folded node.
(b)   There is a singular periodic orbit formed by the slow and fast segments of the reduced and layer problems (slow and fast subsystem) which starts with a fast fiber segment at the folded node. This guarantees that the global return of such singular periodic orbit is within the singular funnel of the folded node.

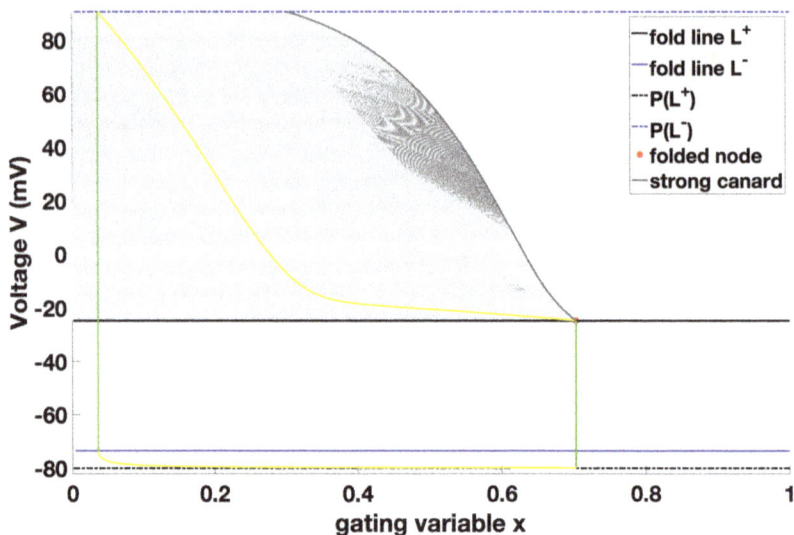

Figure 3. Singular funnel and strong canard on the (x, V)-plane.

Both conditions are satisfied in our situation. This implies that the MMOs are canard-induced and we have canard-induced EADs. Moreover, if $\sqrt{\varepsilon} \ll \mu = \lambda_1/\lambda_2 \approx 0.1141$, then $s_{max} = 4$, cf. (12). Please also note that that system (13) exhibits several types of folded singularities mentioned in (11) for different values of conductance $G_{Ca^{2+}}$. If we increase $G_{Ca^{2+}}$ the folded node will travel on the fold line L^+ to the "right", which means that the values of f and x at the folded node become bigger. Moreover, the ordinary singularity will travel also towards the folded node until both point collide. Then, the folded node becomes a folded saddle. If the folded node travels to the "left", then it will become a folded focus for smaller values of $G_{Ca^{2+}}$. Furthermore, we have a two dimensional fast subsystem, but then this system exhibits only limit point bifurcations and no Andronov-Hopf bifurcations. Hence, there are no Hopf-induced EADs induced by an enhanced calcium current. Thus, we have shown that system (13) exhibits only canard-induced EADs via an enhanced calcium current. Finally, we want to highlight also that system (13) exhibits Hopf-induced EADs provided we consider a fixed singular perturbation parameter ε and $\delta \to 0$. In this case we have a three dimensional fast subsystem with one bifurcation parameter x, but the occurring EADs are then related to a reduced potassium current [1].

2.3. The Study of EADs Using Bifurcation Analysis

In Section 2.2 we established that EADs related to an enhanced calcium current are canard-induced. Here, we did simultaneous the discussion for $G_{Ca^{2+}} = 0.032 \frac{mS}{cm^2}$ and all $C_m = \tau_d \cdot G_{K^+}$, provided $G_{Ca^{2+}} \leq G_{K^+}$. In addition, from (12) we know that $s_{max} = 4$ if $\sqrt{\varepsilon} \ll 0.1141$. Regarding our setting for the relaxation time constant $\tau_d = 20$ ms and the membrane capacity $C_m = 1 \frac{\mu F}{m^2}$ we see that

$\sqrt{\bar{\varepsilon}} = 0.25 \gg 0.1141$ and thus, $s_{max} = 4$ is not satisfied. Moreover, for a setting like $\tau_d = 40$ ms and $C_m = 2\frac{\mu F}{m^2}$ one diverges more from the condition $\sqrt{\bar{\varepsilon}} \ll 0.1141$, since $\sqrt{\bar{\varepsilon}} = 1/\sqrt{2}$. However, the system still exhibits MMOs or EADs but does not satisfy (12), since the condition $\sqrt{\bar{\varepsilon}} \ll \mu$ is barely to fulfil.

Our next step is the study of system (1) using bifurcation analysis. In general, a bifurcation of a dynamical system is a qualitative change in its dynamics produced by varying parameters. Since we investigate the occurrence of EADs induced by an enhancement in the calcium current $I_{Ca^{2+}}$, we will choose the conductance $G_{Ca^{2+}}$ as bifurcation parameter to be able to simulate the decreasing or mainly the increasing of the calcium current. Moreover, we will use our observation from above to analyse the behaviour of system (1). First of all, determining the equilibrium curve of system (1), which is basically the equilibria of this system for different values of $G_{Ca^{2+}}$, yields two stable branches and one unstable branch for all parameter settings. Depending on the parameter setting the equilibrium curve loses or wins stability via a sub- or supercritical Andronov-Hopf bifurcation, cf. also [15]. An Andronov-Hopf bifurcation is characterised by a pair of purely imaginary eigenvalues, where the equilibrium changes stability and a unique limit cycle bifurcates from it, i.e., it is the birth of a limit cycle. The distinction into sub- or supercritical means that an unstable or stable limit cycle, respectively, bifurcates. For the standard setting $\tau_d = 20$ ms system (1)–(4) exhibits two supercritical Andronov-Hopf bifurcations (black dots), cf. Figure 4.

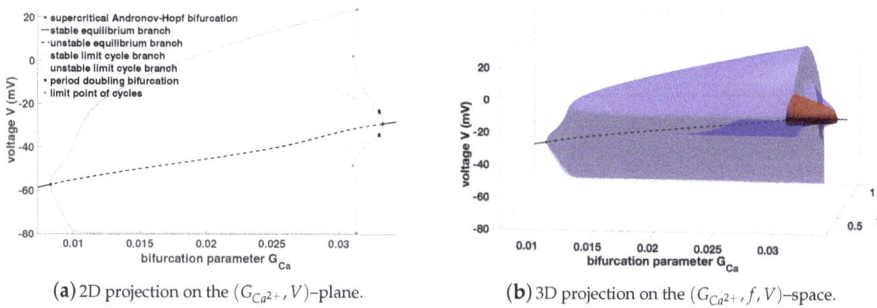

(**a**) 2D projection on the $(G_{Ca^{2+}}, V)$–plane.　　(**b**) 3D projection on the $(G_{Ca^{2+}}, f, V)$–space.

Figure 4. Bifurcation diagram for (1)–(4) with $\tau_d = 20$ ms.

From the first Andronov-Hopf bifurcation ($G_{Ca^{2+}} \approx 0.008253\frac{mS}{cm^2}$) a stable limit cycle branch bifurcates which becomes unstable via a limit point of cycle ($G_{Ca^{2+}} \approx 0.03134055\frac{mS}{cm^2}$) before it wins again stability via a period doubling bifurcation. There is also a second stable limit cycle branch bifurcating from the second Andronov-Hopf bifurcation ($G_{Ca^{2+}} \approx 0.033268\frac{mS}{cm^2}$) which becomes unstable via a period doubling bifurcation (connection of both limit cycle branches), cf. also Figure 5b.

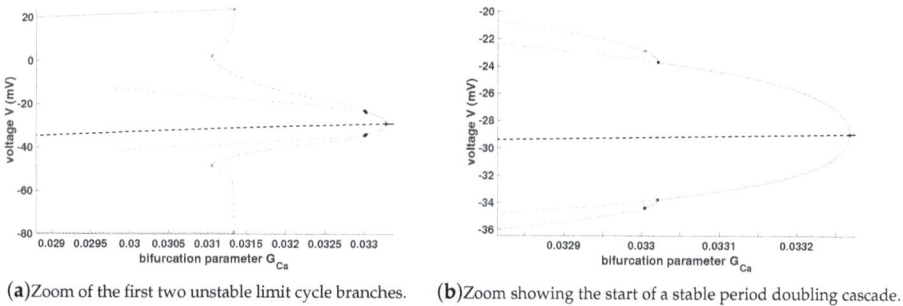

(**a**)Zoom of the first two unstable limit cycle branches.　　(**b**)Zoom showing the start of a stable period doubling cascade.

Figure 5. Zoom of Figure 4a around the second supercritical Andronov-Hopf bifurcation.

This unstable limit cycle branch has of course influence on the system (1) but it does not yields automatically EADs, it also may correspond to an AP. Notice that the limit cycle branches are determined via a continuation algorithm included in MATCONT. The region between the first Andronov-Hopf bifurcation and the first limit point of cycle ($G_{Ca^{2+}} \approx 0.03134055 \frac{mS}{cm^2}$) indicates the region, where no EADs occur, cf. [15]. EADs appear after the first limit point of cycle. In Figure 5a the transient from AP to EADs via the limit point of cycle bifurcation is highlighted, while Figure 5b shows the beginning of a stable period doubling cascade. In Figure 9 we see that this transient might be also via a period doubling bifurcation. Moreover, in Figure 6 we illustrate the limit cycle branches in 3D.

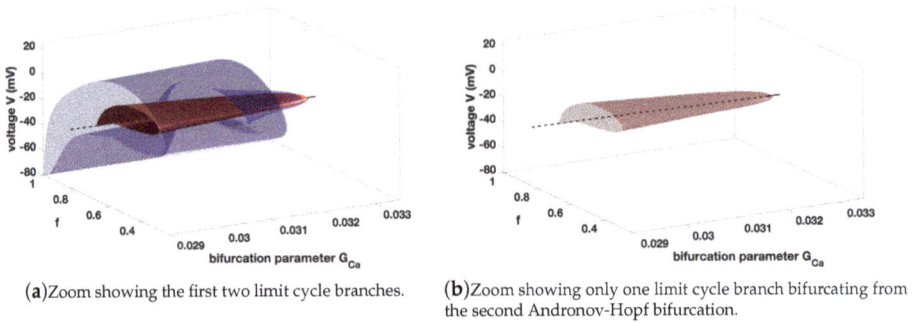

(**a**)Zoom showing the first two limit cycle branches.

(**b**)Zoom showing only one limit cycle branch bifurcating from the second Andronov-Hopf bifurcation.

Figure 6. Zoom of Figure 4b around the second supercritical Andronov-Hopf bifurcation.

For a better understanding we included in Figure 7 also two trajectories, one represents a normal AP, while the other shows an EAD.

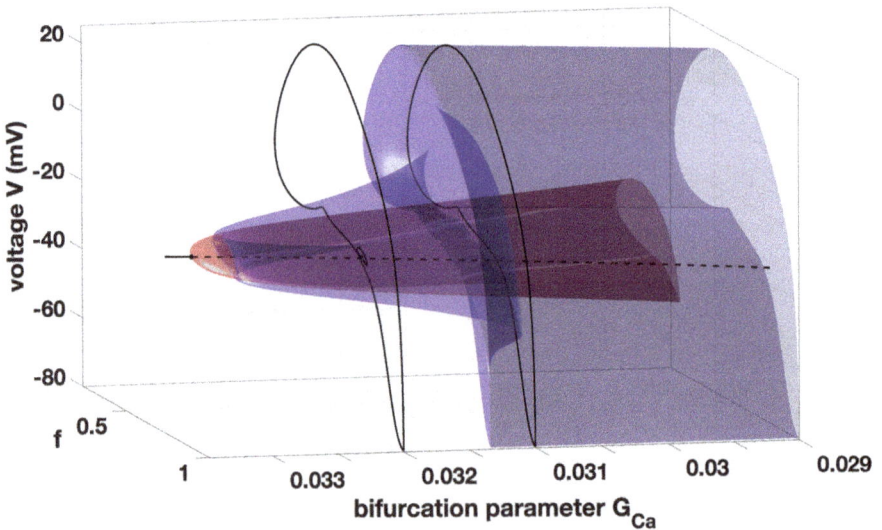

Figure 7. Figure 6a from a different point of view including two trajectories, i.e., one example for a normal action potential (AP) ($G_{Ca^{2+}} = 0.031 \frac{mS}{cm^2}$) and one example for an EAD ($G_{Ca^{2+}} = 0.032 \frac{mS}{cm^2}$).

Notice that we have a four dimensional phase space plus a further dimension for the parameter. Therefore, we have a five dimensional object which we can only plot in 2D or 3D as a projection on a 2D plane or 3D space. This makes the visualisation slightly difficult and it becomes more difficult if the dimension of the system increases. Nevertheless, one gets a good description of the behaviour of the considered system using the bifurcation theory. From the 2D and 3D projection in Figures 4–6 one might get the impression that the limit cycle starting from the second Andronov-Hopf bifurcation is not completed, but this limit cycle terminates at the unstable equilibrium branch, cf. the projection on the $(G_{Ca^{2+}}, x, V)$–space in Figure 8a. In Figure 8b we show for comparison the corresponding bifurcation diagram with $\tilde{\tau}_f = 0.8 \cdot \tau_f$ and $\tilde{\tau}_x = 0.8 \cdot \tau_x$ instead of with τ_f and τ_x, cf. Figure 5b. Here, one sees that the behaviour is different compared to the standard setting, while in the discussion of the GSPT this change has no influence. Thus, it is important to use both approaches for the investigation of such phenomena.

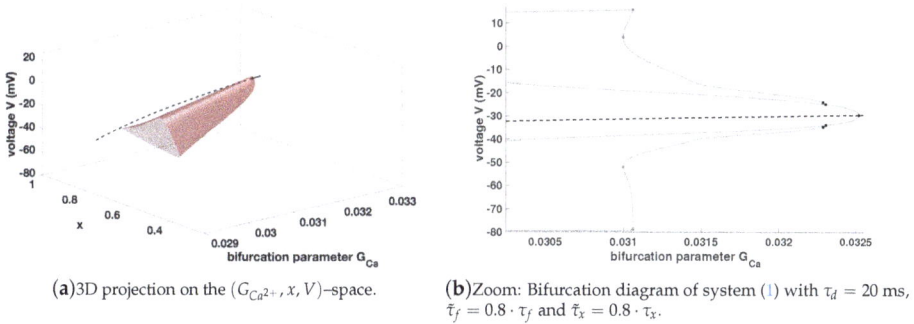

(a)3D projection on the $(G_{Ca^{2+}}, x, V)$–space.

(b)Zoom: Bifurcation diagram of system (1) with $\tau_d = 20$ ms, $\tilde{\tau}_f = 0.8 \cdot \tau_f$ and $\tilde{\tau}_x = 0.8 \cdot \tau_x$.

Figure 8. In (a) a different point of view of Figure 6b is given to illustrate that the limit cycle branch terminates at the unstable equilibrium branch, while in (b) the corresponding bifurcation diagram with $\tilde{\tau}_f = 0.8 \cdot \tau_f$ and $\tilde{\tau}_x = 0.8 \cdot \tau_x$ is stated.

Finally, if we consider the bifurcation diagram of (1)–(3) with $\tau_d = 40$ ms and $C_m = 2\frac{\mu F}{m^2}$ instead of $\tau_d = 20$ ms and $C_m = 1\frac{\mu F}{m^2}$, we see again the importance to consider all these parameters, cf. Figure 9.

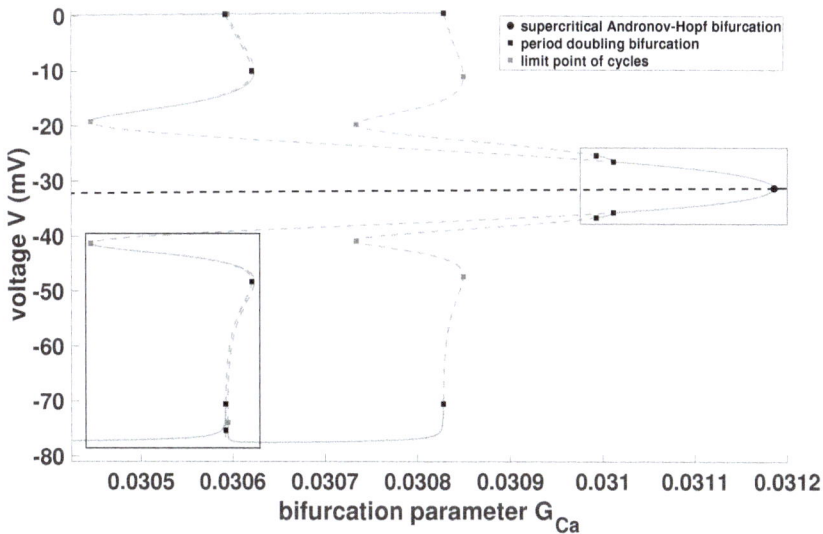

Figure 9. Zoom of bifurcation diagram: $C_m = 2\frac{\mu F}{m^2}$ and $\tau_d = 40$ ms.

In Figures 9 and 10a we see that the system (1) may exhibit different type of MMOs and critical transient regions depending on the choice of the system parameters. Even more, it also shows that

$\varepsilon = 0.5$ in combination with $G_{Ca^{2+}} = 0.032 \frac{mS}{cm^2}$ does not yield MMOs, cf. Figure 9. The reason for this is that the condition $0 < \varepsilon \ll 1$ is not suitable satisfied. Notice that there is no explicit condition how small ε has to be, only it has to be much smaller than 1. Nevertheless, Figure 9 shows the system (1) exhibits MMOs, but for smaller values of $G_{Ca^{2+}}$. For a more suitable visualisation of Figure 9, we present in Figure 10 two zooms of Figure 9. Notice that the system exhibits for this setting two supercritical Andronov-Hopf bifurcations. From the supercritical Andronov-Hopf bifurcation shown in Figure 10b a stable period doubling cascade bifurcates, which is a route to chaos, cf. [29].

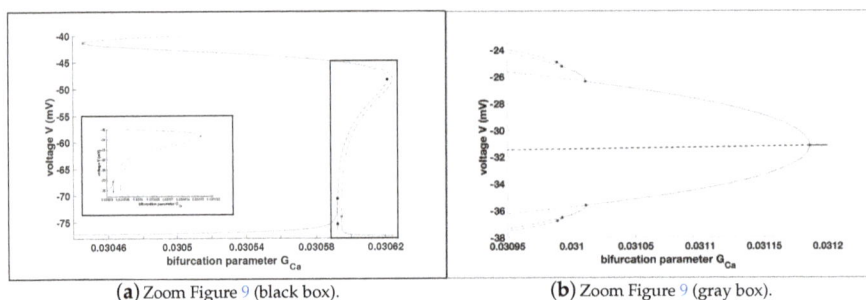

(a) Zoom Figure 9 (black box).   (b) Zoom Figure 9 (gray box).

Figure 10. Zooms of Figure 9 showing a critical transition region and a region around the supercritical Andronov-Hopf bifurcation.

Furthermore, we have shown that the GSPT gives information about the nature of the oscillatory behaviour and even more, one can use the GSPT to determine the important system parameters yielding these oscillations. However, we saw that it is not sufficient to consider only one parameter to analyse the complete dynamics of a dynamical system. Here, we have seen the high relevance for the investigation of MMOs in combination with bifurcation analysis to derive a more detailed understanding of EADs, which one can use to prevent them. In [15] some approaches to control the effect of an enhanced calcium current are established and for this aim a further system parameter is highly interesting, e.g., increasing of τ_d may smooth out this effect yielding EADs. Moreover, these observations, i.e., the system exhibits several time scales and MMOs as in Figure 1, motivate the investigation of system (13) in the sense of the geometric singular perturbation theory.

3. Discussion

In this paper we studied the occurrence of EADs in system (1) related to an enhancement in the calcium current. More precisely, we investigated the sensitivity of the system related to parameter changes. To this aim we used bifurcation theory, numerical bifurcation analysis and GSPT. Moreover, because of the fact that EADs may appear via an enhancement in the calcium current we used the conductance of the calcium current as bifurcation parameter to study the behaviour of system (1) under the influence of an enhanced calcium current. Furthermore, a time scale separation argument motivates to consider further important parameters, cf. (13). Under the assumption that stress, drugs or any diseases have no influence on the steady states of the gating variables, i.e., d_∞, f_∞ and x_∞, we discussed the behaviour of (13) with respect to changes in τ_d, τ_f, τ_x, C_m, $G_{Ca^{2+}}$ and G_{K^+}. Summarising we have shown that system (1) exhibits MMOs or EADs. These MMOs may appear as Hopf-induced MMOs via a reduction of the potassium current or as canard-induced MMOs related to the calcium dynamics of the system. Thus, we pointed out that system (13) may exhibits Hopf-induced EADs only if $\tau_x \to \infty$, which may yield plateau or pseudo-plateau bursting, cf. [30,31]. Furthermore, if $\varepsilon = C_m/(k_t \cdot G) = \tau_d/k_t \to 0$, where k_t is the chosen reference time and G the maximum of the conductances, system (13) may exhibits canard-induced EADs also depending on the choice of $G_{Ca^{2+}}$ and G_{K^+}. Moreover, this shows that EADs may occur via a combination of an enhanced calcium current and a reduced potassium current, cf. [15]. The bifurcation theory in combination with the GSPT and the canard theory [11,32] provides a strategy for the investigation of the complex dynamics of dynamical systems. This strategy yields simultaneously the parameter dependence for the occurrence of complex oscillatory behaviour of the studied system as well as the nature of these oscillations. Further, our approach shows also that the reduction of the system complexity is associated with the loose of information. In addition, if the singular limit is not satisfied, then the GSPT breaks down. However, the GSPT provides a powerful approach to study simpler subsystems and to combine the results of the studies, which yields a better understanding of the original system. This one can use for a specifically targeted examination of the processes, e.g., with the bifurcation theory. The bifurcation theory shows the behaviour of the system nicely with respect to one system parameter. This is also possible for several bifurcation parameters, cf. [15], but it becomes more complicated and time consuming. Moreover, the visualisation becomes more difficult if the phase space and/or the parameter space of the system increase. Therefore, one has to be more careful regarding the interpretation of the result.

Finally, we want to remark that for every new gating variable we have one more system parameter with influence on the appearing of EADs. Even more for each new ion current depending on a specific conductance, there is a further system parameter playing a huge role. Moreover, the investigation of such system using GSPT, yielding on the one hand the important system parameters, as we saw here, on the other hand we have to study 'only' subsystem of a reduced dimension, which is easier to handle. This approach we can use of course to investigate higher dimensional model $C_m \dot{V} = -I_{ion} + I_{stim}$, as in [33] or in [2]. High dimensional system are not only more challenging to study, in fact one has more possibilities to control oscillatory dynamics in such systems. Therefore, it is highly interesting and important to study high dimensional systems in theory as well as in applications. To this aim the GSPT and the bifurcation theory are important components. The numerical efforts will be higher but this will be a acceptable price which is to pay. Finally, we want to emphasise that the future project is the extension from the cellular level to the tissue level, cf. e.g., [21,34–37].

Acknowledgments: The author wishes to thank the anonymous referees for their careful reading of the original manuscript and their comments that eventually led to an improved presentation.

Conflicts of Interest: The author declares no conflict of interest.

References

1. Xie, Y.; Izu, L.T.; Bers, D.M.; Sato, D. Arrhythmogenic Transient Dynamics in Cardiac Myocytes. *Biophys. J.* **2014**, *106*, 1391–1397. [CrossRef] [PubMed]

2. Clayton, R.H.; Bernus, O.; Cherry, E.M.; Dierckx, H.; Fenton, F.H.; Mirabella, L.; Panfilov, A.V.; Sachse, F.B.; Seemann, G.; Zhang, H. Models of cardiac tissue electrophysiology: Progress, challenges and open questions. *Prog. Biophys. Mol. Biol.* **2011**, *104*, 22–48. [CrossRef] [PubMed]

3. Fink, M.; Niederer, S.A.; Cherry, E.M.; Fenton, F.H.; Koivumäki, J.T.; Seemann, G.; Thul, R.; Zhang, H.; Sachse, F.B.; Beard, D.; et al. Cardiac cell modelling: Observations from the heart of the cardiac physiome project. *Prog. Biophys. Mol. Biol.* **2011**, *104*, 2–21. [CrossRef] [PubMed]

4. Landstrom, A.P.; Dobrev, D.; Wehrens, X.H.T. Calcium Signaling and Cardiac Arrhythmias. *Circ. Res.* **2017**, *120*, 1969–1993. [CrossRef] [PubMed]

5. Roden, D.M.; Viswanathan, P.C. Genetics of acquired long QT syndromeg. *J. Clin. Investig.* **2005**, *115*, 2025–2032. [CrossRef] [PubMed]

6. Nieuwenhuyse, E.V.; Seemann, G.; Panfilov, A.V.; Vandersickel, N. Effects of early afterdepolarizations on excitation patterns in an accurate model of the human ventricles. *PLoS ONE* **2017**, *12*, e0188867. [CrossRef] [PubMed]

7. Zimik, S.; Vandersickel, N.; Nayak, A.R.; Panfilov, A.V.; Pandit, R. A Comparative Study of Early Afterdepolarization-Mediated Fibrillation in Two Mathematical Models for Human Ventricular Cells. *PLoS ONE* **2015**, *10*, e0130632. [CrossRef] [PubMed]

8.   Vandersickel, N.; Panfilov, A.V. A study of early afterdepolarizations in human ventricular tissue. In Proceedings of the 2015 Computing in Cardiology Conference (CinC), Nice, France, 6–9 September 2015; pp. 1213–1216.

9.   Sato, D.; Clancy, C.E.; Bers, D.M. Dynamics of sodium current mediated early afterdepolarizations. *J. Clin. Investig.* **2017**, *3*, e00388. [CrossRef] [PubMed]

10.   Bergfeldt, L.; Lundahl, G.; Bergqvist, G.; Vahedi, F.; Gransberg, L. Ventricular repolarization duration and dispersion adaptation after atropine induced rapid heart rate increase in healthy adults. *J. Electrocardiol.* **2017**, *50*, 424–432. [CrossRef]

11.   Desroches, M.; Guckenheimer, J.; Krauskopf, B.; Kuehn, C.; Osinga, H.M.; Wechselberger, M. Mixed-Mode Oscillations with Multiple Time Scales. *SIAM Rev.* **2012**, *54*, 211–288. [CrossRef]

12.   C. Kuehn. *Multiple Time Scale Dynamics*; Applied Mathematical Sciences; Springer: Heidelberg, Germany; New York, NY, USA, 2015; Volume 191.

13.   Kuznetsov, Y.A. *Elements of Applied Bifurcation Theory*; Springer: New York, NY, USA, 1998.

14.   Tsaneva-Atanasova, K.; Shuttleworth, T.J.; Yule, D.I.; Thompson, J.L.; Sneyd, J. Calcium Oscillations and Membrane Transport: The Importance of Two Time Scales. *Multiscale Model. Simul.* **2005**, *3*, 245–264. [CrossRef]

15.   Erhardt, A.H. Bifurcation Analysis of a Certain Hodgkin-Huxley Model Depending on Multiple Bifurcation Parameters. *Mathematics* **2018**, *6*, 103. [CrossRef]

16.   Dhooge, A.; Govaerts, W.; Kuznetsov, Y.A. MATCONT: A MATLAB Package for Numerical Bifurcation Analysis of ODEs. *ACM Trans. Math. Softw.* **2003**, *29*, 141–164. [CrossRef]

17.   Dhooge, A.; Govaerts, W.; Kuznetsov, Y.A.; Meijer, H.G.E.; Sautois, B. New features of the software MatCont for bifurcation analysis of dynamical systems. *Math. Comput. Model. Dyn. Syst.* **2008**, *14*, 147–175. [CrossRef]

18.   Govaerts, W.; Kuznetsov, Y.A.; Dhooge, A. Numerical Continuation of Bifurcations of Limit Cycles in MATLAB. *SIAM J. Sci. Comput.* **2005**, *27*, 231–252. [CrossRef]

19.   Hodgkin, A.L.; Huxley, A.F. A quantitative description of membrane current and its application to conduction and excitation in nerve. *J. Physiol.* **1952**, *117*, 500–544. [CrossRef] [PubMed]

20.   Noble, D. A modification of the Hodgkin-Huxley equations applicable to Purkinje fibre action and pacemaker potentials. *J. Physiol.* **1962**, *160*, 317–352. [CrossRef] [PubMed]

21.   ten Tusscher, K.; Noble, D.; Noble, P.J.; Panfilov, A.V. A model for human ventricular tissue. *Am. J. Physiol. Heart. Circ. Physiol.* **2004**, *286*, H1573–H1589. [CrossRef]

22.   Szmolyan, P.; Wechselberger, M. Canards in $\mathbb{R}^3$. *J. Differ. Equ.* **2001**, *177*, 419–453. [CrossRef]

23.   Vo, T.; Bertram, R.; Tabak, J.; Wechselberger, M. Mixed mode oscillations as a mechanism for pseudo-plateau bursting. *J. Comp. Neurosci.* **2010**, *28*, 443–458. [CrossRef]

24.   Wechselberger, M. Existence and Bifurcation of Canards in $\mathbb{R}^3$ in the Case of a Folded Node. *SIAM J. Appl. Dyn. Syst.* **2005**, *4*, 101–139. [CrossRef]

25.   Rubin, J.; Wechselberger, M. Giant Squid-hidden Canard: The 3D Geometry of the Hodgkin-Huxley Model. *Biol. Cybern.* **2007**, *97*, 5–32. [CrossRef] [PubMed]

26.   Vo, T.; Tabak, J.; Bertram, R.; Wechselberger, M. A geometric understanding of how fast activating potassium channels promote bursting in pituitary cells. *J. Comp. Neurosci.* **2013**, *36*, 259–278. [CrossRef] [PubMed]

27.   Szmolyan, P.; Wechselberger, M. Relaxation oscillations in $\mathbb{R}^3$. *J. Differ. Equ.* **2004**, *200*, 69–104. [CrossRef]

28.   Brøns, M.; Krupa, M.; Wechselberger, M. Mixed Mode Oscillations Due to the Generalized Canard Phenomenon. *Fields Inst. Commun.* **2006**, *49*, 39–63.

29.   Kügler, P.; Bulelzai, M.A.K.; Erhardt, A.H. Period doubling cascades of limit cycles in cardiac action potential models as precursors to chaotic early Afterdepolarizations. *BMC Syst. Biol.* **2017**, *11*, 42. [CrossRef] [PubMed]

30.   Teka, W.; Tsaneva-Atanasova, K.; Bertram, R.; Tabak, J. From Plateau to Pseudo-Plateau Bursting: Making the Transition. *Bull. Math. Biol.* **2011**, *73*, 1292–1311. [CrossRef] [PubMed]

31.   Tsaneva-Atanasova, K.; Osinga, H.M.; Rieb, T.; Sherman, A. Full system bifurcation analysis of endocrine bursting models. *J. Theor. Biol.* **2010**, *264*, 1133–1146. [CrossRef] [PubMed]

32.   Benoit, E.; Callot, J.F.; Diener, F.; Diener, M. Chasse au canard. *Collect. Math.* **1981**, *31–32*, 37–119.

33.   Luo, C.H.; Rudy, Y. A model of the ventricular cardiac action potential. Depolarization, repolarization, and their interaction. *Circ. Res.* **1991**, *68*, 1501–1526. [CrossRef]

34.   Sundnes, J.; Lines, G.T.; Nielsen, B.F.; Mardal, K.A.; Tveito, A. *Computing the Electrical Activity in the Heart*; Springer: Heidelberg, Germany, 2006.

35. Tveito, A.; Jæger, K.H.; Kuchta, M.; Mardal, K.A.; Rognes, M.E. A Cell-Based Framework for Numerical Modeling of Electrical Conduction in Cardiac Tissue. *Front. Phys.* **2017**, *5*, 48. [CrossRef]
36. Vandersickel, N.; Kazbanov, I.V.; Nuitermans, A.; Weise, L.D.; Pandit, R.; Panfilov, A.V. A Study of Early Afterdepolarizations in a Model for Human Ventricular Tissue. *PLoS ONE* **2015**, *9*, e84595. [CrossRef] [PubMed]
37. Quarteroni, A.; Manzoni, A.; Vergara, C. The cardiovascular system: Mathematical modelling, numerical algorithms and clinical applications. *Acta Numer.* **2017**, *26*, 365–590. [CrossRef]

processes

MDPI

Article

An Integrated Mathematical Model of Cellular Cholesterol Biosynthesis and Lipoprotein Metabolism

Frances Pool [1], Peter K. Sweby [2] and Marcus J. Tindall [2,3,*,†]

[1]   Institute of Ophthalmology, University College London, Gower Street, London WC1E 6BT, UK; fran_pool@hotmail.co.uk
[2]   Department of Mathematics and Statistics, University of Reading, Whiteknights, Reading RG6 6AX, UK; p.k.sweby@reading.ac.uk
[3]   Institute of Cardiovascular and Metabolic Research, University of Reading, Whiteknights, Reading RG6 6AA, UK
*   Correspondence: m.tindall@reading.ac.uk; Tel.: +44-118-378-8989
†   Current address: Department of Mathematics and Statistics, University of Reading, Whiteknights, Reading RG6 6AX, UK.

Received: 29 June 2018; Accepted: 10 August 2018; Published: 18 August 2018

check for updates

Abstract: Cholesterol regulation is an important aspect of human health. In this work we bring together and extend two recent mathematical models describing cholesterol biosynthesis and lipoprotein endocytosis to create an integrated model of lipoprotein metabolism in the context of a single hepatocyte. The integrated model includes a description of low density lipoprotein (LDL) receptor and cholesterol synthesis, delipidation of very low density lipoproteins (VLDLs) to LDLs and subsequent lipoprotein endocytosis. Model analysis shows that cholesterol biosynthesis produces the majority of intracellular cholesterol. The availability of free receptors does not greatly effect the concentration of intracellular cholesterol, but has a detrimental effect on extracellular VLDL and LDL levels. We test our model by considering its ability to reproduce the known biology of Familial Hypercholesterolaemia and statin therapy. In each case the model reproduces the known biological behaviour. Quantitative differences in response to statin therapy are discussed in the context of the need to extend the work to a more in vivo setting via the incorporation of more dietary lipoprotein related processes and the need for further testing and parameterisation of in silico models of lipoprotein metabolism.

Keywords: ordinary differential equation; SREBP-2

---

1. Introduction

Cholesterol is an intrinsic part of living cells. Every cell in the human body requires cholesterol in order to produce and maintain a healthy cell membrane. The formation of hormones and of bile acids that assist in the digestion of food, depend on cholesterol. Myelin, which covers nerve axioms to assist the conduction of electrical impulses, facilitating movement, vision, taste and the processing of sensory input is 20%, by weight, cholesterol [1]. This makes cholesterol vital for our nervous system and for memory and learning to take place. All cells have the ability to produce and regulate cholesterol, but the liver is primarily responsible for the metabolism of dietary cholesterol and is the only organ that can remove it from the body via the formation of bile.

Despite being such an important part of cellular health, irregular control of cholesterol homeostasis in hepatocytes (liver cells) can cause the liver to poorly process dietary cholesterol. This in turn can lead to high levels of circulating plasma cholesterol, which is widely known to be a major risk factor for a large number of cardiac diseases, for example coronary artery disease (CAD), cardiovascular disease (CVD) and coronary heart disease (CHD). The chance of developing CAD, CVD or CHD is determined by risk factors, some of which are modifiable. Modifiable risk factors include body weight, blood pressure and blood lipid levels, which are all influenced by exercise levels, smoking and diet. Non-modifiable risk factors include genetic predisposition, age, gender and ethnicity [2]. Whilst we may not be able to control these factors, the effects are sometimes modifiable with pharmaceutical interventions such as statins.

Dietary or exogenous cholesterol however, accounts for merely 20% of the body's cholesterol. The other 80%, endogenous cholesterol, is produced mainly by hepatocytes, but also by cells in the central nervous system and reproductive organs [3]. Each cell is subject to the cholesterol biosynthesis cascade initiated by 3-hydroxy-3-methyl-glutaryl-coenzyme A reductase (HMGCR). This signalling pathway cues the change in rate of production or inhibition of cholesterol in response to declining or increasing cellular cholesterol levels. In this case the transcription factor, sterol regulatory element-binding protein 2 (SREBP-2), is blocked from upregulating mRNA transcription of the HMGCR gene when cellular levels of cholesterol are high, but is free to upregulate transcription when levels are low. This allows the cell to change the rate of cholesterol production according to its needs.

Fats and cholesterol from a normal diet enter the blood stream, through the stomach, having been packed into carrier molecules known as a lipoproteins. Lipids are insoluble and must thus be packed into particles in order to be transported around the body. Lipoproteins are surrounded by phospholipids and apolipoproteins. As well as surrounding lipoprotein molecules, apolipoproteins play a vital role in the binding of lipoproteins to receptors on cell surfaces for removal from circulation. They also act as activators for lipolytic enzymes involved in metabolism.

There are five main classes of lipoprotein: chylomicrons, very low density lipoproteins (VLDL), intermediate density lipoprotein (IDL), low density lipoprotein (LDL) and high density lipoprotein (HDL). They are classed due to their varying triglyceride, cholesterol and apolipoprotein contents. Lipoproteins continuously exchange lipids and proteins with cells and other lipoproteins leading to a reduction in lipid content as the particles vary from being chylomicrons, to chylomicron remnants, VLDL, IDL and subsequently LDL particles.

Carrying fats as an energy source through the blood stream to cells in need, lipoproteins eventually end up in the liver where they are removed from circulation by a process known as receptor mediated endocytosis (RME). The rate of lipoprotein uptake is regulated by the number of available free receptors on the cell surface. Receptors are synthesised by the cell. Newly synthesised receptors are placed on the surface of the cell where they collect in clathrin coated pits. Apolipoproteins attach the lipoprotein to the receptor, after which the clathrin pit encloses around the lipoprotein and pinches off forming endocytotic vesicles which are internalised. Empty pits may also undergo this process. Following internalisation, the clathrin coating is shed and vesicles merge together to form larger endosomes within which the lipoprotein dissociates from the receptor. Some receptors are removed at this point and recycled to the cell surface. The endosomes then combine with lysosomes within the cell and the contents are degraded by lysosomal enzyme hydrolysis releasing amino acids and cholesterol for use in cellular metabolism [4].

Receptors, once synthesised or recycled, insert randomly on the cell surface before diffusing into clathrin coated pits. The concentration of receptors in the pits determine how many lipoproteins can bind and be internalised at any one time. LDL receptor (LDLR) synthesis is governed by SREBP-2. When intracellular cholesterol concentrations are low transcription of LDLR is upregulated, increasing the uptake of lipoproteins. Similarly high levels of cholesterol lead to downregulation of LDLR synthesis, decreasing lipoprotein endocytosis. In high cholesterol concentrations, receptor and cholesterol synthesis is inhibited. RME is the target of drugs, used in cardiovascular therapy, known as statins which inhibit cholesterol biosynthesis and up-regulate receptor synthesis, thus increasing the amount of lipoproteins cleared from the circulation.

Dysregulated cholesterol biosynthesis and lipoprotein metabolism can lead to a number of health conditions. Dyslipidemia, raised levels of LDL in blood plasma and/or reduced levels of high density lipoprotein (HDL), is a major health issue throughout the world [5], which has been linked to increases in dietary fat and sugar intake and sedimentary lifestyles. Hypercholesterolemia, elevated levels of cholesterol in blood plasma, has been linked to cardiovascular and pulmonary inflammation [6] and the overloading of macrophages with cholesterol in vitro has been shown to initiate immune responses. In contrast, unduly lowering the biosynthetic production of cholesterol levels, such as more recently shown via pathogenic infections which may lead to sepsis, can have dramatic adverse results [7,8].

There exists a growing literature on the mathematical modelling of lipoprotein metabolism as recently reviewed in [9]. Such models have generally been formulated using the theory of linear and nonlinear ordinary differential equations and parameterised and tested, to varying degrees, against the experimental literature. The mathematical models reviewed in [9] were tested for their ability to correctly predict the response of each to statin therapy. They found that only a small proportion of models within the literature correctly predicted the well known effect of statins on increasing LDL uptake from the circulation.

In contrast there are few mathematical models of cholesterol biosynthesis. Those that do exist vary in the size of the mathematical models formulated (number of variables and parameters) and complexity. In [10] the authors derived, parameterised and analysed a three variable nonlinear ordinary differential equation (ODE) model of cholesterol biosynthesis via the HMGCR pathway. They demonstrated that whilst the system only exhibited one steady-state, three types of behaviour were possible; monotic, damped and oscillatory. A more recent ODE model of the mevalonate pathway has been formulated and analysed in detail by [11]. This model describes cholesterol biosynthesis via the HMGCR and squalene synthase pathways, demonstrates the effect of the cholesterol-SREBP-2 feedback on the network's temporal responses, whilst more localised positive feedbacks within the network ensure cholesterol levels remained tightly bound should any products within the pathway be adversely increased or decreased.

Limited work has focused on integrating molecular scale cholesterol synthesis with lipoprotein endocytosis and LDLR synthesis. One exception is that of the unpublished work of [12], which integrated a description of LDL endocytosis [13] with that of cholesterol biosynthesis [10]. This was favourably evaluated by [9] in assessing how well the model reproduced the known cellular response to statins, but no mathematical or computational analysis of the model was undertaken.

In this work we present a model of integrated cholesterol biosynthesis, which includes a description of SREBP-2 regulation by cholesterol (as detailed in [14]), a full description of VLDL and LDL uptake (as detailed in [15]) coupled with a description of receptor biosynthesis. The work provides a full account of model formulation, analysis and testing thereof. We are motivated by the following considerations. Firstly, we wish to develop a well-informed integrated mathematical model of the core exogenous and endogenous cholesterol pathways within a hepatocyte. We wish to evaluate whether a simplified model formulated in an in vitro context can capture the known in vivo biological response of the system in respect of Familial Hypercholesterolaemia and statin therapy, without having to complicate the model by accounting for other in vivo aspects (for instance HDL, chylomicron remnants, VLDL hepatocyte recycling). Secondly, we wish to consider how rates of VLDL to LDL delipidation coupled with competition between the two particles for cell membrane level receptors may affect intracellular cholesterol levels. High levels of circulating LDL is an indicator of risk in a number cardiovascular diseases and one of the main sources of circulating LDL is delipidation. Because of this we have included VLDL to LDL delipidation in this model in order to explore it's effects without the complication of a full description of dietary lipoprotein metabolism. Finally, we wish to evaluate any differences that the assumption of a continuum of receptors on the surface of a cell has (as per [15]) versus that of discrete description of receptors bound by differing numbers of VLDL and LDL particles (as per [13]).

Our paper is organised as follows. In Section 1.1 we discuss the main biological features included in our integrated mathematical model before presenting an ODE model of the system in Section 2.1. Details of the model parameterisation are discussed in Section 2.4, which is followed by numerical simulations of the governing system of equations in Section 3.1. Results of computational and mathematical model analysis are presented in Section 3.2, before we investigate the effect of different classes of FH on extracellular levels of LDL and intracellular cholesterol levels. The effect of varying levels of statin therapy are investigated in Section 3.4 before we summarise and discuss our findings in Section 4.

Whilst our model has been formulated and parameterised in an in vitro context, the extrapolation to an in vivo setting is not considerably different given hepatocytes will be surrounded by lipoproteins within the liver. We thus wish to test how well this extrapolation works by testing the model against known in vivo outcomes in respect of Familial Hypercholesterolaemia and response to statin therapy.

1.1. Cholesterol Biosynthesis and Lipoprotein Metabolism

Our work here couples the endocytosis model of VLDL and LDL metabolism in an hepatocyte described in [15] with the description of cholesterol biosynthesis detailed in [10] and extends it with descriptions of LDLR synthesis, VLDL uptake and VLDL to LDL delipidation. An overview of the main processes included in our model is given in Figure 1 with further details on the exact mechanisms provided in Figure 2.

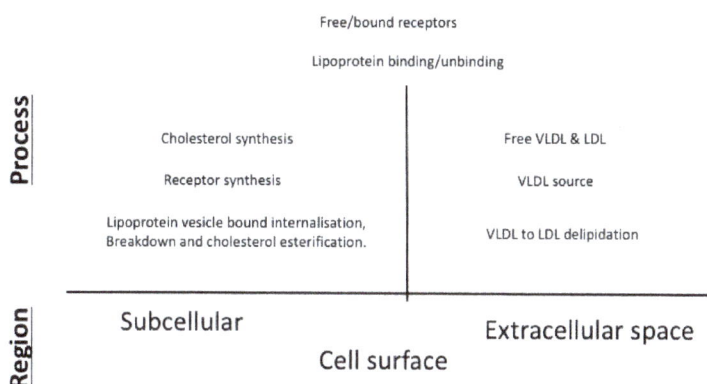

Figure 1. An overview of the main features included in our integrated mathematical model of cholesterol and receptor biosynthesis coupled with lipoprotein (VLDL and LDL) endocytosis.

While the model described in this work extends previous descriptions of in vitro lipoprotein endocytosis, it has not been formulated with a specific in vitro cell experiment in mind, as was the case in [15]. Instead it seeks to describe relevant processes at the subcellular and extracellular scale, which can be found both in vitro and in vivo, thus providing a means of extrapolating between the two. We assume concentrations of VLDL and LDL are fed to hepatocytes in a controlled manner, thereby describing the basic mechanisms of cholesterol synthesis and LDL and VLDL metabolism without the added complexity of describing other dietary lipoprotein metabolism.

These two models have been individually parameterised, analysed and their behaviour tested against published in vitro experimental data as detailed in each publication [10,15]. The integrated model consists of three main compartments: (i) the cell nucleus in which genetic regulation of HMGCR and LDLR occurs; (ii) the cell cytoplasm surrounded by the cell membrane in which all processes related to VLDL and LDL binding and breakdown, cell receptor and cholesterol regulation take place; and (iii) the extracellular space around the cell containing sources of VLDL and LDL, as shown in Figure 2. Whilst full details on each of the mathematical models can be found in each of the respective references, we provide here a summary of the main processes incorporated into the model for completeness, along with descriptions of the additional processes required for an integrated model. To ensure these mechanisms are clear in the context of the mathematical formulation presented in Section 2.1, we define the respective model variables as each mechanism is discussed. Parameter values associated with each process are detailed in Figure 2 and Table 1.

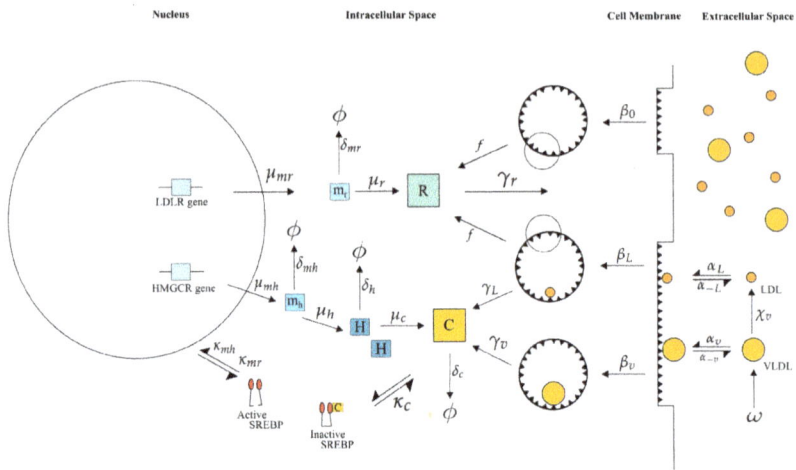

Figure 2. Our integrated model of cholesterol biosynthesis and lipoprotein endocytosis in a hepatocyte. Transcription of HMGCR and LDLR mRNA (at rates μ_{mh} and μ_{mr}, respectively) by SREBP-2 (κ_{mh}, κ_{mr}), leads to the synthesis of HMGCR (μ_h), LDLR (μ_r) and cholesterol (μ_c), which negatively regulates SREBP-2 (κ_c). In the extracellular space, VLDL delipidates to LDL (χ_v) and each bind/unbind to LDLR on the cell surface ($\alpha_v, \alpha_{-v}, \alpha_L, \alpha_{-L}$). Finally VLDL and LDL occupied receptor pits are endocytosed (β_V, β_L) as well as empty ones (β_0), cholesterol extracted from internalised lipoproteins (γ_v and γ_L) and receptors recycled in the intracellular space (f, γ_r). Here m_h represents HMGCR mRNA, m_r LDLR mRNA, H HMGCR, R_I receptors in the internal store, c intracellular cholesterol and ϕ degradation of the respective entity.

At the genetic level SREBP-2 transcribes the HMGCR gene to produce HMGCR mRNA, HMGCR and cholesterol (at rates $\bar{\mu}_{mh}$, $\bar{\mu}_h$ and $\bar{\mu}_c$ respectively) as described in [10]. Our integrated model also requires a description of LDLR synthesis as a result of SREBP-2 transcription which is described by the following biochemical equation

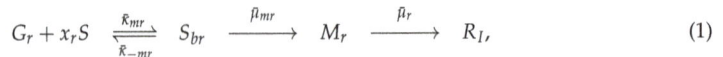

$$G_r + x_r S \underset{\bar{\kappa}_{-mr}}{\overset{\bar{\kappa}_{mr}}{\rightleftharpoons}} S_{br} \xrightarrow{\bar{\mu}_{mr}} M_r \xrightarrow{\bar{\mu}_r} R_I, \tag{1}$$

where unbound free LDLR gene is represented by G_r, free SREBP-2 is represented by S and S_{br} represents SREBP-2 bound to the LDLR gene. Once SREBP-2 and the LDLR gene become bound transcription of LDLR mRNA, M_r is upregulated and accordingly, translation of the mRNA to create receptors, R_I occurs. These are placed in the internal receptor store of the cell, assumed to be in the cell cytoplasm.

The reaction rate constants of $\bar{\kappa}_{mr}$ represent SREBP-2 binding to the LDLR free gene, whilst $\bar{\kappa}_{-mr}$ represents the reaction rate of SREBP-2 unbinding from the LDLR gene. Here x_r is the number of binding sites on the LDLR gene SREBP-2 binds to. The rate of transcription of mRNA responsible for receptor synthesis is $\bar{\mu}_{mr}$ and $\bar{\mu}_r$ is the rate of translation of LDLR from LDLR mRNA.

The incorporated mechanisms of VLDL and LDL endocytosis and their respective dynamics at the cellular level are as follows. LDL particles in the medium surrounding the cell (denoted $\bar{l}_E$) bind to free receptors ($\bar{r}_F$) in clathrin pits on the cell surface at rate $\bar{a}_L$ and unbind at rate $\bar{a}_{-L}$. In binding, LDL particles effectively bind to one receptor, but can occlude up to m_l surrounding receptors. The bound receptor and LDL ($\bar{l}_{RB}$) and the free and occluded receptors in the pit are then internalised (at rate $\bar{\beta}_L$) to become internalised vesicles ($\bar{l}_I$).

Internalised vesicles and their contents subsequently break down whereby the cholesterol $\bar{c}$ contained within the LDL particles is added to the internal cholesterol pool of the cell ($\bar{\gamma}_L$), where it is esterified for other cellular processes at a rate proportional to it's concentration ($\bar{\delta}_c$). A proportion f of the free receptors in the vesicle are placed in the cells internal cell store, ready for recycling back to the cell surface ($\bar{\gamma}_r$). The rest are degraded.

VLDL particle binding, internalisation and subsequent breakdown are governed by a similar series of processes represented by variables denoted with $\bar{v}$ instead of $\bar{l}$, for example $\bar{v}_E$ for the concentration of extracellular VLDL and $\bar{a}_v$ for binding of VLDL to receptors. Pits consisting of P empty receptors (no LDL or VLDL bound) may also be internalised $\bar{\beta}_0$.

Further to these processes, VLDL supply to the serum and VLDL to LDL delipidation is described by

$$\bar{\omega} \quad \longrightarrow \quad V_E \quad \xrightarrow{\bar{\chi}_v} \quad L_E, \tag{2}$$

where $\bar{\omega}$ is a constant supply of VLDL to the serum and $\bar{\chi}_v$ is the respective reaction rate constant of delipidation.

2. Materials and Methods

2.1. Mathematical Formulation

Applying the Law of Mass Action to those mechanisms described in Section 1.1 coupled with the models of [10,15] leads to the system of Equations (3)–(20), that integrate cholesterol and receptor biosynthesis with lipoprotein metabolism metabolism.

Biosynthesis of cholesterol and LDLR via SREBP-2 are described by

$$\frac{d\bar{g}_h}{d\bar{t}} = \left(\bar{\kappa}_{-mh}\bar{s}_{bh} - \bar{\kappa}_{mh}\bar{s}^{x_h}\bar{g}_h\right), \tag{3}$$

$$\frac{d\bar{g}_r}{d\bar{t}} = \left(\bar{\kappa}_{-mr}\bar{s}_{br} - \bar{\kappa}_{mr}\bar{s}^{x_r}\bar{g}_r\right), \tag{4}$$

$$\frac{d\bar{s}}{d\bar{t}} = \left(x_h\bar{\kappa}_{-mh}\bar{s}_{bh} - x_h\bar{\kappa}_{mh}\bar{s}^{x_h}\bar{g}_h + x_r\bar{\kappa}_{-mr}\bar{s}_{br} - x_r\bar{\kappa}_{mr}\bar{s}^{x_r}\bar{g}_r - \bar{\kappa}_c\bar{c}^{x_c}\bar{s} + \bar{\kappa}_{-c}\bar{c}_b\right), \tag{5}$$

$$\frac{d\bar{s}_{bh}}{d\bar{t}} = \left(-\bar{\kappa}_{-mh}\bar{s}_{bh} + \bar{\kappa}_{mh}\bar{s}^{x_h}\bar{g}_h\right), \tag{6}$$

$$\frac{d\bar{s}_{br}}{d\bar{t}} = \left(-\bar{\kappa}_{-mr}\bar{s}_{br} + \bar{\kappa}_{mr}\bar{s}^{x_r}\bar{g}_r\right), \tag{7}$$

$$J\frac{d\bar{m}_h}{d\bar{t}} = \bar{\mu}_{mh}\bar{s}_{bh} - \bar{\delta}_{mh}\bar{m}_h, \tag{8}$$

$$J\frac{d\bar{m}_r}{d\bar{t}} = \bar{\mu}_{mr}\bar{s}_{br} - \bar{\delta}_{mr}\bar{m}_r, \tag{9}$$

$$\frac{d\bar{h}}{d\bar{t}} = \bar{\mu}_h\bar{m}_h - \bar{\delta}_h\bar{h}, \tag{10}$$

where the time dependent variables represent concentrations of each of the respective entities detailed in Figure 2, such that $\bar{g}_h$ is the concentration of free HMGCR gene, $\bar{g}_r$ that of LDLR gene, $\bar{s}$ is free unbound SREBP-2, $\bar{s}_{bh}$ is SREBP-2 bound to the HMGCR gene, $\bar{s}_{br}$ is SREBP-2 bound to the LDLR gene, $\bar{m}_h$ is the concentration of HMGCR mRNA, $\bar{m}_r$ is that of LDLR mRNA and $\bar{h}$ is HMGCR. We note that x_c is the number of binding sites for cholesterol to bind to SREBP-2, x_r is the number of binding sites on the LDLR gene for SREBP to bind to and x_h is the number of binding sites on the HMGCR gene to which SREBP binds.

The mechanisms of LDL and VLDL endocytosis and the subsequent processing of cholesterol and receptors, is given by the following

$$W\frac{d\bar{l}_E}{d\bar{t}} = -\bar{\alpha}_L \bar{r}_f \bar{l}_E + \bar{\alpha}_{-L} \bar{l}_{RB} + W(\bar{\chi}_v \bar{v}_E), \tag{11}$$

$$\frac{d\bar{l}_{RB}}{d\bar{t}} = \bar{\alpha}_L \bar{r}_f \bar{l}_E - \bar{\alpha}_{-L} \bar{l}_{RB} - \bar{\beta}_L \bar{l}_{RB}, \tag{12}$$

$$\frac{d\bar{l}_I}{d\bar{t}} = \bar{\beta}_L \bar{l}_{RB} - \bar{\gamma}_L \bar{l}_I, \tag{13}$$

$$W\frac{d\bar{v}_E}{d\bar{t}} = -\bar{\alpha}_v \bar{r}_f \bar{v}_E + \bar{\alpha}_{-v} \bar{v}_{RB} + W(-\bar{\chi}_v \bar{v}_E + \bar{\omega}), \tag{14}$$

$$\frac{d\bar{v}_{RB}}{d\bar{t}} = \bar{\alpha}_v \bar{r}_f \bar{v}_E - \bar{\alpha}_{-v} \bar{v}_{RB} - \bar{\beta}_v \bar{v}_{RB}, \tag{15}$$

$$\frac{d\bar{v}_I}{d\bar{t}} = \bar{\beta}_v \bar{v}_{RB} - \bar{\gamma}_v \bar{v}_I, \tag{16}$$

$$\frac{d\bar{r}_f}{d\bar{t}} = \bar{\gamma}_r \bar{r}_I - P\bar{\beta}_0 \bar{r}_f^P - \tilde{m}_l \bar{\beta}_L \bar{l}_{RB} - m_l \bar{\alpha}_L \bar{l}_E \bar{r}_f + m_l \bar{\alpha}_{-L} \bar{l}_{RB} - \tilde{m}_v \bar{\beta}_v \bar{v}_{RB} - m_v \bar{\alpha}_v \bar{v}_E \bar{r}_f$$
$$+ m_v \bar{\alpha}_{-v} \bar{v}_{RB}, \tag{17}$$

$$\frac{d\bar{r}_I}{d\bar{t}} = \bar{\mu}_r \tilde{m}_r - \bar{\gamma}_r \bar{r}_I + Pf\bar{\beta}_0 \bar{r}_f^P + f(m_l + \tilde{m}_l)\bar{\beta}_L \bar{l}_{RB} + f(m_v + \tilde{m}_v)\bar{\beta}_v \bar{v}_{RB}, \tag{18}$$

$$\frac{d\bar{c}}{d\bar{t}} = R_L^{chol} \bar{\gamma}_L \bar{l}_I + R_v^{chol} \bar{\gamma}_v \bar{v}_I + \bar{\mu}_c \bar{h} - \bar{\delta}_c \bar{c} + J(x_c \bar{\kappa}_{-c} \bar{c}_b - x_c \bar{\kappa}_c \bar{c}^{x_c} \bar{s}), \tag{19}$$

$$\frac{d\bar{c}_b}{d\bar{t}} = (\bar{\kappa}_c \bar{c}^{x_c} \bar{s} - \bar{\kappa}_{-c} \bar{c}_b), \tag{20}$$

where the concentration of free LDL surrounding the cell is $\bar{l}_E$, $\bar{l}_{RB}$ is receptor bound LDL, $\bar{l}_I$ is internalised LDL, $\bar{v}_E$ is free VLDL surrounding the cell, $\bar{v}_{RB}$ is receptor bound VLDL, $\bar{v}_I$ is internalised VLDL, $\bar{r}_f$ are free unbound receptors, $\bar{r}_I$ are internalised receptors, $\bar{c}$ is cholesterol and $\bar{c}_b$ is the SREBP-2 cholesterol bound complex.

Equations (3)–(9) and (20) describe the synthesis of HMGCR mRNA and LDLR mRNA via SREBP-2 as regulated by cholesterol (Equation (19)). Equation (10) describes the production of HMGCR from HMGCR mRNA while production of LDLR, which is assumed to automatically join the internal receptor store, is accounted for in equation (18). Equations (11)–(20) are essentially those as detailed in [15] with the addition of the respective genetic synthesis terms describing cholesterol and LDLR production. Equations (11) and (14) detail the association of extracellular LDL and VLDL, respectively, to receptors and VLDL to LDL delipidation with a constant source of VLDL. Equations (12) and (15) describe bound LDL and VLDL formation, unbinding and internalisation, respectively, whilst Equations (13) and (16) detail the internalisation of LDL and VLDL bound pits and their respective breakdown. Equations (17) and (18) describe free and internalised receptor dynamics, whilst Equations (19) and (20) detail endogenous and exogenous cholesterol regulation, the latter the effect of cholesterol binding/unbinding to free SREBP-2.

The scaling parameters J and W are ratios representing the difference in volume between the compartments in which the reactions take place. The nucleus of the cell constitutes approximately 10% [16] of the volume of the total cell and thus we set $J = 0.1$. We assume the volume surrounding the cell is considerably greater than that of the cell, which from Jackson et al. [17] gives $W \approx 1.5 \times 10^3$. This accounts for the proportional concentrations between the three compartments: serum, cell and nucleus.

Many of the initial conditions are set equal to zero in order to help understand if the overall system response is in agreement with the known biological behaviour and are given by

$$
\begin{aligned}
&\bar{g}_h(0) = \bar{g}_{h0}, \quad \bar{g}_r(0) = \bar{g}_{r0}, \quad \bar{s}(0) = \bar{s}_0, \quad \bar{s}_{bh}(0) = 0, \quad \bar{s}_{br}(0) = 0 \\
&\bar{m}_h(0) = \bar{m}_{h0}, \quad \bar{m}_r(0) = \bar{m}_{r0}, \quad \bar{h}(0) = \bar{h}_0, \quad \bar{I}_E(0) = \bar{I}_{E0}, \quad \bar{I}_{RB}(0) = 0, \\
&\bar{I}_I(0) = 0, \quad \bar{v}_E(0) = \bar{v}_{E0}, \quad \bar{v}_{RB}(0) = 0, \quad \bar{v}_I(0) = 0, \quad \bar{r}_f(0) = \bar{r}_{f0}, \\
&\qquad\qquad \bar{r}_I(0) = 0, \quad \bar{c}(0) = 0 \quad \text{and} \quad \bar{c}_b(0) = 0.
\end{aligned} \tag{21}
$$

2.2. Model Reduction

Our model consists of 18 coupled linear and non-linear ODEs. We now seek to reduce the number of equations by application of the quasi-steady state approximation and conservation laws as detailed in Appendix A. This reduces the number of equations to 12 such that

$$
J \frac{d\bar{m}_h}{d\bar{t}} = \frac{\bar{\mu}_{mh}^*}{1 + \left(\frac{\bar{\kappa}_{mh}\left(1 + \left(\frac{\bar{c}}{\bar{\kappa}_c}\right)^{x_c}\right)}{\bar{s}_0} \right)^{x_h}} - \bar{\delta}_{mh}\bar{m}_h, \tag{22}
$$

$$
J \frac{d\bar{m}_r}{d\bar{t}} = \frac{\bar{\mu}_{mr}^*}{1 + \left(\frac{\bar{\kappa}_{mr}\left(1 + \left(\frac{\bar{c}}{\bar{\kappa}_c}\right)^{x_c}\right)}{\bar{s}_0} \right)^{x_r}} - \bar{\delta}_{mr}\bar{m}_r, \tag{23}
$$

$$
\frac{d\bar{h}}{d\bar{t}} = \bar{\mu}_h\bar{m}_h - \bar{\delta}_h\bar{h}, \tag{24}
$$

$$
W \frac{d\bar{I}_E}{d\bar{t}} = -\bar{\alpha}_L\bar{r}_f\bar{I}_E + \bar{\alpha}_{-L}\bar{I}_{RB} + W\bar{\chi}_v\bar{v}_E, \tag{25}
$$

$$
\frac{d\bar{I}_{RB}}{d\bar{t}} = \bar{\alpha}_L\bar{r}_f\bar{I}_E - \bar{\alpha}_{-L}\bar{I}_{RB} - \bar{\beta}_L\bar{I}_{RB}, \tag{26}
$$

$$
\frac{d\bar{I}_I}{d\bar{t}} = \bar{\beta}_L\bar{I}_{RB} - \bar{\gamma}_L\bar{I}_I, \tag{27}
$$

$$
W \frac{d\bar{v}_E}{d\bar{t}} = -\bar{\alpha}_v\bar{r}_f\bar{v}_E + \bar{\alpha}_{-v}\bar{v}_{RB} - W\bar{\chi}_v\bar{v}_E + W\bar{\omega}, \tag{28}
$$

$$
\frac{d\bar{v}_{RB}}{d\bar{t}} = \bar{\alpha}_v\bar{r}_f\bar{v}_E - \bar{\alpha}_{-v}\bar{v}_{RB} - \bar{\beta}_v\bar{v}_{RB}, \tag{29}
$$

$$
\frac{d\bar{v}_I}{d\bar{t}} = \bar{\beta}_v\bar{v}_{RB} - \bar{\gamma}_v\bar{v}_I, \tag{30}
$$

$$
\begin{aligned}
\frac{d\bar{r}_f}{dt} =\ & \bar{\gamma}_r\bar{r}_I - P\bar{\beta}_0\bar{r}_f^P - m_l\bar{\beta}_L\left(\frac{\bar{r}_f\bar{I}_{RB}}{\bar{r}_{f0} - \bar{r}_f}\right) - m_l\bar{\alpha}_L\bar{I}_E\bar{r}_f + m_l\bar{\alpha}_{-L}\bar{I}_{RB} \\
& -m_v\bar{\beta}_v\left(\frac{\bar{r}_f\bar{v}_{RB}}{\bar{r}_{f0} - \bar{r}_f}\right) - m_v\bar{\alpha}_v\bar{v}_E\bar{r}_f + m_v\bar{\alpha}_{-v}\bar{v}_{RB},
\end{aligned} \tag{31}
$$

$$
\begin{aligned}
\frac{d\bar{r}_I}{dt} =\ & \bar{\mu}_r\bar{m}_r - \bar{\gamma}_r\bar{r}_I + Pf\bar{\beta}_0\bar{r}_f^P + fm_l\bar{\beta}_L\bar{I}_{RB}\left(1 + \frac{\bar{r}_f}{\bar{r}_{f0} - \bar{r}_f}\right) \\
& + fm_v\bar{\beta}_v\bar{v}_{RB}\left(1 + \frac{\bar{r}_f}{\bar{r}_{f0} - \bar{r}_f}\right),
\end{aligned} \tag{32}
$$

$$
\frac{d\bar{c}}{d\bar{t}} = R_L^{chol}\bar{\gamma}_L\bar{I}_I + R_v^{chol}\bar{\gamma}_v\bar{v}_I + \bar{\mu}_c\bar{h} - \bar{\delta}_c\bar{c}, \tag{33}
$$

with the initial conditions

$$
\begin{aligned}
&\bar{m}_h(0) = \bar{m}_{h0}, \quad \bar{m}_r(0) = \bar{m}_{r0}, \quad \bar{h}(0) = \bar{h}_0, \quad \bar{I}_E(0) = \bar{I}_{E0}, \\
&\bar{I}_{RB}(0) = 0, \quad \bar{I}_I(0) = 0, \quad \bar{v}_E(0) = \bar{v}_{E0}, \quad \bar{v}_{RB}(0) = 0, \\
&\bar{v}_I(0) = 0, \quad \bar{r}_f(0) = \bar{r}_{f0}, \quad \bar{r}_I(0) = 0 \quad \text{and} \quad \bar{c}(0) = 0.
\end{aligned} \tag{34}
$$

2.3. Non-Dimensionalisation

The model was non-dimensionalised according to the following rescalings

$$\bar{t} = \frac{t}{\delta_{mh}}, \quad \bar{m}_h = \bar{s}_0 m_h, \quad \bar{m}_r = \bar{s}_0 m_r, \quad \bar{h} = \bar{s}_0 h, \quad \bar{l}_E = \bar{l}_{E0} l_E,$$
$$\bar{l}_{RB} = \bar{l}_{E0} l_{RB}, \quad \bar{l}_I = \bar{l}_{E0} l_I, \quad \bar{v}_E = \bar{v}_{E0} v_E, \quad \bar{v}_{RB} = \bar{v}_{E0} v_{RB}, \quad \bar{v}_I = \bar{v}_{E0} v_I, \tag{35}$$
$$\bar{r}_f = \bar{r}_{f0} r_f, \quad \bar{r}_I = \bar{r}_{f0} r_I, \quad \bar{c} = \bar{c}_T c,$$

where $c_T = 1.89 \times 10^{19}$ molecules/mL and $s_0 = 8.21 \times 10^{16}$ molecules/mL are the total concentrations of cholesterol and SREBP-2 in a hepatocyte [12,18], respectively. Substituting these rescalings into Equations (22)–(34) leads to

$$J\frac{dm_h}{dt} = \frac{\mu_{mh}}{1 + \left(\kappa_{mh}\left(1 + \left(\frac{c}{\kappa_c}\right)x_c\right)\right)^{x_h}} - \delta_{mh}m_h, \tag{36}$$

$$J\frac{dm_r}{dt} = \frac{\mu_{mr}}{1 + \left(\kappa_{mr}\left(1 + \left(\frac{c}{\kappa_c}\right)x_c\right)\right)^{x_r}} - \delta_{mr}m_r, \tag{37}$$

$$\frac{dh}{dt} = \mu_h m_h - \delta_h h, \tag{38}$$

$$W\frac{dl_E}{dt} = -\alpha_L r_f l_E + \alpha_{-L} l_{RB} + W\chi_v \rho_v v_E, \tag{39}$$

$$\frac{dl_{RB}}{dt} = \alpha_L r_f l_E - \alpha_{-L} l_{RB} - \beta_L l_{RB}, \tag{40}$$

$$\frac{dl_I}{dt} = \beta_L l_{RB} - \gamma_L l_I, \tag{41}$$

$$W\frac{dv_E}{dt} = -\alpha_v r_f v_E + \alpha_{-v} v_{RB} - W\chi_v v_E + W\omega, \tag{42}$$

$$\frac{dv_{RB}}{dt} = \alpha_v r_f v_E - \alpha_{-v} v_{RB} - \beta_v v_{RB}, \tag{43}$$

$$\frac{dv_I}{dt} = \beta_v v_{RB} - \gamma_v v_I, \tag{44}$$

$$\frac{dr_f}{dt} = \gamma_r r_I + \frac{m_l}{\vartheta_l}\left(-\beta_L \frac{r_f l_{RB}}{1-r_f} - \alpha_L l_E r_f + \alpha_{-L} l_{RB}\right)$$
$$- \beta_0 r_f^P + \frac{m_v}{\vartheta_v}\left(-\beta_v \frac{r_f v_{RB}}{1-r_f} - \alpha_v v_E r_f + \alpha_{-v} v_{RB}\right), \tag{45}$$

$$\frac{dr_I}{dt} = -\gamma_r r_I + f\beta_0 r_f^P + f\frac{m_l}{\vartheta_l}\left(1 + \frac{r_f}{1-r_f}\right)\beta_L l_{RB}$$
$$+ f\frac{m_v}{\vartheta_v}\left(1 + \frac{r_f}{1-r_f}\right)\beta_v v_{RB} + \mu_r m_r, \tag{46}$$

$$\frac{dc}{dt} = R_L^{chol}\sigma_l \gamma_L l_I + R_v^{chol}\sigma_v \gamma_v v_I + \mu_c h - \delta_c c, \tag{47}$$

with the initial conditions

$$m_h(0) = m_{h0}, \quad m_r(0) = m_{r0}, \quad h(0) = h_0, \quad l_E(0) = 1,$$
$$l_{RB}(0) = 0, \quad l_I(0) = 0, \quad v_E(0) = v_{E0}, \quad v_{RB}(0) = 0, \tag{48}$$
$$v_I(0) = 0, \quad r_f(0) = r_{f0}, \quad r_I(0) = 0 \quad \text{and} \quad c(0) = 0.$$

All non-dimensional parameters in terms of dimensional ones and their values are shown in Table 2.

2.4. Parameter Estimation

We have utilised parameterisations detailed in [10,15] to inform each of the models, respectively. Table 1 details each dimensional parameter, their value and source. Non-dimensional parameter values and their definitions (in terms of dimensional ones) are given in Table 2.

In the case of the additional reactions involving LDLR synthesis (LDLR transcription, translation, mRNA degradation and the dissociation of SREBP-2 for the LDLR gene) these were determined in a similar manner to that detailed in [10] with further details provided in Appendix B. The rates of VLDL and LDL delipidation and the source of extracellular VLDL were calculated as also shown in Appendix B.

Initially the integrated model was informed with the parameter values detailed in [10,15] and derived in Appendix B. A local sensitivity analysis was then used to determine which parameters required variation such that the integrated model reproduced known intracellular cholesterol concentrations [12]. Where adjustments to parameter values determined in [10,15] where made, this is detailed in Table 1 (denoted "This study" along with either the citing of [10,15]) and Appendix B.

Table 1. Dimensional model parameters. Molec. denotes molecules, r receptors and conc. concentration.

Parameter	Description	Dimensional Value	Units	Reference
$\bar{\mu}_{mh}^*$	Rate of HMGCR mRNA transcription.	5.17×10^5	$\frac{molec.}{mL\ s}$	[19,20]
$\bar{\mu}_{mr}^*$	Rate of receptor mRNA transcription.	4.56×10^6	$\frac{molec.}{mL\ s}$	[19,21]
$\bar{\mu}_h$	Rate of HMGCR translation.	3.32×10^{-2}	$\frac{1}{s}$	[20,22]
$\bar{\mu}_c$	Rate of cholesterol production.	2.16×10^3	$\frac{1}{s}$	[23–25]
$\bar{\mu}_r$	Rate of receptor translation.	5.10×10^{-1}	$\frac{1}{s}$	[26]
$\bar{\delta}_{mh}$	Rate of HMGCR mRNA degradation.	4.48×10^{-5}	$\frac{1}{s}$	[27]
$\bar{\delta}_{mr}$	Rate of receptor mRNA degradation.	4.48×10^{-5}	$\frac{1}{s}$	[28]
$\bar{\delta}_h$	Rate of HMGCR degradation.	6.42×10^{-5}	$\frac{1}{s}$	[29]
$\bar{\delta}_c$	Rate of cholesterol degradation.	1.20×10^{-4}	$\frac{1}{s}$	This study.
$\bar{\kappa}_{mh}$	SREBP-HMGCR gene binding affinity.	8.21×10^{16}	$\frac{molec.}{mL}$	This study.
$\bar{\kappa}_c$	Cholesterol-SREBP-2 dissociation constant.	8.91×10^{18}	$\frac{molec.}{mL}$	This study.
$\bar{\kappa}_{mr}$	LDLR gene-SREBP-2 dissociation constant.	8.21×10^{16}	$\frac{molec.}{mL}$	This study.
$\bar{\alpha}_L$	Rate of LDL-receptor binding.	6.66×10^{-17}	$\frac{mL}{r\ s}$	[30]
$\bar{\alpha}_{-L}$	Rate of LDL-receptor unbinding.	5.90×10^{-4}	$\frac{1}{s}$	[30]
$\bar{\alpha}_v$	Rate of VLDL-receptor binding.	9.32×10^{-16}	$\frac{mL}{r\ s}$	[15,17,31]
$\bar{\alpha}_{-v}$	Rate of VLDL-receptor unbinding.	2.95×10^{-4}	$\frac{1}{s}$	[15,17]
$\bar{\beta}_L$	Rate of LDL internalisation.	2.70×10^{-3}	$\frac{1}{s}$	[30,32,33]
$\bar{\beta}_v$	Rate of VLDL internalisation.	2.70×10^{-3}	$\frac{1}{s}$	[15]
$\bar{\beta}_0$	Rate of free receptor internalisation.	0	$\frac{mL^{(P-1)}}{r^{(P-1)}s}$	[34]
$\bar{\gamma}_L$	Rate of LDL to cholesterol conversion.	3.33×10^{-3}	$\frac{1}{s}$	[33]
$\bar{\gamma}_v$	Rate of VLDL to cholesterol conversion.	3.33×10^{-3}	$\frac{1}{s}$	[15]
$\bar{\gamma}_r$	Rate of receptor recycling.	1.00×10^{-2}	$\frac{1}{s}$	[30]
$\bar{\lambda}_v$	Rate of VLDL-LDL delipidation.	8.7×10^{-6}	$\frac{1}{s}$	[35]
M_l	Receptors covered by bound LDL.	1		[32]
M_v	Receptors covered by bound VLDL.	2		[15]
P	Number of receptors per pit.	180		[15]
f	Fraction of receptors recycled.	0.7		[36]
R_L^{chol}	Average cholesterol content per LDL.	3400		[37]
R_v^{chol}	Average cholesterol content per VLDL.	3100		[17]
J	Nucleus to cell ratio.	0.1		[16]
W	Cell medium to cell volume ratio.	1.50×10^3		[15]
x_c	Molec. of cholesterol to inactivate SREBP-2.	4		[12]
x_h	Number of binding sites for SREBP-2 on HMGCR gene.	3		[12]

Table 1. *Cont.*

Parameter	Description	Dimensional Value	Units	Reference
x_r	Number of binding sites for SREBP-2 on receptor gene.	1		[12]
ω	Influx of extracellular VLDL.	6×10^7	molec./mL	This study.
$\overline{m}_{h0}$	Initial HMGCR mRNA conc.	3.0×10^9	molec./mL	This study [10]
$\overline{m}_{r0}$	Initial LDLR mRNA conc.	5.0×10^9	molec./mL	[38]
$\overline{h}_0$	Initial HMGCR conc.	9.04×10^{11}	molec./mL	This study/[10]
$\overline{l}_{E0}$	Initial extracellular LDL conc.	1.17×10^{13}	molec./mL	[15,17]
$\overline{v}_{E0}$	Initial extracellular VLDL conc.	2.95×10^{12}	molec./mL	[15,17]
$\overline{r}_{f0}$	Initial unbound receptor conc.	3.27×10^{13}	molec./mL	[15,30]

Table 2. Table of non-dimensional parameters including their definition and value.

Parameter	Description	Definition	Non-Dimensional Value
μ_{mh}	Rate of HMGCR mRNA transcription.	$\frac{\mu^*_{mh}}{s_0\delta_{mh}}$	1.406×10^{-7}
μ_{mr}	Rate of receptor mRNA transcription.	$\frac{\mu^*_{mr}}{s_0\delta_{mh}}$	1.240×10^{-6}
μ_h	Rate of HMGCR translation.	$\frac{\overline{\mu}_h}{\delta_{mh}}$	7.4011×10^2
μ_r	Rate of receptor translation.	$\frac{\overline{\mu}_r \overline{s}_0}{\delta_{mh}\overline{r}_{f0}}$	2.876×10^7
μ_c	Rate of cholesterol synthesis.	$\frac{\overline{\mu}_c}{\delta_{mh}}$	2.099×10^5
κ_{mh}	HMGCR DNA-SREBP-2binding affinity.	$\frac{\overline{\kappa}_{mh}}{\overline{s}_0}$	1
κ_{mr}	Receptor DNA-SREBP-2binding affinity.	$\frac{\overline{\kappa}_{mr}}{\overline{s}_0}$	1
κ_c	SREBP-Cholesterol dissociation constant.	$\frac{\overline{\kappa}_c}{\overline{c}_0}$	4.714×10^{-1}
δ_{mh}	Rate of HMGCR mRNA degradation.	$\frac{\overline{\delta}_{mh}}{\delta_{mh}}$	1
δ_{mr}	Rate of receptor mRNA degradation.	$\frac{\overline{\delta}_{mr}}{\delta_{mh}}$	1
δ_h	Rate of HMGCR degradation.	$\frac{\overline{\delta}_h}{\delta_{mh}}$	1.433
δ_c	Rate of cholesterol degradation.	$\frac{\overline{\delta}_c}{\delta_{mh}}$	2.679
α_L	Rate of receptor-LDL binding.	$\frac{\overline{\alpha}_L \overline{r}_0}{\delta_{mh}}$	4.846×10^1
α_{-L}	Rate of receptor-LDL unbinding.	$\frac{\overline{\alpha}_{-L}}{\delta_{mh}}$	1.317×10^1
α_v	Rate of receptor-VLDL binding.	$\frac{\overline{\alpha}_v \overline{r}_0}{\delta_{mh}}$	6.782×10^2
α_{-v}	Rate of receptor-VLDL unbinding.	$\frac{\overline{\alpha}_{-v}}{\delta_{mh}}$	6.585
β_L	Rate of LDL internalisation.	$\frac{\overline{\beta}_L}{\delta_{mh}}$	6.027×10^1
β_v	Rate of VLDL internalisation.	$\frac{\overline{\beta}_v}{\delta_{mh}}$	6.027×10^1
β_0	Rate of empty pit internalisation.	$\frac{P\overline{\beta}_0 \overline{r}_0^{P-1}}{\delta_{mh}}$	0
γ_L	Rate of LDL-cholesterol conversion.	$\frac{\overline{\gamma}_L}{\delta_{mh}}$	7.440×10^1
γ_v	Rate of VLDL-cholesterol conversion.	$\frac{\overline{\gamma}_v}{\delta_{mh}}$	7.440×10^1
γ_r	Rate of receptor recycling.	$\frac{\overline{\gamma}_r}{\delta_{mh}}$	2.232×10^2
χ_v	Rate of VLDL-LDL breakdown.	$\frac{\overline{\chi}_v}{\delta_{mh}}$	1.94×10^{-1}
ω	Influx of extracellular VLDL.	$\frac{\overline{\omega}}{\delta_{mh}\overline{v}_{E0}}$	4.540×10^{-1}
ϑ_l	Ratio of initial free receptors to initial extracellular LDL.	$\frac{\overline{r}_{f0}}{\overline{l}_{E0}}$	2.786
ϑ_v	Ratio of initial free receptors to initial extracellular VLDL.	$\frac{\overline{r}_{f0}}{\overline{v}_{E0}}$	1.105×10^1
σ_l	Ratio of initial extracellular LDL to intracellular cholesterol concentration.	$\frac{\overline{l}_{E0}}{\overline{c}_0}$	6.190×10^{-7}
σ_v	Ratio of initial extracellular VLDL to to intracellular cholesterol concentration.	$\frac{\overline{v}_{E0}}{\overline{c}_0}$	1.561×10^{-7}
ρ_v	Ratio of extracellular VLDL to LDL concentration.	$\frac{\overline{v}_{E0}}{\overline{l}_{E0}}$	2.521×10^{-1}

Table 2. *Cont.*

Parameter	Description	Definition	Non-Dimensional Value
m_{h0}	Initial HMGCR mRNA concentration.	$\frac{m_{h0}}{s_0}$	3.65×10^{-8}
m_{r0}	Initial LDLR mRNA concentration.	$\frac{m_{r0}}{s_0}$	6.09×10^{-8}
h_0	Initial HMGCR concentration.	$\frac{h_0}{s_0}$	1.10×10^{-5}
v_{E0}	Initial extracellular VLDL concentration	$\frac{v_0}{\bar{v}_0}$	1
r_{f0}	Initial free receptor concentration.	$\frac{r_{f0}}{\bar{r}_{f0}}$	0.999

3. Results

3.1. Numerical Simulations

The system of Equations (36)–(48), parameterised with Table 1, was solved using the Matlab stiff differential equation solver ODE15s, given the stiffness coefficient of the system was determined to be $\lambda = 143,451$. A plot of the simulation is shown in Figure 3. We have re-dimensionalised time on the horizontal axis and run the system for approximately 175 h (until steady state) to capture the whole range of behaviours exhibited.

The solutions in Figure 3 show initially (up to 10 h) HMGCR and receptor mRNA increase in response to the initially low cholesterol levels which leads to an increase in HMGCR, internal receptors and cholesterol. VLDL and LDL bind rapidly to free receptors, however VLDL molecules bind more rapidly than LDL due to their greater binding affinity. The rapid binding of lipoproteins leads to an increase in bound LDL and VLDL and hence internalised LDL and VLDL increase. Intracellular cholesterol concentrations increase as cholesterol is extracted from the internalised lipoproteins, and receptors are stored internally and recycled to the cell surface. As intracellular cholesterol concentrations increase, the negative feedbacks from SREBP-2 inhibit HMGCR and LDLR mRNA transcription and hence less HMGCR and LDLR are synthesised. This decrease activates the feedforward/feedback mechanisms and the cell exhibits transient oscillatory type behaviour as a result of the system dynamics.

After this initial period, the molecular components of the system settle to a stable steady-state whilst the longer timescale events of VLDL and LDL endocytosis continue to occur. Eventually each component of this part of the system settles down to a non-zero stable steady-state, a result of the constant influx of VLDL to the system; extracellular VLDL settle before that of extracellular LDL given delipidation and an increased receptor-molecule binding affinity for VLDL than LDL. Internalised receptors tend to a non-zero steady state as the cell is constantly producing receptors to keep in the internal store ready for insertion onto the cell surface.

The concentration of intracellular cholesterol increases initially as a result of the biosynthesis cascade with cholesterol extracted from internalised VLDL and LDL having a significantly less impact on cholesterol levels after the first 10 h.

3.2. Model Analysis

3.2.1. Steady-State Analysis

Given the occurrence of negative and positive feedbacks (genetic and whole cell scale) within the system, we undertook a steady-state analysis of Equations (36)–(48) to understand how many biologically feasible steady-states it may exhibit; more than one real, positive steady-state may indicate more complex underlying system dynamics which have not been previously elucidated experimentally. This analysis was conducted in the absence ($\omega = 0$) and presence ($\omega \neq 0$) of a source of VLDL particles. In the case of $\omega = 0$ we obtained the expected result that all of the extracellular lipoproteins are internalised and esterified, leading to an abundance of free receptors on the cell surface and LDLR and cholesterol being produced via their respective biosynthetic pathways

$$
\begin{bmatrix} m_h^* \\ m_r^* \\ h^* \\ l_E^* \\ l_{RB}^* \\ l_I^* \\ v_E^* \\ v_{RB}^* \\ v_I^* \\ r_f^* \\ r_I^* \\ c^* \end{bmatrix}
=
\begin{bmatrix} \dfrac{\mu_{mh}}{\delta_{mh}\left(1+\left(\kappa_{mh}\left(1+\left(\dfrac{\mu_c\mu_h m_h^*}{\delta_h\delta_c}}{\kappa_c}\right)^4\right)\right)^3\right)} \\ \dfrac{\mu_{mr}}{\delta_{mr}\left(1+\kappa_{mr}\left(1+\frac{c^*}{\kappa_c}\right)^4\right)} \\ \dfrac{\mu_h m_h^*}{\delta_h} \\ 0 \\ 0 \\ 0 \\ 0 \\ 0 \\ 0 \\ \sqrt[P]{\dfrac{\gamma_r r_I^*}{\beta_0}} \\ \dfrac{\mu_r m_r^*}{\gamma_r(1-f)} \\ \dfrac{\mu_c\mu_h m_h^*}{\delta_h\delta_c} \end{bmatrix} ,
\tag{49}
$$

where the * notation indicates steady-state. This result was also verified numerically.

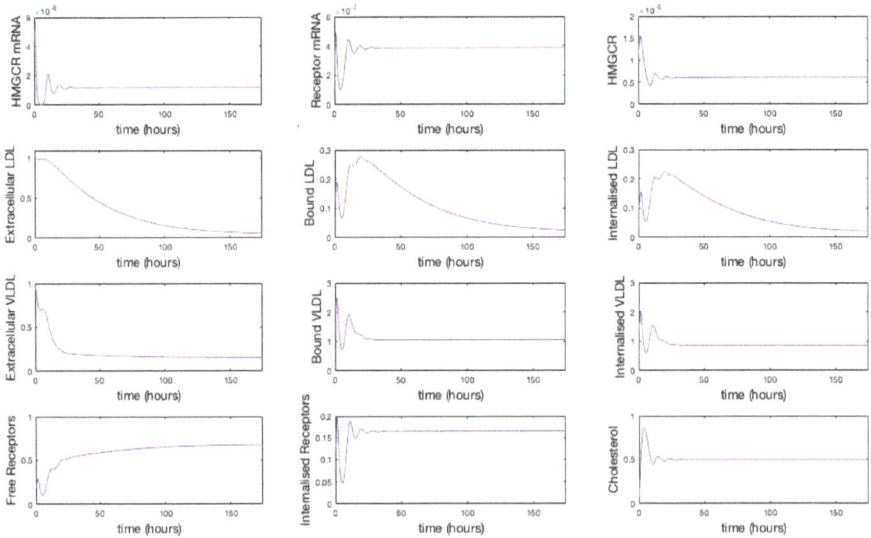

Figure 3. Numerical simulation of Equations (36)–(48). Initially mRNA levels increase in response to zero cholesterol in the system, which leads to an increase in HMGCR and internal receptor levels. VLDL and LDL bind to receptors and are internalised where cholesterol is extracted. The increase in HMGCR and extraction of cholesterol from internalised lipoproteins cause intracellular cholesterol concentrations to increase. Damped oscillations can be seen as HMGCR mRNA responds to changing cholesterol concentrations and the receptor mechanism responds accordingly. Following this initial transient behaviour, the molecular and cholesterol parts of the system settle to a relatively stable steady-state whilst VLDL and LDL continues to be extracted from the extracellular environment until a final steady-state is reached.

In the case of $\omega \neq 0$ the system of equations could only be reduced to the four state system of

$$\left(1 - \frac{1}{f}\right)\gamma_r r_I^* + \frac{m_l}{\vartheta_l}\rho_v(W\omega - \gamma_v v_I^*)\left(1 - (\alpha_{-L} + \beta_L) + \frac{\alpha_{-L}}{\beta_L}\right)$$

$$+\frac{m_v}{\vartheta_v}\gamma_v v_I^*\left(1 - \frac{(\alpha_{-v} + \beta_v)}{\beta_v} + \frac{\alpha_{-v}}{\beta_v}\right) + \frac{\mu_r \mu_{mr} f_2(c^*)}{\delta_{mr}} = 0, \tag{50}$$

$$-\gamma_r r_I^* + \frac{f m_l}{\vartheta_l}\left(1 + \frac{r_f^*}{1 - r_f^*}\right)\rho_v(W\omega - \gamma_v v_I^*) + \frac{f m_v}{\vartheta_v}\left(1 + \frac{r_f^*}{1 - r_f^*}\right)\gamma_v v_I^*$$

$$+\frac{\mu_r \mu_{mr} f_2(c^*)}{\delta_{mr}} = 0, \tag{51}$$

$$\gamma_r r_I^* + \frac{m_l}{\vartheta_l}\rho_v(W\omega - \gamma_v v_I^*)\left(\frac{-r_f^*}{1 - r_f^*} - (\alpha_{-L} + \beta_L) + \frac{\alpha_{-L}}{\beta_L}\right)$$

$$+\frac{m_v}{\vartheta_v}\gamma_v v_I^*\left(\frac{-r_f^*}{1 - r_f^*} - \frac{(\alpha_{-v} + \beta_v)}{\beta_v} + \frac{\alpha_{-v}}{\beta_v}\right) = 0, \tag{52}$$

$$R_L^{chol}\sigma_l\rho_v(W\omega - \gamma_v v_I^*) + R_v^{chol}\sigma_v\gamma_v v_I^* + \frac{\mu_c \mu_h \mu_{mh} f_1(c^*)}{\delta_h \delta_{mh}} - \delta_c c^* = 0, \tag{53}$$

where

$$f_1(c) = \frac{1}{1 + \left(\kappa_{mh}\left(1 + \left(\frac{c^*}{\kappa_c}\right)^{x_c}\right)\right)^{x_h}}$$

and

$$f_2(c) = \frac{1}{1 + \left(\kappa_{mr}\left(1 + \left(\frac{c^*}{\kappa_c}\right)^{x_c}\right)\right)^{x_r}}.$$

This system was then solved numerically in Maple (Version 2016.2) informed by the parameter values given in Table 2. Only one steady-state was determined and the values obtained were found to concur with the steady-states obtained in Figure 3.

3.2.2. Sensitivity analysis

Sensitivity analysis was conducted to determine how the output of our model can be apportioned to the varying sources of input. We conducted local sensitivity analysis, varying each parameter 100-fold above and below its initial value, whilst keeping all the other variables constant as shown in Table 2. We quantitatively measured, primarily, the effect of mechanisms on the steady state intracellular cholesterol concentrations whilst also looking for significant variations in key elements of the system, for example extracellular LDL levels which are an indicator of risk in CVD. What follows is a summary of our findings with more details provided in Appendix C.

Sensitivity analysis of our fully integrated model demonstrated that the respective components (cholesterol biosynthesis and endocytosis) reproduced similar effects on intracellular cholesterol levels as they did when considered in isolation. For instance the integrated model shows the same competition effect between LDL and VLDL for LDLR as detailed in [13,15]; a result of VLDL having a greater receptor binding affinity than LDL. Likewise, this behaviour can be exacerbated by the number of receptors occluded by a bound lipoprotein and by the rate of receptor synthesis. However, where there are limited binding sites on the cell surface, smaller LDL particles are able to bind in spaces that larger VLDL are unable to. We found that variation in parameters affecting LDLR synthesis does not greatly affect cholesterol levels.

We found that biosynthesis of cholesterol has a greater effect on cellular cholesterol levels than the uptake of extracellular lipoproteins, which agrees with the known biology [3]. The concentration of free LDLR does not greatly affect intracellular cholesterol levels, but does have a much greater effect on the concentration of extracellular LDL and VLDL, both of which are major risk factors for health conditions such as CVD. Therefore we know the receptor mechanism plays an important role in lipoprotein clearance which agrees with experimental evidence [39].

The amount of cholesterol within an LDL or VLDL molecule (R_l^{chol}, R_v^{chol}) was found to have very little effect on intracellular cholesterol levels. However, when R_l^{chol} and R_v^{chol} were increased, the concentration of HMGCR mRNA and LDLR mRNA were reduced. For 10-fold increased cholesterol in VLDL, HMGCR mRNA was reduced by nearly 20% and receptor mRNA reduced by around 5%. To induce similar levels of mRNA reduction, the cholesterol content of LDL needs to be increased 100-fold. This is due to the increased amount of cholesterol being brought in to the cell via receptor mediated endocytosis and shows very tight genetic regulatory control of intracellular cholesterol concentrations.

We have also found for certain values that parameters linked with cholesterol and receptor biosynthesis (μ_{mh}, μ_h, μ_c, κ_{mh}, δ_{mh}, μ_{mr}, μ_r, J, x_h and x_c) can produce damped periodic behaviour; the damping a result of the difference in the volumes of the cell nucleus and cytoplasm.

Sensitivity analysis of the VLDL-LDL delipidation parameter, χ_v, provided some interesting results. Increasing the rate of delipidation results in a 40% decrease in extracellular VLDL concentrations, but a 450% increase in extracellular LDL concentrations. Conversely, however, a 10-fold decrease in the delipidation rate produces around an 8% increase in extracellular VLDL concentrations but, significantly, nearly a 90% decrease in extracellular LDL concentrations. Furthermore, there is no other significant changes in the rest of the system due to the perturbation of χ_v. Our model would suggest, then, that delipidation of VLDL to LDL would be a good candidate as a target for LDL reduction therapies.

3.3. Investigating Familial Hypercholesterolaemia (FH)

In this section we investigate whether our model is able to reproduce the known effects of the disease Familial Hypercholesterolaemia (FH), the aetiology of which is well known. By altering specific model parameters we can quantitatively represent the effect of genetic mutations on extracellular VLDL and LDL levels.

FH is a genetic disorder, primarily of the LDL receptor gene and is characterised by high levels of circulating LDL cholesterol. In certain populations (including French and Canadians) 1 in 67 people suffer FH, with an increased risk of heart disease 20 times greater than non-sufferers [40]. Furthermore, almost all people with FH will require plasma cholesterol-lowering drugs.

The gene pertaining to the LDL receptor is located on chromosome 19 and a number of mutations have been identified in the DNA of individuals affected by this disorder [41]. Hobbs et al. [42] identified five categories of LDL receptor defects, which are listed in Table 3 with a description of the biological traits and parameters in our model that are affected. We will model each case by multiplying each relevant parameter by 0, 0.2, 0.4 and 1. We are unable to model type II since our model does not, with the exception of the nucleus and cytoplasm, sub-compartmentalise the cell.

Table 3. Familial hypercholesterolaemia class types and their relation to parameters in our model.

Class	Description	Parameter Affected
I	LDLR not synthesised.	μ_{mr}
II	LDLR not transported to the golgi apparatus.	
III	LDL-LDLR binding ineffective.	α_L
IV	Bound LDL not internalised properly.	β_L
V	Receptors not recycled effectively.	f

3.3.1. Class I FH

Let us first consider Class I where LDL receptors are not synthesised. In this case the associated parameter is μ_{mr}, which we vary in order to investigate the model response. The results illustrated in Figure A1 show how the inhibition of receptor synthesis prolongs uptake of extracellular LDL, which would equate to higher circulating plasma LDL levels.

As expected, with $\mu_{mr} = 0$, receptor numbers deplete and lipoproteins are unable to be internalised. Levels of extracellular VLDL increase because of continuous influx, as do LDL as VLDL are broken down into LDL. This increase in extracellular LDL concentration biologically would increase the risk of health problems. When varying μ_{mr} successively we find that even a small increase in the number of receptors synthesised decreases the levels of extracellular LDL and VLDL. For instance increasing the value from 0 to 20% of normal function halves the concentration of circulating VLDL and LDL. Increasing the value from 0 to 40% of normal function decreases the concentration of circulating VLDL and LDL by 80–90%, respectively.

3.3.2. Class III FH

We now consider Class III where binding of LDL and receptors is ineffective. In this case the associated parameter is α_l (LDL receptor binding), which we vary in order to investigate the model response. The results are illustrated in Figure A2 and show that the amount of extracellular LDL is affected significantly by the inability of LDL to bind to LDLR on the cell surface. We can see that increasing LDL-LDLR binding from 0 to just 20% of normal function decreases extracellular LDL concentrations by 67%. Subsequently restoring normal function reduces extracellular LDL concentrations by nearly 97%.

3.3.3. Class IV FH

Here LDL bound to receptors on the cell surface are not internalised properly. In this case we vary the associated parameter, β_l and investigate the model response. The results illustrated in Figure A3 demonstrate the concentration of extracellular LDL is significantly affected if $\beta_l = 0$, but is only marginally altered if β_l is increased. There is also a significant difference between the amount of bound LDL when $\beta_l = 0$ and when β_l is increased, however this does not appear to significantly affect extracellular VLDL concentrations. Aside from when $\beta_l = 0$, the system is fairly robust to changes in the internalisation rate of bound LDL particles.

3.3.4. Class V FH

Finally we consider Class V where LDL receptors are not recycled effectively. To investigate this case we vary the associated parameter f. The results in Figure A4 show the number of free and internalised receptors declines significantly for reduced receptor recycling which causes an increase in extracellular concentrations of LDL and VLDL. However the number of bound and internalised VLDL is not affected as significantly as the number of bound and internalised LDL. This is because VLDL have a greater binding affinity and so are more successful in binding competition. We also see a reduction by more than half in both free and internalised receptor concentrations between the usual value $f = 0.7$ and altered values $f \leq 0.7$.

3.3.5. Individual Class FH Summary

Having explored the effects of different classes of FH, we have found that Class I has the greatest effect on extracellular LDL and Class IV the least.

The lack of variation in intracellular cholesterol, HMGCR mRNA and HMGCR levels suggest that despite the effects of FH, the cell is able to maintain intracellular cholesterol levels genetically. This makes sense as without this control the concentration of cholesterol may decline to cytotoxic levels and the cell perish. However we know that humans are able to survive with FH indicating their cells do not perish.

We now wish to simulate the effects of being afflicted with a combination of all four cases.

3.3.6. Combined FH

We used Latin Hypercube Sampling [43], to consider the possible outcomes for a range of combined effects of FH Class types I, III, IV and V. Latin Hypercube Sampling generates a sample of plausible collections of parameter values from a multidimensional distribution. The method takes the midpoint of each quartile for parameters selected and randomly samples the combination of the effects of the four FH class types. In this case this leads to the four hypothetical combined FH cases detailed below.

FH Combined Case 1—(62.5% of μ_{mr}, 12.5% of α_L, 87.5% of β_L, 87.5% of f),
FH Combined Case 2—(87.5% of μ_{mr}, 62.5% of α_L, 37.5% of β_L, 12.5% of f),
FH Combined Case 3—(12.5% of μ_{mr}, 37.5% of α_L, 62.5% of β_L, 37.5% of f),
FH Combined Case 4—(37.5% of μ_{mr}, 87.5% of α_L, 12.5% of β_L, 62.5% of f).

Figure 4 provides a summary of the effects of each of the combined cases of FH on HMGCR mRNA, HMGCR, LDLR mRNA, extracellular LDL and VLDL and cellular cholesterol levels. For completeness full model results are provided in Appendix E . These lead to a disruption in receptor production, free receptors, extracellular levels of VLDL and LDL and the binding and internalisation of VLDL and LDL that we would expect to see as a result of the disease. Our model shows that despite lipoprotein uptake being significantly reduced, the cell will keep intracellular cholesterol levels within a tightly controlled range as a result of genetic regulation via the SREBP-2 cholesterol feedback. This feedback ensures the cell responds to low levels of cholesterol by upregulating cholesterol biosynthesis, allowing it to produce around 80% of the cholesterol the cell needs, in spite of disruptions to receptor function. Whilst direct comparison with experimental values of intracellular cholesterol is not possible due to a lack of reported values in the literature, we postulate here that this effect could be tested experimentally via a series of VLDL and LDL uptake experiments. For populations of cells each affected by the different FH classes, the relative difference in the uptake of the lipoproteins could be compared to that of a control group of cells, in order to discern the differences detailed here.

From these model results we can infer the increased susceptibility to CVD events, as a result of FH leading to increased plasma LDL levels concurs with the known biology. Furthermore, we can see from the samplings taken, Case 3 leads to the greatest rise in plasma LDL levels, due to low receptor synthesis and recycling combined with that of low LDL receptor binding affinity.

3.4. Modelling Statin Therapy

We can also consider if our model produces the known biological response to statins, globally the most commonly used pharmaceutical treatment for lowering plasma cholesterol levels. These drugs competitively bind to HMGCR preventing binding with HMGCoA and so inhibiting cholesterol biosynthesis. This reduces intracellular cholesterol concentrations thereby up-regulating receptor synthesis which clears more lipoproteins from the circulation. In this model, the effect of taking statins can be modelled by modifying the transcription of HMGCR mRNA, μ_{mh}. We here show the numerical results for an idealised statin that instantaneously halts transcription of HMGCR mRNA, for 11 doses over a period of 7 days such that

$$\mu_{mh} = \begin{cases} 0, & \text{for approx. } 9\,\text{h } 45\,\text{m} + n \times 14\,\text{h } 45\,\text{m} \leq t \leq 23\,\text{h} + n \times 14\,\text{h } 45\,\text{m,} \\ 1.406 \times 10^{-7}, & \text{otherwise,} \end{cases}$$

where n is the number of dosage periods. Although this is a dramatic change in μ_{mh} it is sufficient to show that the model replicates the expected dynamical behaviour. We have run the model to steady state and begun our simulation from that point. At time $t = 9$ h 45 m we set $\mu_{mh} = 0$ for the equivalent of approximately 13 h. Dosing in this way, the solutions give a 17.6% reduction in extracellular LDL. No further discernible differences in extracellular LDL levels were perceived after this period.

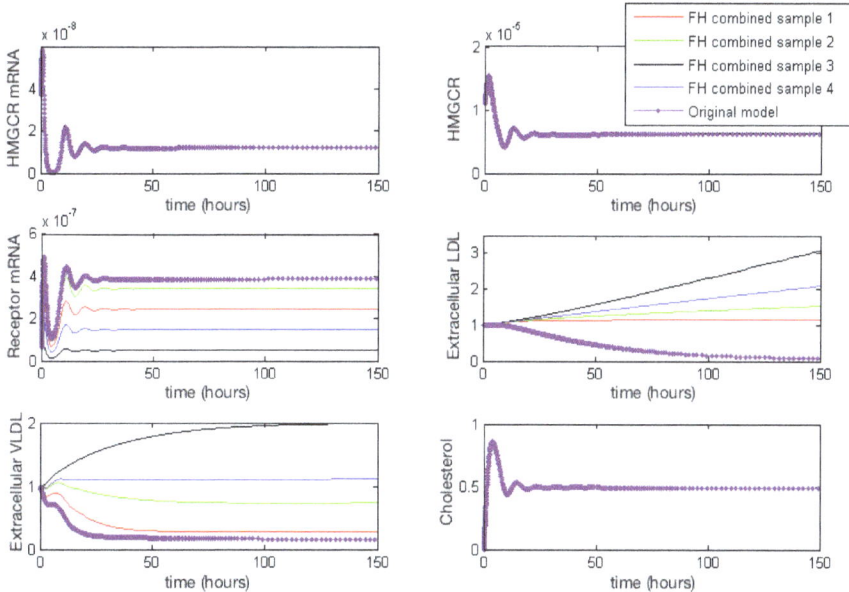

Figure 4. Familial Hypercholesterolaemia combined effects on HMGCR mRNA, HMGCR, LDLR mRNA, extracellular LDL and VLDL and cellular cholesterol levels. Results for all model variables are given in Appendix E.

It is indicated that in general statins cause a 25%–55% decrease in LDL-cholesterol [44]. We were able to achieve a 25% reduction in extracellular concentrations by setting the rate of receptor mRNA transcription $\mu_{mr}^{statin} = 0.3 \times \mu_{mr}$, with the same 11 doses over 7 days. The solutions in Figure 5 show that upon receiving a statin dose, levels of HMGCR mRNA and HMGCR decline to zero as transcription is inhibited. Solutions for all model variables are provided in Appendix F. Due to the lack of biosynthesis, cholesterol levels also decline dramatically which in turn up-regulates the transcription of receptor mRNA; a response by the cell to bring more cholesterol in to maintain healthy levels. Subsequently we see a rise in internal receptor levels and hence free receptors on the cell surface. Extracellular LDL and VLDL decrease as they bind to the abundant free receptors and are endocytosed. We see that with each statin dose extracellular LDL concentrations gradually decline for 7 days until they level out at a 25% decrease. Whilst of the same order of magnitude as that observed clinically we believe differences are a result of the short term duration of our statin application versus the longer term scale of measurements taken in patients (e.g. weeks or months). Furthermore, our model does not contain a detailed description of other elements of lipoprotein metabolism, for instance chylomicrons, or that of VLDL production by the hepatocyte. LDL levels are also directly linked to those of VLDL at present, whereas in vivo it is known they do not vary as much as other lipoproteins postprandially [45].

After 168 h (7 days) we allow μ_{mh} to return to its steady-state value of 1.406×10^{-7}. In doing so concentrations of each of the two biosynthesis pathways (cholesterol and LDLR) exhibit periodic overshoot type behaviour as the system returns to its pre-stimulus steady-state; a result of the homoclinic Hopf bifurcation behaviour that the cholesterol biosynthesis pathway exhibits [10].

Those entities directly affected by this change, for example receptor synthesis, also exhibit such behaviour, but this is dampened in the case of bound VLDL and LDL and considerably more so in the case of their extracellular levels.

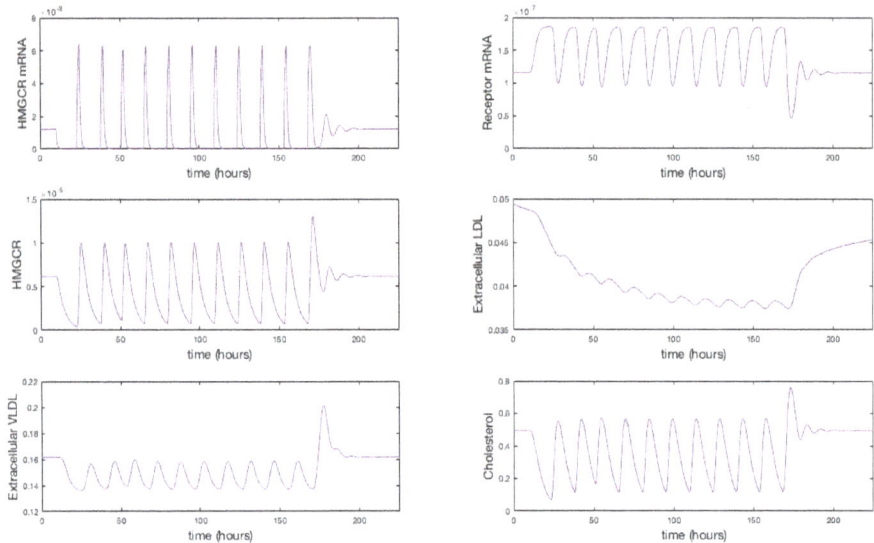

Figure 5. A simulation showing the effect of statin therapy on our integrated model of cholesterol metabolism. Here 11 doses are applied over a seven day period starting at approximately $t = 10$ h. The inhibition of HMGCR mRNA transcription subsequently decreases HMGCR and cholesterol biosynthesis which up-regulates receptor mRNA transcription and receptor synthesis, which leads to a reduction in extracellular LDL and VLDL levels. A complete set of model solutions are provided in Appendix F.

4. Discussion and Conclusions

In this paper we have formulated, solved and analysed a nonlinear deterministic ODE model describing the key mechanisms of hepatocyte endocytosis of VLDL and LDL coupled with a description of cholesterol and receptor biosynthesis via the HMGCR pathway.

Parameterised with data from the relevant literature, the model was solved using the Matlab stiff differential equation solver ODE15s. Solutions showed the system synthesising cholesterol and receptors in response to mRNA transcription and translation of each entity, and uptake of extracellular VLDL and LDL as a result of receptor synthesis and the extraction of cholesterol from internalised lipoproteins.

Sensitivity analysis showed that the model qualitatively reproduced the known biology of lipoprotein uptake and receptor and cholesterol regulation [17,30,33]. It highlighted the competition effect between LDL and VLDL when binding to free receptors on the cell surface. VLDL have a greater binding affinity than LDL and so are removed from the lipoprotein rich medium faster than LDL. However, we found that when receptor numbers were reduced, LDL particles are able to bind in spaces that VLDL are not able to, given their differences in sizes, and thus extracellular LDL levels decrease faster than VLDL ones. We found the concentration of free receptors on the cell surface for lipoproteins to bind to have little effect on intracellular cholesterol levels but greatly impacted concentrations of extracellular LDL and VLDL which is a major risk factor for health problems such as CVD. Sensitivity analysis also demonstrated that periodic behaviour exhibited by the cholesterol biosynthesis system [12] becomes damped as the signal dissipates from transcription in the cell nucleus to translation in the cell cytoplasm.

Our work has highlighted that small changes in the rate of VLDL to LDL delipidation had significant effects on extracellular LDL concentrations but not the rest of the system. Hence this process could be a good candidate target for LDL reductive therapies. While modelling processes in this way is helpful to identifying possible new methods of treatments, the enzymes responsible for this reaction (lipoprotein lipase, hepatic lipase and cholesteryl ester transfer protein) have dual functions which are not represented by our model, so it is hard to predict the full extent of including this change without further work.

Our model demonstrates that cholesterol biosynthesis is the dominant source of cholesterol for the cell. Thus any major disruption of this pathway is likely to have a detrimental effect on human health. Thus whilst any therapy targeted at reducing intracellular cholesterol, such as statins, will favourably decrease plasma levels of circulating lipoproteins in time, our model results suggest the counteracting of any decrease in intracellular cholesterol is limited by the number of receptors available on the cell surface (a function of the cell size and receptor internationalisation and recycling) and their rate of internalisation. Hence in cases where cholesterol biosynthesis is compromised we speculate it will be difficult to supplement intracellular cholesterol levels via increases in plasma cholesterol levels alone.

Although the mathematical formulation of lipoprotein endocytosis used here assumes the surface of the cell is covered in a continuum of LDLR [15], the main outcome of the model (total cholesterol content) does not greatly differ to that of [12] in respect of predicting each class of Familial Hypercholesterolaemia and statin therapy. We, however, note the differences between each of [13,15] at the lower level detail of receptor occupancy levels and rate of lipoprotein uptake require further investigation.

We have been able to explore the effects of the genetic disease Familial Hypercholesterolaemia and statins using our integrated model of cholesterol metabolism. We found that Case I FH has the greatest effect on extracellular LDL concentrations and Case IV the least. The effects of combined cases were shown to affect receptor mRNA, free and internalised receptor levels and the extracellular concentrations of LDL and VLDL. We also found the model replicates the qualitative effects of statins very closely.

Quantitatively the model produced a 25% reduction in extracellular LDL levels for repeated statin dosing over a seven day period when the receptor mRNA transcription rate was decreased by 70%. Whilst our the model able to reproduce the clinically reported lower bound of extracellular LDL reductions following statin therapy, we believe improvement in this result could be made by including a more thorough description of in vivo lipoprotein metabolism (e.g., chylomicrons, HDL) and longer dosing periods. Thus we believe further model extensions coupled with clinically informed parameterisation of this work are required to fully capture the quantitative regulation of lipoproteins and their responses to statin therapy.

In conclusion, our work has demonstrated that mathematical modelling can provide a useful tool for understanding the cellular (lipoprotein endocytosis) and subcellular (biosynthesis and genetic regulation of cholesterol and receptors) processes that occur during lipoprotein metabolism. Whilst the level of abstraction of our mathematical model is quite high, this work demonstrates that such simplifications of a complex system can still reproduce the known biology of disease states and therapeutic interventions. Future work and extensions to the model presented here is thus needed to consider the effect of other aspects of the overall system, occurring at the subcellular and tissue level. There thus remains scope for further testing and application of such models and their extension to contexts that include a description of other dietary lipoproteins.

Author Contributions: F.P. undertook research and wrote the paper, P.K.S. and M.J.T. both supervised the research and wrote the paper.

Funding: F.P. acknowledges the support of an Engineering Physical Sciences Research Council (EPSRC) UK CASE studentship in collaboration with Syngenta (EP/P505682/1 & EP/J500501/1). M.J.T. is grateful for the support of a Research Council UK Fellowship (EP/C508777/1) during parts of the period in which this work was undertaken.

Conflicts of Interest: The atuhors declare no conflict of interest.

Appendix A. Model Reduction

Equations (3)–(20) were reduced as follows. We first observe that the total number of genes with a cell remains constant whether bound to SREBP-2 or otherwise. Adding Equations (3) and (6) for HMGCR synthesis, and (4) and (6) for LDLR synthesis, respectively, and integrating with respect to time leads to

$$\bar{g}_h + \bar{s}_{bh} = \bar{g}_{h0} \quad \text{and} \quad \bar{g}_r + \bar{s}_{br} = \bar{g}_{r0}. \tag{A1}$$

The total amount of SREBP-2 in a cell is also constant such that

$$\bar{s} + \bar{s}_{bh} + \bar{s}_{br} + \bar{c}_b = \bar{S}_0 \approx \bar{s} + \bar{c}_b, \tag{A2}$$

since $\bar{s}_{bh} + \bar{s}_{br} \ll \bar{s} + \bar{c}_b$.

We further reduce the system by assuming DNA-transcription factor binding is rapid in comparison to the reaction events in the cell [10,46] and so reaches a steady state faster than the rest of the system. This rapid equilibrium approximation applied to Equation (6) leads to

$$\bar{s}_{bh} \approx \frac{\bar{g}_{h0}\bar{s}^{x_h}}{\bar{s}^{x_h} + \bar{K}_{mh}^{x_h}}, \tag{A3}$$

where $\bar{K}_{mh} = \left(\frac{\bar{k}_{-mh}}{\bar{k}_{mh}}\right)^{\frac{1}{x_h}}$.

Applying the same approximation to Equation (7) leads to

$$\bar{s}_{br} \approx \frac{\bar{g}_{r0}\bar{s}^{x_r}}{\bar{s}^{x_r} + \bar{K}_{mr}^{x_r}}, \tag{A4}$$

where $\bar{K}_{mr} = \left(\frac{\bar{k}_{-mr}}{\bar{k}_{mr}}\right)^{\frac{1}{x_r}}$.

Substitution of result (A3) into Equation (8) gives

$$J\frac{d\bar{m}_h}{d\bar{t}} = \frac{\bar{\mu}_{mh}\bar{g}_{h0}\bar{s}^{x_h}}{\bar{s}^{x_h} + \bar{K}_{mh}^{x_h}} - \bar{\delta}_{mh}\bar{m}_h, \tag{5.19a}$$

and similarly for result (A4) into Equation (9) we find

$$J\frac{d\bar{m}_r}{d\bar{t}} = \frac{\bar{\mu}_{mr}\bar{g}_{r0}\bar{s}^{x_r}}{\bar{s}^{x_r} + \bar{K}_{mr}^{x_r}} - \bar{\delta}_{mr}\bar{m}_r. \tag{5.20a}$$

We can substitute both results (A3) and (A4) into Equation (5) to give, after cancelling,

$$\frac{d\bar{s}}{d\bar{t}} = \bar{k}_{-c}(\bar{s}_0 - \bar{s}) - \bar{k}_c\bar{c}^{x_c}\bar{s}. \tag{5.16d}$$

Finally we assume that cholesterol-SREBP-2binding is rapid in comparison to other reaction events in the cell [10] such that

$$\bar{k}_{-c}(\bar{s}_0 - \bar{s}) - \bar{k}_c\bar{c}^{x_c}\bar{s} \approx 0,$$

which upon rearranging yields

$$\bar{s} \approx \frac{\bar{K}_c^{x_c} \bar{s}_0}{\bar{c}^{x_c} + \bar{K}_c^{x_c}} = \frac{\bar{s}_0}{1 + (\frac{\bar{c}}{\bar{K}_c})^{x_c}}, \tag{A5}$$

with $\bar{K}_c = \left(\frac{\bar{R}_{-c}}{\bar{R}_c}\right)^{\frac{1}{x_c}}$.

Using result (A5) we can express Equation (5.19a) in terms of c such that

$$J\frac{d\bar{m}_h}{d\bar{t}} = \frac{\bar{\mu}_{mh}^*}{1 + \left(\frac{\bar{R}_{mh}(1+(\frac{\bar{c}}{\bar{K}_{mh}})^{x_c}))}{\bar{s}_0}\right)^{x_h}} - \bar{\delta}_{mh}\bar{m}_h. \tag{5.19b}$$

with $\bar{\mu}_{mh}^* = \bar{\mu}_{mh}\bar{g}_{h0}$.

Similarly for (5.20a) we have,

$$J\frac{d\bar{m}_r}{d\bar{t}} = \frac{\bar{\mu}_{mr}^*}{1 + \left(\frac{\bar{R}_{mr}(1+(\frac{\bar{c}}{\bar{K}_c})^{x_c}))}{\bar{s}_0}\right)^{x_r}} - \bar{\delta}_{mr}\bar{m}_r, \tag{5.20b}$$

with $\bar{\mu}_{mr}^* = \bar{\mu}_{mr}\bar{g}_{r0}$.

Following the work in Tindall et al. [15] we seek to define the number of bound and occluded receptors. Quantities of $\bar{m}_{l,v}$ are dependent on the average occupancy of surface receptors. In high levels of extracellular LDL and VLDL, average occupancy will be high and free receptors will be low, hence fewer free receptors will be internalised with each pit and $\bar{m}_{l,v}$ will have a small value. Intuitively then, in low concentrations of extracellular LDL and VLDL, $\bar{m}_{l,v}$ will have a larger value. To simplify, we can assume the total number of receptors is approximately constant over shorter time-scales, such as that of pit internalisation, then we can say the total number of receptors on the cell surface is

$$\bar{r}_b + \bar{r}_f = m_l \bar{l}_{RB} + m_v \bar{v}_{RB} + \bar{r}_f \approx \bar{r}_k, \tag{A6}$$

where $\bar{r}_b$ represents bound receptors.

We can also assume, as pits are internalised, a fraction of all receptors are internalised also, given by

$$\bar{r}_b + \bar{r}_f = m_l \bar{l}_{RB} + m_v \bar{v}_{RB} + \bar{m}_l(\bar{r}_f)\bar{l}_{RB} + \bar{m}_v(\bar{r}_f)\bar{v}_{RB}. \tag{A7}$$

We assume the contribution of free receptor internalised with each particle is divided equitably between all bound particles, that is $\bar{m}_l(\bar{r}_f) = m_l T(\bar{r}_f)$ and $\bar{m}_v(\bar{r}_f) = m_v T(\bar{r}_f)$ with $T(\bar{r}_f)$ being the total density of free receptors. We can find $T(\bar{r}_f)$ by calculating the difference between Equations(A6) and (A7) giving

$$T(\bar{r}_f) = \frac{\bar{r}_f}{m_l \bar{l}_{RB} + m_v \bar{v}_{RB}} = \frac{\bar{r}_f}{\bar{r}_{f0} - \bar{r}_f}, \tag{A8}$$

where , since the number of internalised receptors is small, a suitable value for $\bar{r}_k$ is $\bar{r}_{f0}$

Appendix B. Parameter Estimation

$\tilde{m}_{r0}$—**Initial value of receptor mRNA**: Rudling et al. (2002) [38] details copy numbers of mRNA found in human liver cells under basal conditions. We take a value of 48 copies of receptor mRNA per cell, i.e. per 10^{-9} mL which gives 48 molecules/10^{-9} mL = 4.8×10^{10} molecules/ml. This value was then refined using sensitivity analysis to give $\tilde{m}_{r0} = 5.0 \times 10^9$ molecules/mL.

$\tilde{\mu}_{mr}$ —**Rate of receptor mRNA transcription**: As for $\tilde{\mu}_{mh}$, one human LDLR mRNA transcript is 5265 bases long [21]. To transcribe one molecule of LDLR mRNA, from one gene, at a rate of 12 bases per second, takes

$$\frac{5265 \text{ bases}}{12 \text{ bases/s}} = 438.75 \text{ s}.$$

So per gene we have 1/438.75 s = 2.28×10^{-3} molecules/s. There are two genes in a liver cell so 4.56×10^{-3} LDLR mRNA molecules are synthesised per cell per second. With a cell volume equal to 10^{-9} mL this gives 4.56×10^{-3} molecules/s / 1×10^{-9} mL giving $\tilde{\mu}_{mr} = 4.56 \times 10^6$ molecules/mL/s.

$\tilde{\mu}_r$—**Rate of receptor translation**: Soutar and Knight (1990) [19] tell us a human LDLR mRNA transcript contains 839 amino acids. For one ribosome to transcribe one molecule of LDLR protein, from one LDLR mRNA, at 6 amino acids per second, it takes 839 amino acids / 6 amino acids/s = 139.83 s. Then per ribosome this gives

$$\frac{1 \text{ molecule}}{139.83 \text{ s}} = 7.15 \times 10^{-3} \text{ molecules/s}.$$

The coding region of LDLR mRNA is 839 amino acids $\times$ 3 = 2517 nucleotides long, and a ribosome can attach every 35 nucleotides, so we have 71.91 ribosomes per mRNA molecule. Finally, 7.15×10^{-3} molecules/s/ribosome $\times$ 71.91 ribosomes gives $\tilde{\mu}_r = 0.51$ molecules/s.

$\tilde{\delta}_{mr}$—**Rate of receptor mRNA degradation**: In the absence of further details we assume the rate of receptor mRNA degradation is equivalent to that HMGCR mRNA degradation and so take $\tilde{\delta}_{mr} = 4.48 \times 10^{-5}$ /s.

$\tilde{\kappa}_{mr}$—**Dissociation of SREBP-2 for receptor gene**: Yang and Swartz (2011) [47] quantify DNA binding affinities to other transcription factors at 54.2 nmol. We convert this value into units of molecules/mL by the use of Avogadro's constant, such that

$$\frac{100 \times 10^{-9} \text{ moles}}{1000 \text{ mL}} \times (6.022 \times 10^{23} \text{ molecules/mol}) = 3.26 \times 10^{13} \text{ molecules/mL},$$

as an estimate we took $\tilde{K}_2 = O(10^{13})$. This value was then refined using sensitivity analysis to give $\tilde{\kappa}_{mr} = 8.21 \times 10^{16}$ molecules/mL.

$\tilde{\omega}$—**Rate of VLDL synthesis**: We estimated a value for the rate of VLDL synthesis as $\tilde{\omega} = 6.00 \times 10^7$ particles/mL/s. This value was derived by a local sensitivity analysis such that model reproduced known intracellular cholesterol concentrations.

x_r—**Number of binding sites on LDLR gene for SREBP-2**: The number of binding sites on the receptor gene available to SREBP-2 [12] is taken as being **1**.

Appendix C. Sensitivity Analysis

Results of the local sensitivity analysis of Equations (36)–(48) showing the relative percentage change from steady-state as each parameter was varied 100-fold below and above the values indicated in Table 2. Here "mRNA H" is HMGCR mRNA, "LE" is extracellular LDL, "LRB" is receptor bound LDL, "LI" is intenalised LDL, "VE" is extracellular VLDL, "VRB" is receptor bound VLDL, "VI" is internalised VLDL, "RF" is free receptors and "RI" is internalised receptors. Green indicates a less than

10% change from the steady-state value found when the model was solved using the values in Table 2, yellow greater than 10% but less than 50% and pink greater than 50%.

Para	Value	mRNA H	mRNA LDLR	HMGR	LE	LRB	LI	VE	VRB	VI	RF	RI	Cholesterol
μ_{mh}	1.41E-05	-14.59	0.00	63.48	790.00	740.21	741.92	6.15	0.00	0.00	-5.79	0.00	51.85
	1.41E-06	-14.54	0.00	32.27	38.43	38.12	38.22	0.30	0.00	0.00	-0.30	0.00	2.70
	1.41E-07	0.00	0.00	0.00	0.00	0.00	0.00	0.00	0.00	0.00	0.00	0.00	0.00
	1.41E-08	179.17	0.00	-40.10	-5.54	-5.51	-5.52	-0.04	0.00	0.00	0.04	0.00	-0.39
	1.41E-09	243.25	0.00	36.54	-5.88	-5.85	-5.87	-0.05	0.00	0.00	0.05	0.00	-0.42
μ_h	7.41E+04	49.70	0.00	86.68	813.33	760.77	762.53	6.33	0.00	0.00	-5.95	0.00	53.27
	7.41E+03	37.02	0.00	34.19	39.30	38.98	39.08	0.31	0.00	0.00	-0.31	0.00	2.76
	7.41E+02	0.00	0.00	0.00	0.00	0.00	0.00	0.00	0.00	0.00	0.00	0.00	0.00
	7.41E+01	-48.08	0.00	-49.00	-5.54	-5.51	-5.53	-0.04	0.00	0.00	0.04	0.00	-0.39
	7.41E+00	110.31	0.00	-94.29	-5.88	-5.85	-5.87	-0.05	0.00	0.00	0.05	0.00	-0.42
μ_{mr}	1.24E-04	-88.68	9900.00	-1.62	-39.23	-23.96	-23.98	-19.18	1.14	1.14	25.15	9900.00	-0.53
	1.24E-05	-88.00	900.00	-1.76	-36.86	-22.44	-22.46	-17.74	1.06	1.06	22.85	900.00	-0.51
	1.24E-06	0.00	0.00	0.00	0.00	0.00	0.00	0.00	0.00	0.00	0.00	0.00	0.00
	1.24E-07	-45.36	-90.00	-2.20	13354.46	314.37	314.24	1085.45	-63.48	-63.47	-96.92	-90.00	-38.95
	1.24E-08	-34.64	-99.00	1.06	18855.76	-58.38	-58.40	1613.23	-96.24	-96.24	-99.78	-99.00	-93.78
μ_r	2.87E+09	-53.04	0.00	0.19	-39.23	-23.96	-23.98	-19.18	1.14	1.14	25.15	9900.00	-0.53
	2.87E+08	-18.04	0.00	0.49	-36.86	-22.44	-22.46	-17.74	1.06	1.06	22.85	900.00	-0.51
	2.87E+07	0.00	0.00	0.00	0.00	0.00	0.00	0.00	0.00	0.00	0.00	0.00	0.00
	2.87E+06	150.02	0.00	1.95	13367.47	314.64	314.50	1085.58	-63.48	-63.48	-96.92	-90.00	-38.94
	2.87E+05	19.59	0.00	2.24	18871.66	-58.36	-58.37	1613.25	-96.24	-96.24	-99.78	-99.00	-93.77
μ_c	2.10E+07	-50.40	0.00	-98.12	869.15	809.66	811.51	6.76	0.00	0.00	-6.33	0.00	56.66
	2.10E+06	-35.63	0.00	-86.47	42.32	41.97	42.08	0.33	0.00	0.00	-0.33	0.00	2.97
	2.10E+05	0.00	0.00	0.00	0.00	0.00	0.00	0.00	0.00	0.00	0.00	0.00	0.00
	2.10E+04	41.99	0.00	413.68	-5.55	-5.52	-5.54	-0.04	0.00	0.00	0.04	0.00	-0.39
	2.10E+03	-18.98	0.00	487.28	-5.89	-5.86	-5.88	-0.05	0.00	0.00	0.05	0.00	-0.42

Para	Value	mRNA H	mRNA LDLR	HMGR	LE	LRB	LI	VE	VRB	VI	RF	RI	Cholesterol
W	1.50E+04	64.45	-0.02	-9.58	19043.49	3461.80	3460.55	1622.10	220.54	220.56	-81.39	-0.02	430.68
	1.50E+03	108.51	-0.02	11.47	14329.49	3323.55	3322.44	1161.44	204.15	204.17	-76.27	-0.02	406.55
	1.50E+02	0.00	0.00	0.00	0.00	0.00	0.00	0.00	0.00	0.00	0.00	0.00	0.00
	1.50E+01	136.40	0.00	3.11	-99.34	-99.19	-99.19	-91.41	-89.45	-89.45	22.80	0.00	-90.09
	1.50E+00	-4.43	0.00	2.94	-99.99	-99.99	-99.99	-99.15	-98.94	-98.94	25.14	0.00	-99.00
X_h	3.00E+02	62.47	0.00	-64.37	-5.83	-5.80	-5.82	-0.05	0.00	0.00	0.05	0.00	-0.41
	3.00E+01	161.29	0.00	-49.10	-5.47	-5.44	-5.46	-0.04	0.00	0.00	0.04	0.00	-0.39
	3.00E+00	0.00	0.00	0.00	0.00	0.00	0.00	0.00	0.00	0.00	0.00	0.00	0.00
	3.00E-01	33.14	0.00	170.75	5238.03	3704.46	3709.08	39.61	-0.63	-0.63	-28.82	-0.01	250.42
	3.00E-02	-1.72	-0.01	436.04	7241.67	4615.96	4620.49	54.10	-1.11	-1.12	-35.83	-0.01	309.32
X_c	4.00E+02	19.26	0.00	42.02	1031.05	949.25	951.37	8.01	-0.01	-0.01	-7.42	0.00	66.29
	4.00E+01	170.34	0.00	50.47	397.91	383.91	384.86	3.10	0.01	0.01	-3.00	0.00	27.03
	4.00E+00	0.00	0.00	0.00	0.00	0.00	0.00	0.00	0.00	0.00	0.00	0.00	0.00
	4.00E-01	200.24	-10.45	18.98	5.39	2.29	2.28	2.85	-0.17	-0.17	-2.94	-10.45	-0.01
	4.00E-02	304.02	-30.18	37.08	19.30	6.39	6.38	11.36	-0.68	-0.68	-10.81	-30.18	-0.24
X_r	1.00E+02	122.24	-0.72	3.60	7474.77	4688.80	4693.25	56.50	-1.16	-1.16	-36.84	-0.73	313.70
	1.00E+01	143.86	-0.06	3.79	6672.82	4374.36	4378.94	50.07	-0.97	-0.97	-34.01	-0.06	293.75
	1.00E+00	0.00	0.00	0.00	0.00	0.00	0.00	0.00	0.00	0.00	0.00	0.00	0.00
	1.00E-01	226.69	0.00	-0.02	-5.58	-5.55	-5.57	-0.04	0.00	0.00	0.04	0.00	-0.39
	1.00E-02	29.61	0.00	-0.97	-5.86	-5.83	-5.85	-0.05	0.00	0.00	0.05	0.00	-0.41

Para	Value	mRNA H	mRNA LDLR	HMGR	LE	LRB	LI	VE	VRB	VI	RF	RI	Cholesterol
m_v	2.00E+02	94.57	0.00	2.61	18829.28	-43.45	-43.47	1594.10	-94.94	-94.94	-99.70	0.00	-91.61
	2.00E+01	27.95	0.00	0.50	12252.49	468.43	468.23	926.62	-52.74	-52.74	-95.40	0.00	-19.02
	2.00E+00	0.00	0.00	0.00	0.00	0.00	0.00	0.00	0.00	0.00	0.00	0.00	0.00
	2.00E-01	300.96	0.00	2.92	-36.26	-22.06	-22.07	-17.39	1.04	1.04	22.30	0.00	-0.50
	2.00E-02	-32.32	0.00	1.41	-38.63	-23.57	-23.58	-18.81	1.12	1.12	24.55	0.00	-0.52
f	7.00E+02	1981486.40	-89.19	258594236.41	2980.92	232.01	-88.13	654.77	-25.53	-97.34	6368.88	17221463.13	-79.62
	7.00E+00	2166716.10	-78.87	262317013.75	2972.12	2225.69	830.47	620.39	425.43	110.17	6368.88	90066830.82	150.11
	7.00E-00	0.00	0.00	0.00	0.00	0.00	0.00	0.00	0.00	0.00	0.00	0.00	0.00
	7.00E-02	136.80	0.00	-3.41	2203.53	718.10	718.35	158.52	-8.22	-8.22	-64.50	-67.74	39.73
	7.00E-03	23.39	0.00	-3.56	2871.97	782.24	782.40	201.29	-10.58	-10.58	-70.32	-69.79	41.54
R_i^{chol}	3.40E+05	218.79	-0.11	-23.98	10.45	10.36	10.38	0.11	0.00	0.00	-0.11	-0.11	721.10
	3.40E+04	-6.15	0.00	-0.35	0.16	0.16	0.16	0.00	0.00	0.00	0.00	0.00	59.01
	3.40E+03	0.00	0.00	0.00	0.00	0.00	0.00	0.00	0.00	0.00	0.00	0.00	0.00
	3.40E-02	-10.68	0.00	1.34	-0.01	-0.01	-0.01	0.00	0.00	0.00	0.00	0.00	-5.89
	3.40E-01	245.86	0.00	2.65	-0.02	-0.02	-0.02	0.00	0.00	0.00	0.00	0.00	-6.48
R_e^{chol}	3.10E+05	61.72	-81.43	38.20	4461.90	419.50	419.39	510.32	-30.48	-30.48	-88.61	-81.43	6430.31
	3.10E+04	269.08	-0.19	-15.01	1.39	1.33	1.34	0.06	0.00	0.00	-0.06	-0.19	841.11
	3.10E+03	0.00	0.00	0.00	0.00	0.00	0.00	0.00	0.00	0.00	0.00	0.00	0.00
	3.10E+02	119.35	0.00	2.90	-0.11	-0.11	-0.11	0.00	0.00	0.00	0.00	0.00	-84.11
	3.10E+01	147.91	0.00	3.39	-0.12	-0.12	-0.12	0.00	0.00	0.00	0.00	0.00	-92.52
J	1.00E+01	207151.59	-22.70	56666311.27	66.46	53.26	53.35	8.50	-0.16	-0.16	-7.98	-22.76	1592.54
	1.00E+00	17.14	-0.01	21371.97	2.14	2.12	2.13	0.02	0.00	0.00	-0.02	-0.01	1.06
	1.00E-01	0.00	0.00	0.00	0.00	0.00	0.00	0.00	0.00	0.00	0.00	0.00	0.00
	1.00E-02	283.21	0.00	-12.81	-0.23	-0.22	-0.22	0.00	0.00	0.00	0.00	0.00	-0.02
	1.00E-03	56.01	0.00	-14.09	-0.25	-0.25	-0.25	0.00	0.00	0.00	0.00	0.00	-0.02

Para	Value	mRNA H	mRNA LDLR	HMGR	LE	LRB	LI	VE	VRB	VI	RF	RI	Cholesterol
γ_v	7.44E+03	-31.69	0.00	-0.38	0.00	0.00	0.00	0.00	0.00	-99.00	0.00	0.00	0.00
	7.44E+02	-30.38	0.00	-0.36	0.00	0.00	0.00	0.00	0.00	-90.00	0.00	0.00	0.00
	7.44E+01	0.00	0.00	0.00	0.00	0.00	0.00	0.00	0.00	0.00	0.00	0.00	0.00
	7.44E+00	-5.47	0.00	0.94	0.00	0.00	0.00	0.00	0.00	900.00	0.00	0.00	0.00
	7.44E-01	-61.79	0.00	-0.65	0.00	0.00	0.00	0.00	0.00	9900.38	0.00	0.00	0.00
γ_r	2.23E+04	13.95	0.00	0.73	-0.02	-0.02	-0.02	0.00	0.00	0.00	0.00	-99.00	0.00
	2.23E+03	-67.91	0.00	-0.73	-0.02	-0.02	-0.02	0.00	0.00	0.00	0.00	-90.00	0.00
	2.23E+02	0.00	0.00	0.00	0.00	0.00	0.00	0.00	0.00	0.00	0.00	0.00	0.00
	2.23E+01	-33.40	0.00	-1.00	0.17	0.17	0.17	0.00	0.00	0.00	0.00	900.00	0.01
	2.23E+00	-81.06	0.00	-0.78	2.13	2.10	2.10	0.04	0.00	0.00	-0.04	9892.10	0.15
χ_v	1.94E+01	-23.39	0.00	1.24	1195.55	1327.24	1327.05	-85.02	-83.50	-83.50	10.18	0.00	8.46
	1.94E+00	180.78	0.00	1.43	505.52	529.18	529.13	-35.25	-32.71	-32.71	3.92	0.00	3.96
	1.94E-01	0.00	0.00	0.00	0.00	0.00	0.00	0.00	0.00	0.00	0.00	0.00	0.00
	1.94E-02	293.28	0.00	1.38	-88.69	-88.75	-88.75	6.00	5.34	5.34	-0.62	0.00	-0.82
	1.94E-03	118.91	0.00	0.97	-98.27	-98.28	-98.27	6.64	5.91	5.91	-0.69	0.00	-0.91
ω	4.54E+01	231.06	-0.03	-9.83	1528070.00	3436.35	3434.94	170915.88	295.94	295.96	-99.77	-0.03	499.41
	4.54E+00	-27.72	-0.03	-12.23	105010.86	3398.30	3397.01	11480.20	284.58	284.60	-96.67	-0.03	486.43
	4.54E-01	0.00	0.00	0.00	0.00	0.00	0.00	0.00	0.00	0.00	0.00	0.00	0.00
	4.54E-02	140.23	0.00	3.15	-93.67	-92.22	-92.23	-91.77	-89.89	-89.89	22.85	0.00	-90.04
	4.54E-03	46.60	0.00	3.06	-99.38	-99.22	-99.23	-99.19	-98.99	-98.99	25.15	0.00	-99.00
m_l	1.00E-02	-13.76	0.00	3.69	15822.29	54.48	54.45	1420.21	-85.24	-85.24	-99.03	0.00	-76.07
	1.00E-01	-15.96	0.00	-0.04	705.70	400.61	400.91	58.03	-1.87	-1.88	-37.90	0.00	24.93
	1.00E-00	0.00	0.00	0.00	0.00	0.00	0.00	0.00	0.00	0.00	0.00	0.00	0.00
	1.00E-01	16.12	0.00	0.29	-2.09	-1.41	-1.41	-0.65	0.04	0.04	0.70	0.00	-0.06
	1.00E-02	70.79	0.00	1.36	-2.28	-1.54	-1.54	-0.72	0.04	0.04	0.76	0.00	-0.07

Para	Value	mRNA H	mRNA LDLR	HMGR	LE	LRB	LI	VE	VRB	VI	RF	RI	Cholesterol
α_v	6.58E+02	261.65	0.00	1.92	536.89	562.72	562.68	578.01	-34.37	-34.37	4.06	0.00	4.63
	6.58E+01	19.66	0.00	1.29	76.89	77.86	77.86	78.80	-4.70	-4.70	0.55	0.00	0.70
	6.58E+00	0.00	0.00	0.00	0.00	0.00	0.00	0.00	0.00	0.00	0.00	0.00	0.00
	6.58E-01	191.14	0.00	1.45	-8.22	-8.28	-8.28	-8.36	0.50	0.50	-0.06	0.00	-0.08
	6.58E-02	130.45	0.00	1.59	-9.05	-9.11	-9.11	-9.20	0.55	0.55	-0.06	0.00	-0.08
β_r	6.03E+03	93.47	0.00	1.37	-19.70	-99.02	-2.37	-0.02	0.00	0.00	0.02	0.00	-0.16
	6.03E+02	186.37	0.00	1.91	-17.96	-90.22	-2.18	-0.02	0.00	0.00	0.02	0.00	-0.15
	6.03E+01	0.00	0.00	0.00	0.00	0.00	0.00	0.00	0.00	0.00	0.00	0.00	0.00
	6.03E+00	257.39	0.00	1.72	309.39	1461.86	56.21	0.42	-0.02	-0.02	-0.43	0.01	3.73
	6.03E-01	-15.99	0.00	0.69	2764.08	15113.46	52.12	0.37	-0.02	-0.02	-0.40	0.02	3.35
β_v	6.03E+03	72.98	0.00	0.84	-9.05	-9.11	-9.11	-9.20	-98.99	0.55	-0.06	0.00	-0.08
	6.03E+02	0.73	0.00	1.11	-8.23	-8.28	-8.28	-8.36	-89.95	0.50	-0.06	0.00	-0.08
	6.03E+01	0.00	0.00	0.00	0.00	0.00	0.00	0.00	0.00	0.00	0.00	0.00	0.00
	6.03E+00	249.85	0.00	1.15	76.91	77.88	77.88	78.80	852.98	-4.70	0.55	0.00	0.70
	6.03E-01	23.10	0.00	0.85	535.73	561.55	561.51	578.32	6466.30	-34.34	4.07	0.02	4.58
β_o	0.00E+00	0.00	0.00	0.00	0.00	0.00	0.00	0.00	0.00	0.00	0.00	0.00	0.00
	0.00E+00	0.00	0.00	0.00	0.00	0.00	0.00	0.00	0.00	0.00	0.00	0.00	0.00
	0.00E+00	0.00	0.00	0.00	0.00	0.00	0.00	0.00	0.00	0.00	0.00	0.00	0.00
	0.00E+00	0.00	0.00	0.00	0.00	0.00	0.00	0.00	0.00	0.00	0.00	0.00	0.00
	0.00E+00	0.00	0.00	0.00	0.00	0.00	0.00	0.00	0.00	0.00	0.00	0.00	0.00
γ_l	7.44E+03	86.54	0.00	1.04	0.00	0.00	-99.00	0.00	0.00	0.00	0.00	0.00	0.00
	7.44E+02	-29.32	0.00	0.51	0.00	0.00	-90.00	0.00	0.00	0.00	0.00	0.00	0.00
	7.44E+01	0.00	0.00	0.00	0.00	0.00	0.00	0.00	0.00	0.00	0.00	0.00	0.00
	7.44E+00	169.64	0.00	0.90	0.00	0.00	901.56	0.00	0.00	0.00	0.00	0.00	0.01
	7.44E-01	108.18	0.00	1.66	0.00	0.00	10126.88	0.00	0.00	0.00	0.00	0.00	0.16

Para	Value	mRNA H	mRNA LDLR	HMGR	LE	LRB	LI	VE	VRB	VI	RF	RI	Cholesterol
							Percent difference in						
δ_{mr}	1.00E+02	28.30	-99.00	2.25	18870.04	-58.36	-58.37	1613.17	-96.24	-96.24	-99.78	-99.00	-93.77
	1.00E+01	36.23	-90.00	-1.39	13353.42	314.48	314.35	1084.90	-63.48	-63.48	-96.92	-90.00	-38.94
	1.00E+00	0.00	0.00	0.00	0.00	0.00	0.00	0.00	0.00	0.00	0.00	0.00	0.00
	1.00E-01	108.86	899.97	1.42	-36.82	-22.40	-22.41	-17.74	1.06	1.06	22.85	899.97	-0.50
	1.00E-02	-44.97	8382.18	0.51	-39.17	-23.91	-23.93	-19.15	1.14	1.14	25.10	8379.91	-0.52
δ_c	2.68E+02	-43.95	0.00	490.03	-5.89	-5.86	-5.88	-0.05	0.00	0.00	0.05	0.00	-99.00
	2.68E+01	-60.20	0.00	419.68	-5.60	-5.57	-5.59	-0.04	0.00	0.00	0.04	0.00	-90.04
	2.68E+00	0.00	0.00	0.00	0.00	0.00	0.00	0.00	0.00	0.00	0.00	0.00	0.00
	2.68E-01	-43.01	-1.25	-88.80	116.21	113.84	114.12	1.27	0.02	0.02	-1.23	-1.26	1402.67
	2.68E-02	-85.94	-91.79	50.83	17151.21	268.01	267.68	1272.80	-70.70	-70.71	-97.86	-91.81	8143.84
α_{li}	4.85E+03	185.09	0.00	1.10	-99.06	-5.99	-6.01	-0.05	0.00	0.00	0.05	0.00	-0.42
	4.85E+02	193.49	0.00	0.64	-90.60	-5.99	-6.01	-0.05	0.00	0.00	0.05	0.00	-0.42
	4.85E+01	0.00	0.00	0.00	0.00	0.00	0.00	0.00	0.00	0.00	0.00	0.00	0.00
	4.85E+00	108.70	0.00	1.34	1966.79	105.00	105.00	0.77	-0.04	-0.04	-0.81	0.00	6.82
	4.85E-01	-2.65	0.00	1.43	3923.74	-59.59	-59.59	-0.44	0.02	0.02	0.46	0.00	-3.90
α^{-}_{li}	1.32E+03	29.92	0.00	1.25	2792.61	53.58	53.56	0.39	-0.02	-0.02	-0.41	0.00	3.45
	1.32E+02	-44.78	0.00	0.74	307.60	55.26	55.30	0.41	-0.02	-0.02	-0.43	0.00	3.67
	1.32E+01	0.00	0.00	0.00	0.00	0.00	0.00	0.00	0.00	0.00	0.00	0.00	0.00
	1.32E+00	14.28	0.00	0.38	-17.96	-2.16	-2.16	-0.02	0.00	0.00	0.02	0.00	-0.15
	1.32E-01	-0.96	0.00	0.26	-19.70	-2.35	-2.35	-0.02	0.00	0.00	0.02	0.00	-0.16
α_v	6.78E+04	191.13	0.00	1.43	-98.22	-98.23	-98.23	-98.93	5.91	5.91	-0.69	0.00	-0.90
	6.78E+03	103.27	0.00	0.78	-88.68	-88.74	-88.74	-89.40	5.34	5.34	-0.62	0.00	-0.82
	6.78E+02	0.00	0.00	0.00	0.00	0.00	0.00	0.00	0.00	0.00	0.00	0.00	0.00
	6.78E+01	-10.30	0.00	0.48	511.16	534.67	534.63	548.65	-32.63	-32.63	3.85	0.00	4.43
	6.78E+00	87.87	0.00	0.71	1199.83	1331.45	1331.27	1400.12	-83.48	-83.48	10.14	0.00	8.76

Para	Value	mRNA H	mRNA LDLR	HMGR	LE	LRB	LI	VE	VRB	VI	RF	RI	Cholesterol
							Percent difference in						
K_{mh}	1.00E+02	-96.67	0.00	6.55	-5.88	-5.85	-5.87	-0.05	0.00	0.00	0.05	0.00	-0.42
	1.00E+01	57.94	0.00	41.23	-5.88	-5.85	-5.87	-0.05	0.00	0.00	0.05	0.00	-0.42
	1.00E+00	0.00	0.00	0.00	0.00	0.00	0.00	0.00	0.00	0.00	0.00	0.00	0.00
	1.00E-01	201.51	0.00	95.53	2420.31	2025.58	2029.25	18.63	-0.13	-0.13	-15.81	0.00	139.50
	1.00E-02	174.92	-0.01	235.08	6299.35	4210.53	4215.15	47.33	-0.87	-0.88	-32.71	-0.01	283.22
K_{mr}	1.00E+02	-23.26	-98.02	1.60	18503.21	-16.90	-16.93	1558.39	-92.59	-92.59	-99.55	-98.02	-87.67
	1.00E+01	22.40	-81.82	4.61	10429.32	707.37	707.15	698.57	-38.71	-38.71	-92.33	-81.82	9.77
	1.00E+00	0.00	0.00	0.00	0.00	0.00	0.00	0.00	0.00	0.00	0.00	0.00	0.00
	1.00E-01	133.89	81.82	3.46	-23.50	-14.78	-14.80	-9.73	0.58	0.58	11.42	81.82	-0.45
	1.00E-02	308.40	98.02	0.67	-25.23	-15.86	-15.87	-10.60	0.63	0.63	12.56	98.02	-0.48
K_c	1.09E+04	131.06	0.00	0.03	0.00	0.00	0.00	0.00	0.00	0.00	0.00	0.00	0.00
	1.09E+03	565.54	0.00	0.30	0.00	0.00	0.00	0.00	0.00	0.00	0.00	0.00	0.00
	1.09E+02	0.00	0.00	0.00	0.00	0.00	0.00	0.00	0.00	0.00	0.00	0.00	0.00
	1.09E+01	506.88	0.00	-0.45	0.02	0.02	0.02	0.00	0.00	0.00	0.00	0.00	0.18
	1.09E+00	529.08	0.00	-65.42	3.08	3.06	3.07	0.02	0.00	0.00	-0.02	0.00	0.18
δ_{mh}	1.00E+02	143.29	0.00	-94.40	-5.88	-5.85	-5.87	-0.05	0.00	0.00	0.05	0.00	-0.42
	1.00E+01	146.62	0.00	-54.53	-5.55	-5.52	-5.53	-0.04	0.00	0.00	0.04	0.00	-0.39
	1.00E+00	0.00	0.00	0.00	0.00	0.00	0.00	0.00	0.00	0.00	0.00	0.00	0.00
	1.00E-01	114.28	0.00	28040.02	55.79	55.26	55.40	0.44	0.00	0.00	-0.43	0.00	5.11
	1.00E-02	335267.30	-14.60	92142922.03	4455.16	2993.42	2995.44	48.52	0.80	0.81	-32.13	-14.78	2774.66
δ_h	1.43E+02	-26.70	0.00	-99.34	-5.89	-5.86	-5.88	-0.05	0.00	0.00	0.05	0.00	-0.42
	1.43E+01	-68.82	0.00	-99.87	-5.57	-5.54	-5.56	-0.04	0.00	0.00	0.04	0.00	-0.40
	1.43E+00	0.00	0.00	0.00	0.00	0.00	0.00	0.00	0.00	0.00	0.00	0.00	0.00
	1.43E-01	-64.87	-0.98	45120714.59	80.40	79.00	79.20	0.93	0.03	0.04	-0.89	-0.99	1302.52
	1.43E-02	18.72	-84.81	249942832.67	14210.11	604.66	603.93	948.20	-48.36	-48.38	-95.07	-84.86	6828.68

Appendix D. Familial Hypercholesterolaemia Analysis

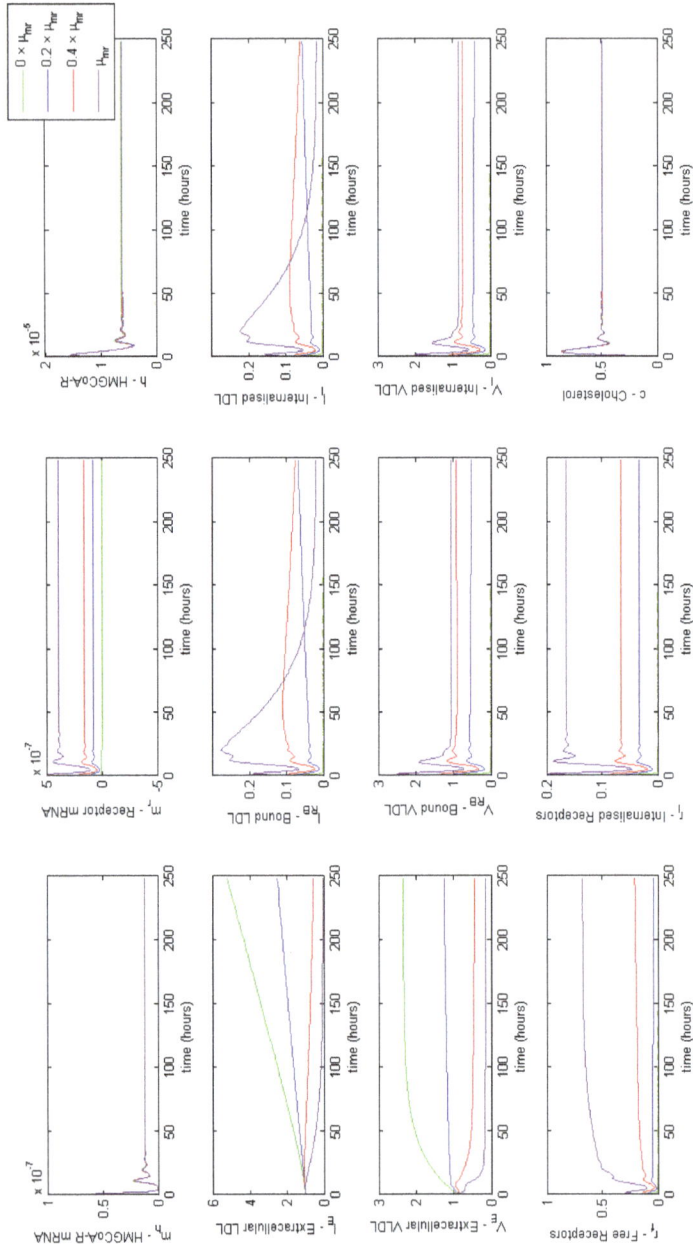

Figure A1. Familial Hypercholesterolaemia Class I; LDLR are not synthesised correctly, affecting parameter μ_{mr}. Results show a lack of receptors has a significant impact on extracellular LDL and VLDL concentrations.

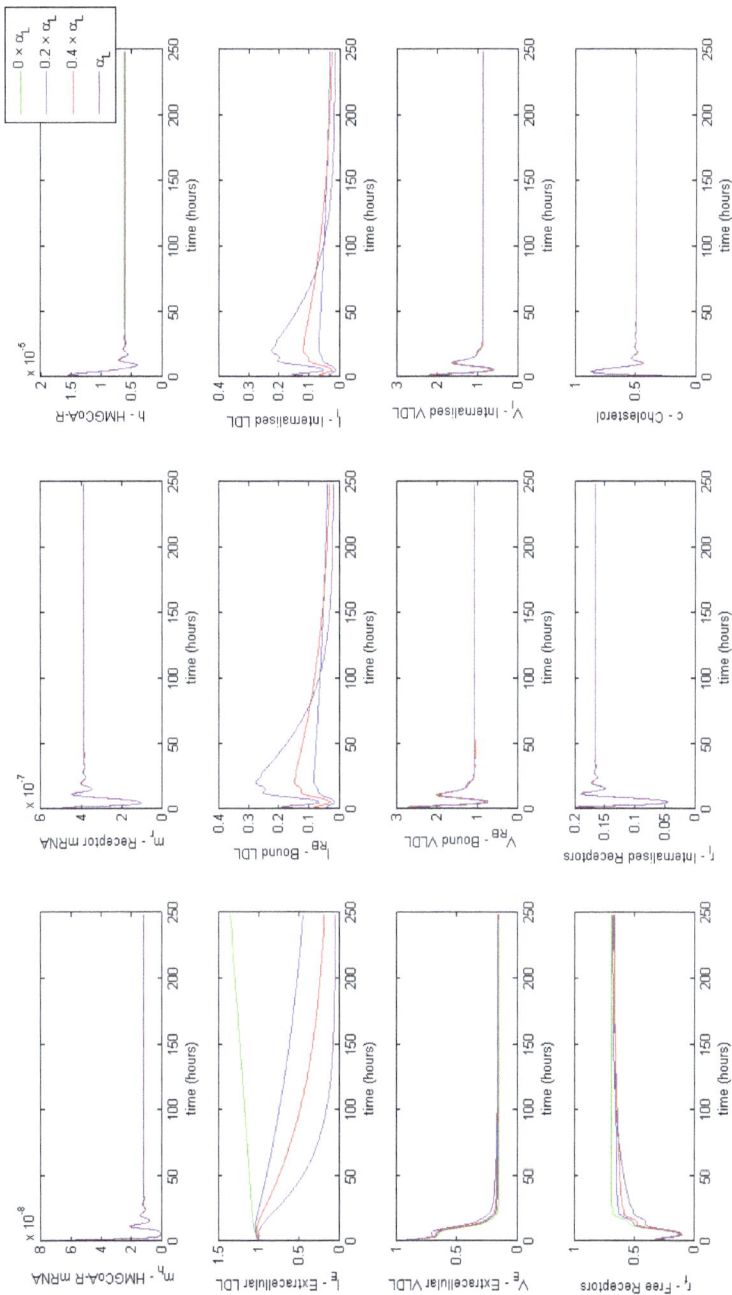

Figure A2. Familial Hypercholesterolaemia Class III; binding of LDL to LDL receptors is ineffective, affecting parameter α_L. Results show a lack of LDL binding prevents LDL being endocytosed.

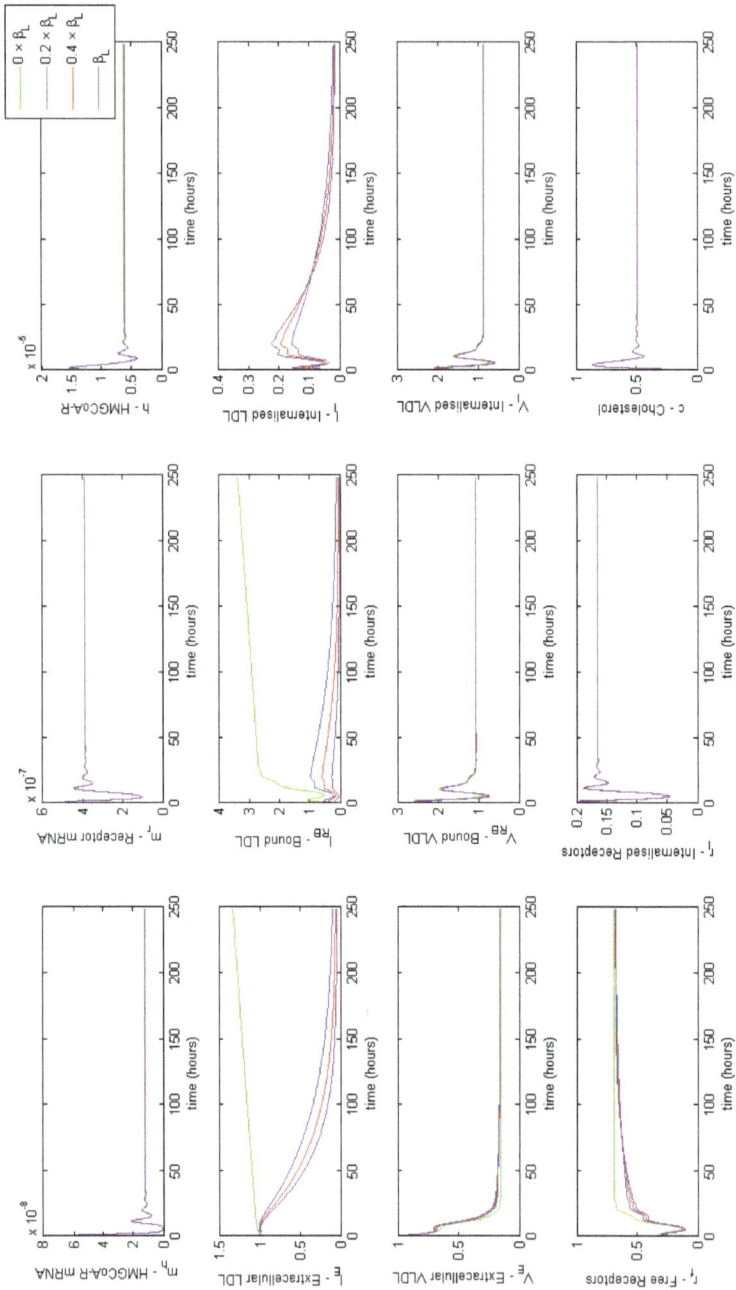

Figure A3. Familial Hypercholesterolaemia Class IV; Bound LDL not internalised properly, affecting parameter β_L. Results show the lack of internalisation of bound LDL particles blocking LDL binding.

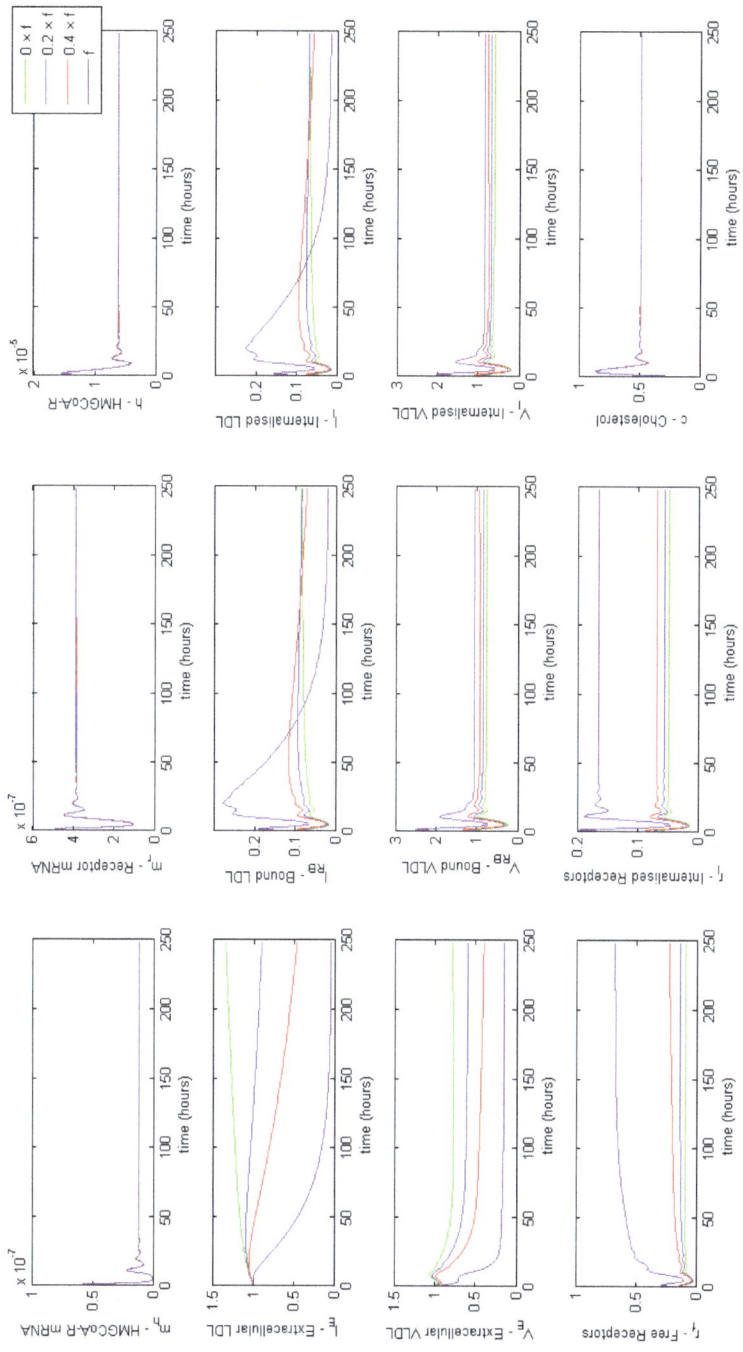

Figure A4. Familial Hypercholesterolaemia Class V; receptors are not recycled properly, affecting model parameter f. Results show a decrease in free and internalised receptors, significantly affecting extracellular concentrations of both LDL and VLDL.

Appendix E. Full Model Results for Combined Cases of Familial Hypercholesterolaemia

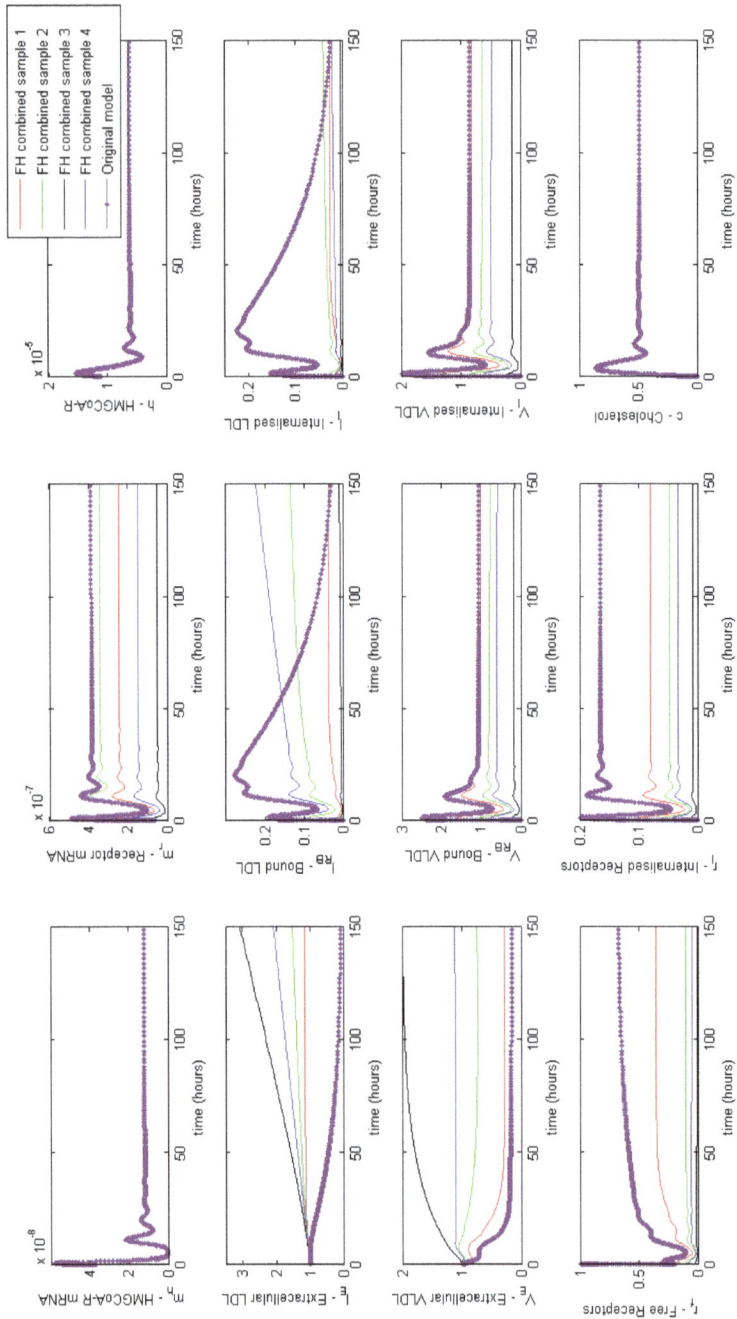

Figure A5. Familial Hypercholesterolaemia combined effects. Showing a combination of reduced synthesis production and recycling and ineffective LDL binding and internalisation can dramatically increase levels of circulating LDL and VLDL.

Appendix F. Full Model Results for Statin Application

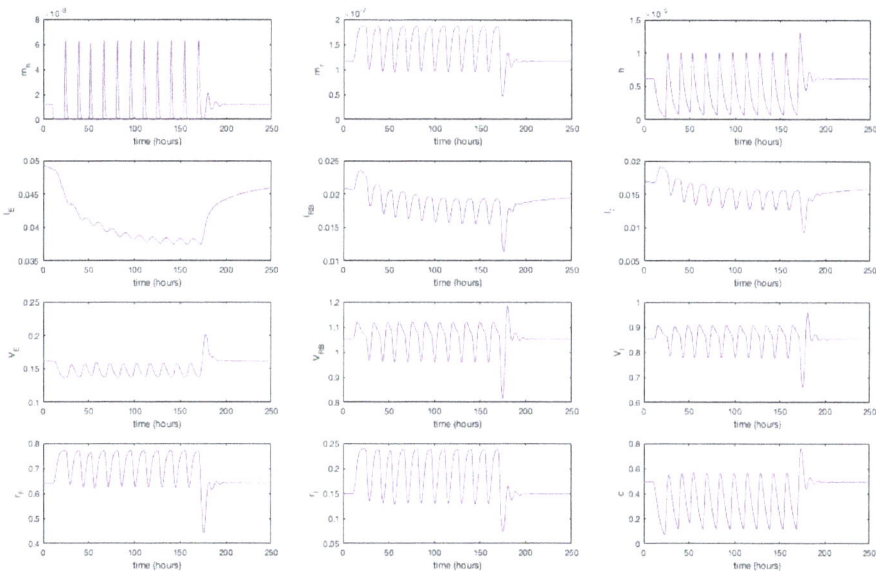

Figure A6. A simulation showing the effect of statin therapy on our integrated model of cholesterol metabolism. Here 11 doses are applied over a seven day period starting at approximately $t = 10$ h.

References

1. O'Brien, J.; Sampson, E. Lipid composition of the normal human brain: Gray matter, white matter, and myelin. *J. Lipid Res.* **1965**, *6*, 537–544. [PubMed]
2. Jousilahti, P.; Vartiainen, E.; Tuomilehto, J.; Puska, P. Sex, age, cardiovascular risk factors, and coronary heart disease A prospective follow-up study of 14,786 middle-aged men and women in Finland. *Circulation* **1999**, *99*, 1165–1172. [CrossRef] [PubMed]
3. Turley, S.; Dietschy, J. The intestinal absorption of biliary and dietary cholesterol as a drug target for lowering the plasma cholesterol level. *Prev. Cardiol.* **2003**, *6*, 29–64. [CrossRef] [PubMed]
4. Goldstein, J.; Brown, M.; Anderson, R.; Russell, D.; Schneider, W. Receptor-mediated endocytosis: Concepts emerging from the LDL receptor system. *Annu. Rev. Cell Biol.* **1985**, *1*, 1–39. [CrossRef] [PubMed]
5. Grundy, S. Inflammation, metabolic syndrome, and diet responsiveness. *Circulation* **2003**, *108*, 126–128. [CrossRef] [PubMed]
6. Gowdy, K.; Fessler, M. Emerging roles for cholesterol and lipoproteins in lung disease. *Pulm. Pharmacol. Ther.* **2013**, *26*, 430–437. [CrossRef] [PubMed]
7. Bermudes, A.; de Carvalho, W.; Dietist, P.; Muramoto, G.; Maranhão, R.; Delgado, A. Changes in lipid metabolism in pediatric patients with severe sepsis and septic shock. *Nutrition* **2018**, *47*, 104–109. [CrossRef] [PubMed]
8. Lagrost, G.; Grosjean, S.; Masson, D.; Deckert, V.; Gautier, T.; Debomy, F.; Vinault, S.; Jeannin, A.; Labbé, J; Bonithon-Kopp, C. Low preoperative cholesterol level is a risk factor of sepsis and poor clinical outcome in patients undergoing cardiac surgery with cardiopulmonary bypass. *Crit. Care Med.* **2014**, *42*, 1065–1073. [CrossRef] [PubMed]
9. Paalvast, Y.; Kuivenhoven, J.; Groen, A. Evaluating computational models of cholesterol metabolism. *Biochim. Biophys. Acta* **2015**, *1851*, 1360–1376. [CrossRef] [PubMed]

10. Bhattacharya, B.; Sweby, P.; Minihane, A.M.; Jackson, K.; Tindall, M. A mathematical model of the sterol regulatory element binding protein 2 cholesterol biosynthesis pathway. *J. Theor. Biol.* **2014**, *349*, 150–162. [CrossRef] [PubMed]

11. Pool, F.; Currie, R.; Sweby, P.; Salazar, D.; Tindall, M. A mathematical model of the mevalonate cholesterol biosynthesis pathway. *J. Theor. Biol.* **2018**, *443*, 157–176. [CrossRef] [PubMed]

12. Bhattacharya, B. Mathematical Modelling of Low Density Lipoprotein Metabolism. Intracellular Cholesterol Regulation. Ph.D. Thesis, University of Reading, Reading, UK, 2011.

13. Wattis, J.; O'Malley, B.; Blackburn, H.; Pickersgill, L.; Panovska, J.; Byrne, H.; Jackson, K. Mathematical model for low density lipoprotein (LDL) endocytosis by hepatocytes. *Bull. Math. Biol.* **2008**, *70*, 2303–2333. [CrossRef] [PubMed]

14. Cromwell, W.; Otvos, J.; Keyes, M.; Pencina, M.; Sullivan, L.; Vasan, R.; Wilson, P.; D'Agostino, R. LDL particle number and risk of future cardiovascular disease in the Framingham Offspring Study implications for LDL management. *J. Clin. Lipidol.* **2007**, *1*, 583–592. [CrossRef] [PubMed]

15. Tindall, M.; Wattis, J.; O'Malley, B.; Pickersgill, L.; Jackson, K. A continuum receptor model of hepatic lipoprotein metabolism. *J. Theor. Biol.* **2009**, *257*, 371–384. [CrossRef] [PubMed]

16. Alberts, B.; Johnson, A.; Lewis, J.; Raff, M.; Roberts, K.; Walter, P. The Cell Cycle. In *Molecular Biology of the Cell*, 5th ed.; Garland Science: New York, NY, USA, 2008; pp. 1053–1114.

17. Jackson, K.; Maitin, V.; Leake, D.; Yaqoob, P.; Williams, C. Saturated fat-induced changes in Sf 60–400 particle composition reduces uptake of LDL by HepG2 cells. *J. Lipid Res.* **2006**, *47*, 393–403. [CrossRef] [PubMed]

18. Lange, Y.; Ye, J.; Rigney, M.; Steck, T. Regulation of endoplasmic reticulum cholesterol by plasma membrane cholesterol. *J. Lipid Res.* **1999**, *40*, 2264–2270. [PubMed]

19. Darzacq, X.; Shav-Tal, Y.; de Turris, V.; Brody, Y.; Shenoy, S.; Phair, R.; Singer, R. In vivo dynamics of RNA polymerase II transcription. *Nat. Struct. Mol. Biol.* **2007**, *14*, 796–806. [CrossRef] [PubMed]

20. Goldstein, J.; Brown, M. Progress in understanding the LDL receptor and HMG-CoA reductase, two membrane proteins that regulate the plasma cholesterol. *J. Lipid Res.* **1984**, *25*, 1450–1461. [PubMed]

21. National Center for Biotechnology Information. Genome. 2015. Available online: http://www.ncbi.nlm.nih.gov/genome (accessed on 31 July 2015).

22. Trachsel, H. *Translation in Eukaryotes*; CRC Press: Boca Raton, FL, USA, 1991; pp. 149–176.

23. Istvan, E.; Palnitkar, M.; Buchanan, S.; Deisenhofer, J. Crystal structure of the catalytic portion of human HMG-CoA reductase: Insights into regulation of activity and catalysis. *EMBO J.* **2000**, *19*, 819–830. [CrossRef] [PubMed]

24. Segal, I. *Enzyme Kinetics Behaviour and Analysis of Rapid Equilibrium and Steady-State Enzyme Systems*; A Wiley-Interscience Publication; Wiley: New York, NY, USA, 1975; pp. 60–100.

25. Tanaka, R.; Edwards, P.; Lan, S.; Knöppel, E.; Fogelman, A. Purification of 3-hydroxy-3-methylglutaryl coenzyme A reductase from human liver. *J. Lipid Res.* **1982**, *23*, 523–530. [PubMed]

26. Soutar, A.; Knight, B. Structure and regulation of the LDL-receptor and its gene. *Br. Med. Bull.* **1990**, *46*, 891–916. [CrossRef] [PubMed]

27. Wilson, G.; Deeley, R. An episomal expression vector system for monitoring sequence-specific effects on mRNA stability in human cell lines. *Plasmid* **1995**, *33*, 198–207. [CrossRef] [PubMed]

28. Vargas, N.; Brewer, B.; Rogers, T.; Wilson, G. Protein kinase C activation stabilizes LDL receptor mRNA via the JNK pathway in HepG2 cells. *J. Lipid Res.* **2009**, *50*, 386–397. [CrossRef] [PubMed]

29. Brown, M.; Dana, S.; Goldstein, J. Regulation of 3-Hydroxy-3-methylglutaryl Coenzyme A Reductase Activity in Cultured Human Fibroblasts comparison of cells from a normal subject and from a patient with homozygous familial hypercholesterolemia. *J. Biol. Chem.* **1974**, *249*, 789–796. [PubMed]

30. Harwood, J.R.H.; Pellarin, L. Kinetics of low-density lipoprotein receptor activity in Hep-G2 cells: Derivation and validation of a Briggs–Haldane-based kinetic model for evaluating receptor-mediated endocytotic processes in which receptors recycle. *Biochem. J.* **1997**, *323*, 649–659. [CrossRef] [PubMed]

31. Mahley, R.; Innerarity, T. Lipoprotein receptors and cholesterol homeostasis. *Biochim. Biophys. Acta-Rev. Biomembr.* **1983**, *737*, 197–222. [CrossRef]

32. Goldstein, J.; Anderson, R.; Brown, M. Coated pits, coated vesicles, and receptor-mediated endocytosis. *Nature* **1979**, *279*, 679–685. [CrossRef] [PubMed]

33. Brown, M.; Goldstein, J. Receptor-mediated endocytosis: Insights from the lipoprotein receptor system. *Proc. Natl. Acad. Sci. USA* **1979**, *76*, 3330–3337. [CrossRef] [PubMed]

34. Basu, S.; Goldstein, J.; Brown, M. Characterization of the low density lipoprotein receptor in membranes prepared from human fibroblasts. *J. Biol. Chem.* **1978**, *253*, 3852–3856. [PubMed]
35. Shearer, K. The regulation and uptake of dietary fats. Ph.D. Thesis, University of Reading, Reading, UK, 2010.
36. Dunn, K.; McGraw, T.; Maxfield, F. Iterative fractionation of recycling receptors from lysosomally destined ligands in an early sorting endosome. *J. Cell Biol.* **1989**, *109*, 3303–3314. [CrossRef] [PubMed]
37. Panovska, J.; Tindall, M.; Wattis, J.; Byrne, H. Mathematical models of hepatic lipoprotein metabolism. In Proceedings of the 5th Mathematics in Medicine Study Group, Oxford, UK, 12–16 September 2005.
38. Rudling, M.; Angelin, B.; Ståhle, L.; Reihnér, E.; Sahlin, S.; Olivecrona, H.; Björkhem, I.; Einarsson, C. Regulation of hepatic low-density lipoprotein receptor, 3-hydroxy-3-methylglutaryl coenzyme A reductase, and cholesterol 7α-hydroxylase mRNAs in human liver. *J. Clin. Endocrinol. Metab.* **2002**, *87*, 4307–4313. [CrossRef] [PubMed]
39. Go, G.; Mani, A. Low-density lipoprotein receptor (LDLR) family orchestrates cholesterol homeostasis. *Yale J. Biol. Med.* **2012**, *85*, 19–28. [PubMed]
40. Foundation, F.H. Familial Hypercholesterolaemia. 2018. Available online: https://thefhfoundation.org (accessed on 6 June 2018).
41. Dammerman, M.; Breslow, J. Genetic basis of lipoprotein disorders. *Circulation* **1995**, *91*, 505–512. [CrossRef] [PubMed]
42. Hobbs, H.; Russell, D.; Brown, M.; Goldstein, J. The LDL receptor locus in familial hypercholesterolemia: Mutational analysis of a membrane protein. *Annu. Rev. Genet.* **1990**, *24*, 133–170. [CrossRef] [PubMed]
43. Iman, R. Latin Hypercube Sampling. In *Encyclopedia of Quantitative Risk Analysis and Assessment*; John Wiley & Sons, Inc.: Hoboken, NJ, USA, 2008.
44. Golan, D.; Tashjian, A.; Armstrong, E. *Principles of Pharmacology: The Pathophysiologic Basis of Drug Therapy*; Lippincott Williams & Wilkins: Philadelphia, PA, USA, 2011.
45. Balasubramaniam, S.; Szanto, A.; Roach, P. Circadian rhythm in hepatic low-density-lipoprotein (LDL)-receptor expression and plasma LDL levels. *Biochem. J.* **1994**, *298*, 39–43. [CrossRef] [PubMed]
46. Lickwar, C.R.; Mueller, F.; Hanlon, S.E.; McNally, J.G.; Lieb, J.D. Genome-wide protein-DNA binding dynamics suggest a molecular clutch for transcription factor function. *Nature* **2012**, *484*, 251–255. [CrossRef] [PubMed]
47. Yang, W.; Swartz, J. A filter microplate assay for quantitative analysis of DNA binding proteins using fluorescent DNA. *Anal. Biochem.* **2011**, *415*, 168–174. [CrossRef] [PubMed]

processes

MDPI

Article

A Framework for the Development of Integrated and Computationally Feasible Models of Large-Scale Mammalian Cell Bioreactors

Parham Farzan and Marianthi G. Ierapetritou *

Department of Chemical and Biochemical Engineering, Rutgers, The State University of New Jersey,
98 Brett Road, Piscataway, NJ 08854, USA; parham.farzan@rutgers.edu
* Correspondence: marianth@soemail.rutgers.edu; Tel.: +1-848-445-2971

Received: 5 May 2018; Accepted: 26 June 2018; Published: 29 June 2018

check for
updates

Abstract: Industrialization of bioreactors has been achieved by applying several core concepts of science and engineering. Modeling has deepened the understanding of biological and physical phenomena. In this paper, the state of existing cell culture models is summarized. A framework for development of dynamic and computationally feasible models that capture the interactions of hydrodynamics and cellular activities is proposed. Operating conditions are described by impeller rotation speed, gas sparging flowrate, and liquid fill level. A set of admissible operating states is defined over discretized process parameters. The burden on a dynamic solver is reduced by assuming hydrodynamics at its fully developed state and implementation of compartmental modeling. A change in the conditions of operation is followed by hydrodynamics switching instantaneously to the steady state that would be reached under new conditions. Finally, coupling the model with optimization solvers leads to improvements in operation.

Keywords: bioreactor integrated modeling; CFD simulation; compartmental modeling; reduced-order model; bioreactor operation optimization

1. Introduction

1.1. The Importance of Reliable Unit Operation Models

Global sales of biopharmaceutical products reached \$228 billion in 2016 [1]. Compared to microbial and yeast-based production systems, mammalian cells possess the cellular machinery to manufacture and secrete large proteins with the necessary post-translational modifications. Mammalian cell cultures are responsible for half of the revenue generated by the biotechnology industry, which is expected to grow by 15% annually [2].

The average commercial-scale titer for mammalian-expressed products has increased by 10-fold since the early 1990s and reached 2.5 g/L in recent years [3]. This has mainly been achieved via clone selection, cell line screening, host cell engineering, improving vectors, and gene amplification. At the same time, the demand and regulations for production of therapeutic proteins from mammalian cells have also been increasing. As a result, the delivery of affordable products at consistent quality is still a challenge for the industry. The industry has responded to this challenge by expanding the manufacturing capacity and improving the operational agility and efficiency of the supply chain. Bioreactors as large as 25,000 L are used, which has increased the capacity of manufacturing sites up to 200,000 L [4]. However, scale-up methodologies have remained based on the overall characterization of the components of the system such as aeration and agitation mechanisms. Considering the total cost of launching an asset and net cash inflow an asset is forecast to deliver, the return on innovation, as the engine of the industry, has been declining. Deloitte Center for Health Solutions monitored the

Research and Development (R&D) performances of 12 major biopharmaceutical companies. The rate of return declined from 10.1% in 2010 to 3.7% in 2016 [5]. The response of the industry has been mainly strategic [6]. The industry has tried to increase the value of the drug pipeline through mergers and acquisitions and the identification of priority customers and emerging markets. It also has pursued a reduction in the cost of launching new assets by developing explicit therapy area focus, balancing in-house and outsourced activities, defining specific missions for the enterprise, utilization of new technologies and advanced analytics in R&D, and data exploitation.

Fulfilling strategic plans depends on the execution of tactical decisions [7]. For example, strategic R&D management involves project selection, budgeting, and commercialization, while tactical R&D addresses the scheduling and resource management necessary for the accomplishment of the project. Integrated decision-making as a means for efficient use of data and knowledge has been highlighted in the "National Strategic Plan for Advanced Manufacturing" published by the Executive Office of the President in 2012. Enterprise-wide decision-making demands the integration of operational activities of planning, scheduling, and control [8]. Control of unit operations needs mechanistic models. These models capture complex reactions and transport phenomena and may demand many CPU hours due to heterogeneity and dynamics of certain systems. Maintaining computational feasibility has been approached by substituting the original detailed model with a reduced model. Reduced models have been developed either by reduction of the order of the model based on evaluation of the significance of its components or development of surrogate models using data obtained through careful experimental design.

1.2. Bioreactor

The goal in the operation of a bioreactor is to enhance the growth, viability, and productivity of organisms by adjusting their environment. Figure 1 presents a cause and effect diagram of bioreactor operation. The main operational parameters are impeller rotation speed, gas sparging flow rate, pH, and temperature. Inefficient mixing creates spatial gradients in mechanical shear, volume fraction of gas phase, dissolved oxygen (DO), carbon dioxide and metabolites concentrations, pH, and temperature. As organisms move inside the reactor they experience fluctuations in environmental conditions that affect metabolism, yield, and quality of product [9]. Due to low solubility in water, high densities of organisms quickly consume all the oxygen in a saturated culture and produce enough carbon dioxide to have inhibitory effects. Therefore, the addition and removal of gases are inevitable parts of fermenter operation. Bioreactor systems can operate under different modes: batch, fed-batch, continuous, and perfusion [10]. Considerations regarding the set-up, operation, and control of bioreactors can be found in the literature [11–15].

The environment refers to the physical and chemical stimuli acting on organisms. Physical stimuli of bubble size distribution and dissipation rate of energy directly affect the viability of cells. Hydrodynamic forces acting on an organism can be quantified using shear stress and energy dissipation rate. Some of the effects of mechanical stress are lysis, change in distribution and number of surface receptors, alteration in production of specific proteins, rate of DNA synthesis, rate of certain metabolic processes, and induction of apoptosis [16]. Chemical stimuli include concentrations of metabolites (e.g., glucose, glutamine, lactate, and ammonia), dissolved oxygen concentration, and pH. Concentrations of other supplements that have functions such as species transport enhancement, growth stimulation, shear protection, and surface charge modification can also be considered. The dynamics of organisms and environment interaction should be modeled based on an understanding of the response times of biological and physical mechanisms. Prokop compared the characteristic times of physical and biological mechanisms [17]. Characteristic time is a measure of the time needed by the mechanism to smooth out a change to a certain extent. It was observed that physical mechanisms of mixing, oxygen transfer to the liquid phase, and diffusion are faster compared to the biological mechanisms of oxygen and substrate consumption.

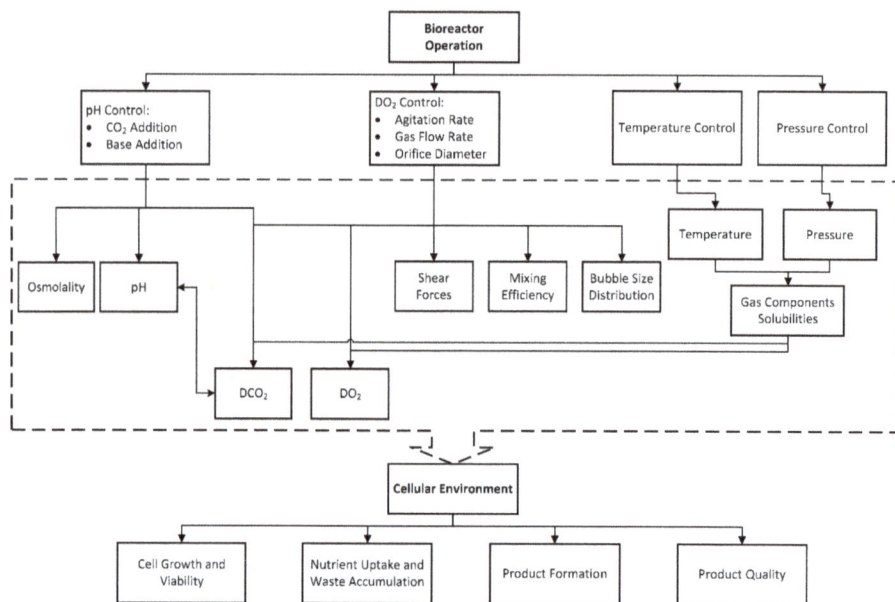

Figure 1. Cause and effect diagram for the operation of a bioreactor.

1.3. Existing Culture Models and Their Need for Improvement

Robust mechanistic bioreactor models facilitate improvement in equipment utilization, medium design, and feeding strategy, and explain causes of scale-up problems such as productivity and byproduct formation [18]. Assuming spatial homogeneity and neglecting the effects of physical environmental parameters have been the basis of most culture models [19–29]. This reduces the problem of simulating relevant portions of cellular activities that control the production of a product of interest. Metabolic models are divided into two main groups based on whether the cell mass composition is considered variable (structured) or fixed (unstructured). Structured models are based on a set of biochemical reactions, which create stoichiometric relations to represent the metabolism of the organism. The reactions contain both extracellular components and intracellular components. The flux of the intracellular metabolites can be determined by assuming a pseudo-steady state inside the cell. Unstructured models utilize a reduced number of reactions to capture metabolism macroscopically. Unstructured models do not have the ability to capture the effects of the growth condition on cellular composition, and as a result they cannot accurately predict balanced growth for a culture in transient condition. In our earlier work examples of different metabolic models and their capabilities have been reviewed [30]. On the unreliability of existing cell culture models, it should be noted that their predictions are independent of whether a shake flask or a large-scale bioreactor is used for the cultivation. These models, in which hydrodynamics is often excluded, usually consider all causes of cell loss in one parameter, e.g., growth rate. The performance of the model deteriorates when the process conditions change as the parameters of unstructured models are highly dependent on strain, cultivation medium, and fermentation conditions [31]. Consequently, existing models do not possess satisfactory predictive power, especially when used outside the calibration range, and literature data can only be used for qualitative studies. In addition to narrow confidence intervals, achieving a satisfactory fit of experimental data often requires considering extra terms or assumptions, which are theoretically difficult to explain. Overall, despite the practical and commercial applications of animal cells, there are only a few reports on their kinetics of growth and production. More specifically, there is no literature report on the kinetic parameters for Chinese hamster ovary (CHO) cell batch culture related to mAb production [32]. Therefore, in this work, instead of final protein concentration, the time-integrated value of biomass is the subject of maximization. The specific attributes of large-scale mammalian cell cultures, e.g., high level of spatial heterogeneity and sensitivity of organisms to physical environmental stimuli, demand a modeling framework that captures both the biology and the hydrodynamics of the system and also their interactions. This work aims to improve on the current state of lumped-parameter bioreactor modeling by capturing the interactions of system components. The selection of components is based on experimental observations. The formulation is devised in order to take into account the inherent dynamics of the system, maintain computational tractability, and couple the model with optimization solvers. In the next section, the modeling and integration of hydrodynamics and biological processes are discussed and the application of compartmental modeling for maintaining computational feasibility is explained. In Section 3, the integrated model is used to find a near-optimal operation scenario. Section 4 is a discussion of the challenges and potential in the area of modeling and the optimization of a bioreactor operation.

2. Development of a Dynamic, Integrated, and Computationally Feasible Bioreactor Model

This work seeks to improve the reliability of bioreactor models by capturing the effects of hydrodynamics on the performance of a bioreactor. Computational Fluid Dynamics (CFD) simulation is employed to calculate the attributes of flow required for this purpose. Biological processes are inherently dynamic; therefore, solving the problem requires dynamic CFD simulations. Dynamic CFD simulations demand the discretization of time using step sizes in the range of 0.01 to 0.1 s [33–37], even for a small reactor. A CPU time of 12 s per CPU for each time step has been reported [34], and 0.5 to 2 s per CPU has been observed for every iteration [36]. Considering the reported CPU times and the fact that bioreactors are usually operated in fed-batch mode for up to two weeks, dynamic CFD simulation of the entire operation is computationally unfeasible. To tackle this problem, in our earlier work a two-step framework was developed [38]. This method is based on the assumption that the effects of metabolic activities on hydrodynamics are negligible [18,39]. The simulation is run in two steady-state and dynamic steps. First the steady state of the two-phase flow inside the bioreactor for specific values of impeller rotation speed and gas sparging flowrate is calculated. At this stage cells are considered merely components of the liquid phase without any biological function. Then the problem is solved dynamically to obtain the evolution of biophase over time. In the dynamic phase of the simulation only species conservation is considered. During the dynamic run, the process parameters of impeller rotation speed and gas sparging flowrate are fixed and the flow remains at its fully developed steady-state condition. Therefore the values of velocity, gas volume fraction, kinetic energy, and energy dissipation rate remain approximately constant. Obtaining the solution in two stages replaces the problem with two smaller systems of equations. This allows for the specification of larger time step sizes and improves the convergence of the simulation.

Deconstructing the problem into steady-state flow and dynamic metabolism problems and solving them sequentially seems to successfully address the batch operation with constant process parameters [38]. Although this simulation procedure exploits off-the-shelf software packages, it is still limited to simple case studies. The improvement in computational feasibility in order to simulate fed-batch operation and study effects of composition and schedule of feeding on the performance of bioreactor is achieved through the development of a compartmental model. Compartmental modeling facilitates time–space decomposition, which significantly reduces computation. Therefore, it has been widely used for modeling the hydrodynamics of stirred tanks [40–43]. In this methodology the reactor is divided into well-mixed zones that do not contain segregated regions. Then fluxes between the zones are calculated based on the expected flow patterns resulting from CFD simulations or experimental data [18,40,41,44]. A number of states of operation are defined by discretizing the process parameters of impeller rotation speed, gas sparging flow rate, and operating volume. Data on fully developed steady state flow under all pre-defined operation states are obtained and stored in flow matrices.

Flow matrices contain information on inlet and outlet fluxes, gas volume fraction, dissipation rate of mechanical energy, and gas superficial velocity of compartments and characterize the flow inside the reactor under specific operating conditions. The change in operating conditions is simulated by replacing the flow matrices of the current state of operation with those associated with the new state. This is based on the assumption that the time for the flow to reach a new steady state as a result of a change in the operational conditions is negligible compared to the total processing time. It should be noted that compartmental modeling provides an approximation of the solution and as such predicts more homogenous distribution for species since each compartment is homogeneous, and does not account for diffusive mass transfer. On the other hand, it makes it possible to take into account hydrodynamics in dynamic analysis of reactor performance and couple the model with optimization solvers. In this work, integration of hydrodynamics with metabolism refers to capturing the effects of dissolved oxygen (DO) concentration, bubbles, and turbulent eddies on the metabolic activities and viability of cells. Biological processes are captured through unstructured modeling. This utilizes a reduced number of reactions to macroscopically capture cellular kinetics.

2.1. Development of CFD Simulations

Computational fluid dynamics (CFD) simulations are developed in ANSYS® Fluent® 15.0.7 for the prediction of spatial variations of environmental parameters. Conservation laws of mass, momentum, and energy are usually used to describe a single phase flow, gas or liquid. If the thermodynamic, transport, and chemical properties of a component need to be specified, the field equations may be accompanied by the constitutive equations of state, stress, chemical reactions, etc. The presence of interfacial surface in a multi-phase flow complicates the mathematical formulation of the problem. To derive the field and constitutive equations of a multi-phase flow, such as inside a bioreactor, local characteristics have to be considered. This is not straightforward due to unknown motions of multiple deformable interfaces, variable fluctuations due to turbulence and moving interfaces, and discontinuity of properties at the interface. Obtaining local mean values of flow properties has been shown to be an efficient way to eliminate instantaneous fluctuations. Three averaging methodologies have been developed: Eulerian, Lagrangian, and Boltzmann statistical averaging. In the Eulerian approach, time and space coordinates are independent and other variables are expressed with respect to them. In the Lagrangian averaging methodology, particle coordinates replace spatial coordinates. If the purpose of modeling is studying the group behavior of particles, the Eulerian approach is preferred. However, if the behavior of individual particles is of interest, the Lagrangian description has a clear advantage [45]. Tracking individual bubbles increases the computation. Additionally, it would only improve model predictive power if the extent of the interactions between individual bubbles and the liquid phase could be quantified. These interactions involve growth, breakage, and agglomeration of bubbles and energy dissipation due to bubble rupture. Therefore, in this study gas and liquid phases are treated as continua and Eulerian averaging is used. The Eulerian multiphase model creates sets of momentum and continuity equations for each phase and couples them through exchanging pressure and interphase coefficients [46]. Turbulence of flow is calculated using the k-ε viscosity model, which has been widely used for stirred tanks [47]. It is a robust model that gives reasonably accurate results for a wide range of turbulent flows [48]. A k-ε model consists of two transport equations, one each for the turbulent kinetic energy (k) and the energy dissipation rate (ε). The motion of the impeller is captured using a multiple reference frame (MRF). To implement the MRF model, the geometry is broken up into stationary and moving zones. The MRF model approximates the flow in the moving zone around the impeller by freezing the motion of the moving part in a specific position and observing the instantaneous flow field. To use flow variables of one zone for calculation of fluxes at the boundary of the adjacent zone, a local reference frame transformation is performed at the interface between cell zones. In the absence of large-scale transient effects due to weak impeller–wall interactions, the MRF approach provides a reasonable approximation of the flow [48].

2.2. Development of the Integrated Model

The state of operation is defined by operating volume, impeller rotation speed, and gas sparging flowrate. For every admissible state a separate CFD simulation is developed to obtain the steady-state flow under associated operating conditions. After the solution is converged, the flow information of computational cells is extracted. Compartments are formed through agglomeration of computational cells. Inlet and outlet fluxes, gas volume fraction, dissipation rate of mechanical energy, and gas superficial velocity of compartments are calculated using data obtained from CFD simulations. The surface between two neighbor compartments is composed of faces of computational cells. The mass flowrates reported for these faces are summed to calculate the flowrates between neighbor compartments. The net flow between two neighbors is not necessarily zero, but the net flow for each compartment should be zero in order to maintain mass conservation. Fluent does not report flowrates for faces at the boundaries. Boundary faces are located at the surface of meshing zones. Although very small, these missing values cause mass imbalance and introduce errors. These errors are smoothed out by making the smallest possible changes to the calculated flowrates. It is assumed that the calculated value for mass flowrate from compartment j to i, F_{ij}, has error ε_{ij}. The sum of squared errors is minimized by solving the problem explained by Equations (1) to (3). The second constraint makes sure that zero elements will remain zero, i.e., flowrates between non-neighbor compartments remain zero. The effect of agitation on distribution of cells between compartments is only understood if the sedimentation of cells is captured. The rate of sedimentation is estimated as $r^2/4$ (mm/h) and r is cell radius in µm [49]. Cell radius is estimated assuming spherical shape, density equal to that of water, and average mass of 1.1165×10^{-6} mg [20].

$$\underset{\varepsilon_{ij}}{\text{minimize}} \sum_i \sum_j \varepsilon_{ij}^2 \qquad (1)$$

subject to:

$$\sum_i F_{ij} - \varepsilon_{ij} = \sum_i F_{ji} - \varepsilon_{ji} \qquad (2)$$

$$-F_{ij} \le \varepsilon_{ij} \le F_{ij} \qquad (3)$$

It has been reported that cells can be damaged when the power input is greater than 22,500 W·m^{-3} [50,51]. Power input is calculated by multiplying the density of the liquid phase (kg/m^3) by the turbulent energy dissipation rate (m^2/s^3). The turbulent energy dissipation rate is the rate of absorption of kinetic energy that breaks up large eddies. This is then converted to heat by viscous forces [52]. A rate of cell damage of 3.4% min^{-1} has been reported for cells in high shear regions [51]. The volume fraction of high shear region in compartments is calculated using the data obtained from CFD simulations. Since cells are assumed to be homogenously distributed inside compartments, the volume fraction of the high shear region is equal to the fraction of cells exposed to shear beyond their tolerable threshold. Therefore, the rate of loss of viable cells under operating condition op and in compartment c is calculated using Equation (4):

$$k_{d,shear\,op,c} = 0.034.\,volume\;fraction\;of\;high\;shear\;region_{op,c}. \qquad (4)$$

Interaction with bubbles has also been reported as one of the sources of cell loss. Cells attach to bubbles, rise with them to the surface, become trapped in the foam layer, and perish. Also, the maximum energy dissipated due to bubble rupture is two or three orders of magnitude higher than the tolerable threshold for cells [51]. No value has been reported in the literature for the rate of cell loss due to interaction with bubbles. One motive for integrated modeling is to capture uncertainty where it occurs. The rate of cell loss due to interaction with bubbles is estimated by assuming an interaction vicinity around bubbles. It is assumed that a fraction of the cells in a compartment that are in the vicinity of bubbles are lost over the average lifespan of a bubble in the compartment. The volume of bubble vicinity is calculated using the reported average bubble diameter of 0.00289 m [53] and assuming a particular value for critical distance from the surface of bubbles. The critical distance is assumed to be equal to the cell radius. The number of bubbles is calculated using gas holdup in the compartment obtained from CFD simulations and the reported value for average bubble diameter. Average bubble lifespan is calculated using the gas holdup and air sparging flowrate. Equation (5) shows the estimated rate of loss of cells due to interactions with bubbles under operating condition *op* and in compartment *c*:

$$k_{d,bubbleop,c} = \frac{\ln\left(1 - \frac{volume\ of\ the\ interaction\ vicinity \times number\ of\ bubbles_{op,c}}{volume_c}\right)}{bubble\ lifespan_{op,c}}. \tag{5}$$

The integrated model predicts viable cell density (VCD) in compartment *c* using Equation (6). μ and μ_d are the metabolic rates of growth and death. They are calculated as functions of metabolites' concentrations using the metabolic model. F^{sed} is the rate of sedimentation. c' and c'' are compartments below and above *c*. *N* is the total number of compartments. It can be seen that the hydrodynamic part of Equation (6) is a linear system of independent ODEs of rank *N*. The discretization of space provided by compartmental modeling allows for the application of proper orthogonal decomposition (POD) for the development of a reduced-order model (ROM) [54]. In order to achieve this, the full rank system is first solved to create time-series snapshots of distribution of cells over compartments under a specific operating condition. Then a set of orthonormal bases are generated through eigen-decomposition [55]. Bases with no significant impact on the solution profile are truncated to obtain the reduced rank model. A system with *N* = 16 was used to evaluate the performance of the ROM developed with this methodology. The ROM showed satisfactory performance, while the reduction in the rank of the system was small. The impact of the initial condition used for generating snapshots became apparent when fewer basis functions were used for approximation. For large reactors with a greater number of compartments, however, it is recommended to investigate the application of this methodology.

$$\frac{dX_c}{dt} = \left(\mu - \mu_d - k_{d,shearop,c} + k_{d,bubbleop,c} - F^{sed}_{c',c} - \frac{\sum_{i=1}^{N} F_{i,c}}{\rho_l.volume_c}\right)X_c$$
$$+F^{sed}_{c,c''}X_{c''} + \sum_{i=1}^{N}\frac{F_{c,i}}{\rho_l.volume_c}X_i \tag{6}$$

Contrary to cells, metabolites are assumed to be homogenously distributed at all times. This is due to the fact that fast diffusion dominates mass transfer in the small reactor considered for case study. For larger bioreactors, local diffusive mass transfer has lower importance relative to convection, so the flux matrix may be used for calculation of distribution of metabolites [18]. The dissolved oxygen (DO) concentration is also assumed to be homogenously distributed. In the calculation of DO concentration, mass transfer and cellular uptake are considered. The ratio of rates of oxygen uptake to carbon dioxide production by cells has been reported to vary within a narrow range around 1 [56]. So the oxygen uptake rate (OUR) is assumed to be 0.35 pmol·cell^{-1}·h^{-1}, which has been reported for carbon dioxide production [57]. The overall volumetric mass transfer coefficient, k_La, is calculated using Equation (7), in which $U_{Gop,q}$ is superficial gas velocity (m/s) for computational cell *q* under operating condition *op* [33]. Volumetric mass transfer coefficient is the product of liquid phase mass transfer coefficient; k_L (m/s) and specific interfacial area; a (m^2·m^{-3}). Mass transfer stops after reaching the saturation concentration at 37 °C. The reported saturation mass fraction of oxygen is 3.43 × 10^{-5} for [58]. The unstructured model is assumed to predict metabolite uptake and production rates when the culture is oxygen-saturated. Experimental data show the dependence of these rates on the concentration of dissolved oxygen [59]. The reported data are used to calculate correction factors for metabolites' uptake and production rates at different concentrations of DO (Figure 2). Uptake and production rates of metabolites predicted by the unstructured model are multiplied by correction factors to take into account the effects of mass transfer mechanism on metabolism of cells.

$$(k_L a)_{op} = \frac{\sum_q 0.412 U_{Gop,q}{}^{0.809} \cdot gas\ volume\ fraction_{op,q} \cdot volume_q}{\sum_q volume_q} \tag{7}$$

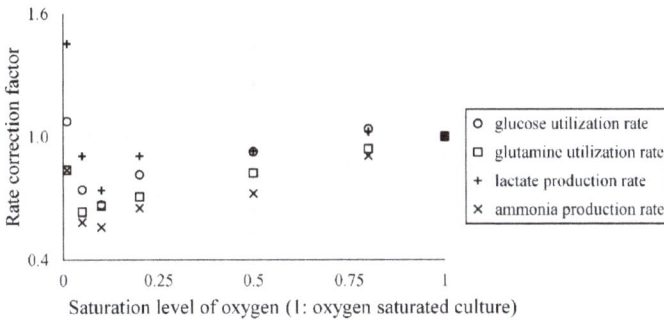

Figure 2. The effect of DO concentration on metabolites' uptake and production rates.

The behavior of the bio-phase is computed through the incorporation of an unstructured metabolic model and the consideration of the effects of environmental parameters on viable cell density. For the purpose of this study, a metabolic model developed in the literature is adapted to represent cellular growth and death rates as functions of the concentrations of metabolites. As discussed, due to the lumped nature of unstructured models, the estimated values of their parameters have narrow confidence intervals. Moreover, these values lose their meanings when the model is used to predict the dynamics of a different bioreactor system. An important group of parameters are threshold metabolites' concentrations, which separate growth and death rates into different regimes. Threshold concentrations are determined by observing how viable cell density reacts to concentrations of metabolites. Capturing the dynamic behavior of the system sometimes requires considering multiple phases for cellular growth and death, during which cells react differently to environmental stimuli. Xing et al. [60] assumed the death phase begins after viable cell density declines by 10% from its peak value. It was assumed that cellular growth did not happen during this phase. The integration with hydrodynamics incorporates additional sources of cell loss into the model, which impacts the viable cell density profile. For these reasons, the metabolic model is merely adapted to explain the integration of physical and biological processes. This paper proposes a framework for capturing the interaction of system components and uncertainty where it occurs. The results presented in this paper are only meant to demonstrate the capabilities of the modeling framework.

2.3. Coupling the Model with Nonlinear Solvers

Maximization of bioreactor yield is achieved through manipulation of process parameters based on the optimal operating policy. Reduction of the order of the model through compartmental modeling provides the formulation and modeling environment necessary for coupling the model with nonlinear or mixed-integer nonlinear programming solvers. Optimization algorithms that use reduced models are categorized based on their level of dependence on the original detailed model for the calculation of gradients [61]. In the proposed formulation, reduced models are developed before calling the optimization algorithm since minimal communication between the algorithm and the original model is necessary to reduce the computational complexity. Advanced nonlinear optimization algorithms are capable of handling numerous decision variables and constraints. However, the inherent dynamics of this problem makes it challenging. The classical approach to dynamic optimization problems takes advantage of Pontryagin's maximum principle and maximizes the control Hamiltonian over the set of all admissible controls [62]. The application of this approach becomes difficult for larger systems with state constraints [63], so direct approaches based on parameterization of variables have been preferred [64,65]. The method of collocation has been proposed for parameterization of variables [66,67]. The stiffness of the system of ordinary differential equations (ODE) is evaluated for different initial values. The stiffness ratio is calculated using the eigenvalues of the Jacobian matrix at different points in time [68]. It is observed that the order of magnitude of the stiffness ratio varies between 6 and 28 throughout the integration. Therefore, an approach based on the discretization of time using fixed step sizes does not provide a good approximation of the solution unless the step size is very small, i.e., less than 10^{-9} h. Instead, integration is carried out using appropriate solvers for stiff, nonlinear ODEs and the problem is formulated for the application of the interior point method [69]. Equations (8)–(11) represent the solution to the mathematical optimization problem. T_i and C_i are the time and composition of the ith feeding. In addition to initial nutrient concentrations, schedule, and composition of feeding, it also finds optimal criteria for setting aeration and agitation rates. Aeration is stopped or started based on the DO level. For the adjustment of impeller rotation speed, a measure of homogeneity is defined based on the relative standard deviation (RSD) of distribution of cells over compartments.

$$\underset{C_0, T_i, C_i, \text{RSD}_{cri}, \text{DO}_{cri}}{\text{maximize}} \int_0^{t_f} \text{biomass} \, dt \tag{8}$$

Subject to:

Process Model: Equation (6)

Control Bounds:

$$0 < \ldots < T_{i-1} < T_i < T_{i+1} < \ldots < t_f \tag{9}$$

$$C_{imin} \leq C_i \leq C_{imax} \tag{10}$$

$$0\% \leq \text{RSD}_{cri}, \text{DO}_{cri} \leq 100\%. \tag{11}$$

3. Case Study

A 3 L bioreactor with Rushton impeller and sparger is considered. The operation lasts for two weeks. The dimensions of the reactor are shown in Figure 3. It is operated at liquid fill levels of 130, 155, 180, and 205 mm. Feeding is simulated by change of liquid fill level. So, overall, three steps of feeding are allowed. For the discretization of space into computational cells, the space is divided into several meshing zones. This helps Fluent to generate hexahedral cells where possible. It also reduces the total number of computational cells and improves the convergence of the solution. Meshing for the highest operating volume results in 363,175 computational cells, 947,988 faces, and 226,971 nodes. Agitation rates of 150, 225, and 300 RPM and two options for aeration are considered: sparging air at 0.01 vessel volume per minute (vvm), and no aeration. Table 1 shows the physical properties calculated from CFD simulations. The reactor is divided into compartments as shown in Figure 4. The data exported from CFD simulations are used to calculate flow matrices for all states of operation. The unstructured model developed by Xing et al. [60] is adapted to predict the behavior of the bio-phase. The metabolic model captures the effects of concentrations of glucose, glutamine, lactate, and ammonia on cellular rates of growth and death in a CHO culture. The rate of utilization of glutamine for essential metabolic functions, i.e., maintenance, is calculated from Equation (12). The values of the model parameters are shown in Table 2.

$$m_{Gln} = \frac{a_1[Gln]}{a_2 + [Gln]} \tag{12}$$

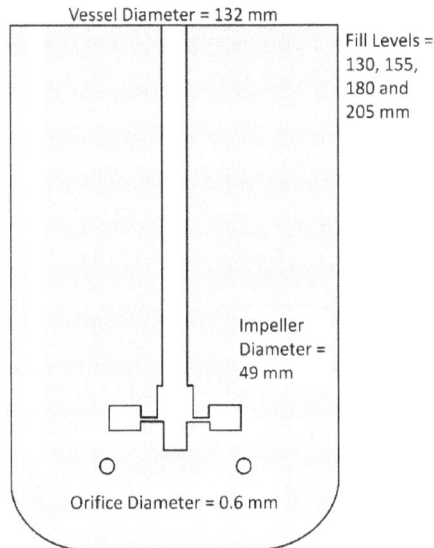

Figure 3. Geometry of the vessel considered for the CFD simulations.

Table 1. Physical properties calculated from CFD simulations.

	No Aeration			Aerated System		
Fill Level (mm)	Impeller Rotation Speed (RPM)	Power Input (W·m^{-3})	Fill Level (mm)	Impeller Rotation Speed (RPM)	Power Input (W·m^{-3})	Volumetric Mass Transfer (h^{-1})
130	150	3.8	130	150	3.5	12.8
	225	11.0		225	11.1	16.6
	300	24.5		300	26.5	14.9
155	150	3.2	155	150	3.1	10.9
	225	9.5		225	9.7	11.3
	300	21.6		300	23.6	16.1
180	150	2.7	180	150	2.4	14.4
	225	8.4		225	7.9	14.8
	300	19.2		300	19.6	20.4
205	150	2.3	205	150	2.7	11.9
	225	7.0		225	6.5	9.6
	300	16.1		300	15.5	13.7

Cellular rates of growth and death are calculated by knowing the Monod constants ($K_{metabolite}$) and using Equations (13) and (14), respectively. Growth and death rates are then used in Equation (6) to calculate viable the cell densities in compartments:

$$\mu = \mu_{max} \frac{[Glc]}{K_{Glc} + [Glc]} \frac{[Gln]}{K_{Gln} + [Gln]} \frac{KI_{Lac}}{KI_{Lac} + [Lac]} \frac{KI_{Amm}}{KI_{Amm} + [Amm]} \tag{13}$$

$$\mu_d = \mu_{dmax} \frac{[Lac]}{KD_{Lac} + [Lac]} \frac{[Amm]}{KD_{Amm} + [Amm]}. \tag{14}$$

The estimation of overall volumetric mass transfer coefficient under present operation conditions is used in Equation (15) to calculate the concentration of DO, where X is the viable cell density.

$$\frac{d[DO]}{dt} = (k_La)_{op}(DO_{eq} - [DO]) - X.OUR \tag{15}$$

Concentrations of metabolites are calculated from Equations (16)–(19) by knowing the values of yield parameters (Y's). The impact of DO on uptake and production rates ($DO_{metabolite}$) is estimated through spline interpolation of the experimental data shown in Figure 2. The model also captures the chemical degradation of glutamine.

$$\frac{d[Glc]}{dt} = -DO_{Glc}\left(\frac{\mu - \mu_d}{Y_{X/Glc}} + m_{Glc}\right)X \tag{16}$$

$$\frac{d[Gln]}{dt} = -DO_{Gln}\left(\frac{\mu - \mu_d}{Y_{X/Gln}} + m_{Gln}\right)X - d_{Gln}[Gln] \tag{17}$$

$$\frac{d[Lac]}{dt} = DO_{Lac}\cdot Y_{Lac/Glc}\left(\frac{\mu - \mu_d}{Y_{X/Glc}} + m_{Glc}\right)X \tag{18}$$

$$\frac{d[Amm]}{dt} = DO_{Amm}\cdot Y_{Amm/Gln}\frac{\mu - \mu_d}{Y_{X/Gln}}X \tag{19}$$

The definition of states of operation creates a finite set of admissible actions for every point in time. Depending on the present operational conditions, it is possible to increase or decrease impeller rotation speed, stop or start air sparging, and feed or not feed. It is also possible to continue under the current conditions. Changing the state of operation is limited to once every 2 h. The concentrations of glucose and glutamine, initially and in feed, are constrained to be under 100 and 10 mM, respectively. The solid lines in Figure 5 show the system operated under the policy obtained from solving the optimization problem described in Equations (8)–(11) (system 1). Although the existence of multiple local optima cannot be ruled out, the solution shows improvement in the yield of operation through manipulation of feeding schedule and composition, agitation, and aeration rates. Nutrient concentrations stay at their upper bounds due to the limited number of feeding steps. The criterion for aeration is 47% of saturated DO concentration, i.e., aeration starts if DO concentration falls below this value and stops otherwise. The obtained criterion for agitation is 0.02%. The impeller rotation speed is increased if the RSD of distribution of cells over compartments is greater than the agitation criterion. In the opposite case, where RSD has a smaller value, the lower agitation rate is selected for the next 2 h. For comparison, the dashed lines in Figure 5 show the system operating with the same initial and feed compositions but a uniform feeding schedule (system 2). System 2 is consistently aerated and agitated at 300 RPM, i.e., the criteria for aeration and agitation are 100% DO saturation and RSD of 0%, respectively. Schedule of feeding should be determined with consideration of capacity of the reactor and duration of operation. Early addition of feed, despite increasing cellular population, causes accumulation of lactate and ammonia in the system, which further inhibits growth. Late feeding, on the other hand, results in poor utilization of nutrients and reactor capacity. Solving the optimization problem results in feeding times of 215, 265, and 300 h after the start of cultivation. Even with a limited number of feeding steps, through the manipulation of the feeding schedule, the cells in system 1 are provided with enough nutrients to maintain growth throughout the operation. Inclusion of mass transfer mechanism in the model leads to an improvement in aeration. The DO concentration drops quickly when aeration stops because of fast consumption by cells. Stopping and starting aeration according to the near-optimal policy prevents loss of viable cells due to unnecessary aeration while guaranteeing that DO is not depleted. The impeller rotation speed switches between 150 and 225 RPM for most of the operation to maintain RSD of distribution of cells at 0.02%. The three sharp peaks in RSD values show disturbances in spatial homogeneity of cells caused by feeding. The declining trend in net growth that happens toward the end of the process is due to the fact that the cellular growth rate, which is reduced by the inhibitory effects of metabolites, cannot compete with the cell loss due to the effects of hydrodynamics. The results demonstrate that integrated modeling is able to capture the behavior of the system using a mechanistic understanding of the reactor without the need for unnecessary assumptions.

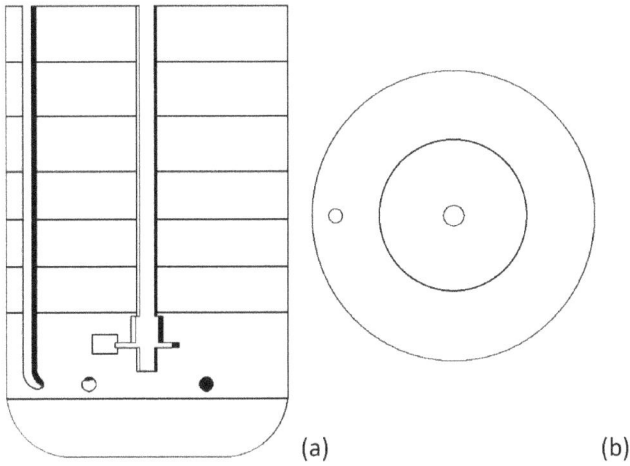

Figure 4. Arrangement of compartments; side (**a**) and top (**b**) views.

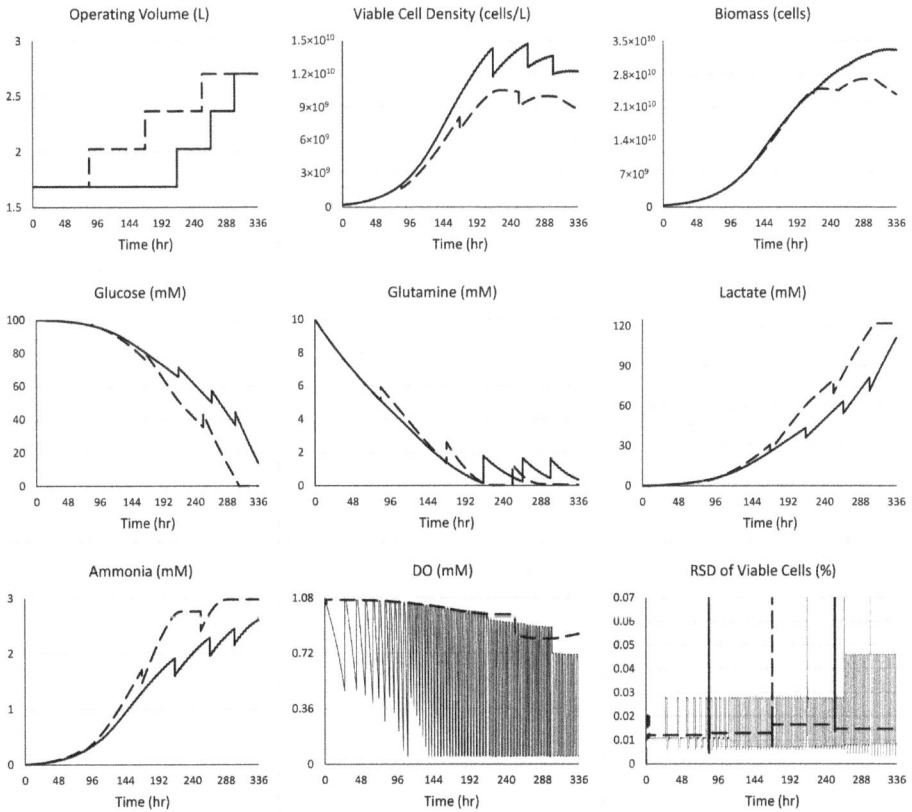

Figure 5. Comparison of two operational polices: near-optimal policy (solid lines); and alternative policy with uniform feeding schedule (dashed lines).

Table 2. Unstructured model parameters.

Parameter	Value	Unit
m_{Glc}	6.92×10^{-11}	mmol·cell^{-1}·h^{-1}
a_1	3.2×10^{-12}	mmol·cell^{-1}·h^{-1}
a_2	2.1	mM
μ_{max}	0.029	h^{-1}
μ_{dmax}	0.016	h^{-1}
K_{Glc}	0.084	mM
K_{Gln}	0.047	mM
KI_{Lac}	43	mM
KI_{Amm}	6.51	mM
KD_{Lac}	45.8	mM
KD_{Amm}	6.51	mM
d_{Gln}	7.2×10^{-3}	h^{-1}
$Y_{X/Glc}$	1.69×10^8	cell·mmol^{-1}
$Y_{X/Gln}$	9.74×10^8	cell·mmol^{-1}
$Y_{Lac/Glc}$	1.23	mmol·mmol^{-1}
$Y_{Amm/Gln}$	0.67	mmol·mmol^{-1}
DO_{eq}	1.0699	mM
OUR	3.5×10^{-10}	mmol·cell^{-1}·h^{-1}

4. Conclusions and Future Directions

The implementation of mechanistic models in the biotechnology industry has been hindered by a lack of universality of cell culture models. Integrated modeling, followed by experimental design and parameter estimation, can lead to quantification of the extent of effects of mechanical shear and bubble interactions on the viability of cells. It also gives better estimations of the cellular rates of growth and death as functions of metabolite concentrations. A model that represents the system well can be linked to proper optimization algorithms to recommend low-cost improvements for the operation of a bioreactor. Furthermore, computationally feasible unit operation models facilitate the integration of control, scheduling, and planning as a leading step toward integrated decision-making.

The incorporation of a buffer system into the model results in better representation of the system. It makes it possible to calculate pH and capture its effects on biological processes. This is achieved at the cost of increasing the nonlinearity and rank of the ODE system. The discrete space of admissible operating conditions can be expanded through finer discretization of process parameters. It can also be replaced by a continuous space through multi-scale surrogate modeling. Data obtained from CFD simulations for computational cells can be used for the development of surrogate models for hydrodynamics [70]. The main challenge in this area is to devise an efficient algorithm to explore the large sampling space. To select a subset of computational cells, Zhao et al. proposed a sampling method based on Latin hypercube designs (LHDs) [71]. After decomposing the complete data into disjoint equally spaced blocks, a subsample is obtained by collecting blocks according to a randomly generated LHD. This method is called LHD-based block bootstrap. It takes into account the spatial dependency and therefore improves the accuracy of estimations. Integration with hydrodynamics introduces new parameters to the cell culture model, i.e., sedimentation rate, tolerable shear threshold, rate of cell damage due to shear, bubble radius, interaction distance with bubbles, and mass transfer coefficients. The uncertainties in the values of these parameters can be included in the dynamic analysis of the operation. The sensitivity of the solution profile to model parameters can be determined and the uncertainty in sensitive model parameters can be considered while solving for the optimal operating policy. Although the proposed approach cannot guarantee that the global solution is obtained, since a local optimization algorithm is utilized, the convergence to the global optimal solution can be improved using initialization strategies in the interior point method.

Processes **2018**, *6*, 82

Author Contributions: Under supervision of M.I., P.F. was responsible for development of research methodology and modeling framework, model computation and analysis of the results.

Funding: This research received no external funding.

Acknowledgments: Financial support from the School of Engineering at Rutgers University is gratefully acknowledged.

Conflicts of Interest: The authors declare no conflicts of interest.

References

1. Moorkens, E.; Meuwissen, N.; Huys, I.; Declerck, P.; Vulto, A.G.; Simoens, S. The Market of Biopharmaceutical Medicines: A Snapshot of a Diverse Industrial Landscape. *Front. Pharmacol.* **2017**, *8*, 314. [CrossRef] [PubMed]
2. Davidson, A.; Farid, S.S. Innovation in Biopharmaceutical Manufacture. *BioProcess Int.* **2014**, *12*, 12–19.
3. Rader, R.A.; Langer, E.S. 30 years of upstream productivity improvements. *BioProcess Int.* **2015**, *13*, 10–15.
4. Farid, S.S. Process economics of industrial monoclonal antibody manufacture. *J. Chromatogr. B-Anal. Technol. Biomed. Life Sci.* **2007**, *848*, 8–18. [CrossRef] [PubMed]
5. Terry, C.; Lesser, N. *Balancing the R&D Equation*; Deloitte Center for Health Solutions: London, UK, 2017.
6. Gyurjyan, G.; Thaker, S.; Westhues, K.; Zwaanstra, C. *Rethinking Pharma Productivity*; McKinsey & Company: New York, NY, USA, 2017.
7. Varma, V.A.; Reklaitis, G.V.; Blau, G.E.; Pekny, J.F. Enterprise-wide modeling & optimization—An overview of emerging research challenges and opportunities. *Comput. Chem. Eng.* **2007**, *31*, 692–711.
8. Grossmann, I.E. Advances in mathematical programming models for enterprise-wide optimization. *Comput. Chem. Eng.* **2012**, *47*, 2–18. [CrossRef]
9. Lara, A.R.; Galindo, E.; Ramírez, O.T.; Palomares, L.A. Living with heterogeneities in bioreactors. *Mol. Biotechnol.* **2006**, *34*, 355–381. [CrossRef]
10. Xie, L.; Zhou, W.; Robinson, D. Protein production by large-scale mammalian cell culture. *New Compr. Biochem.* **2003**, *38*, 605–623.
11. Spier, R.E. Encyclopedia of Cell Technology. In *Wiley Biotechnology Encyclopedias*; Wiley-Interscience: Hoboken, NJ, USA, 2000.
12. Panda, T. *Bioreactors: Analysis and Design*; Tata McGraw-Hill Education Private Limited: New York, NY, USA, 2011.
13. Ho, C.S.; Wang, D.I.C. *Animal Cell Bioreactors*; Biotechnology Series; Davies, J.E., Ed.; Butterworth-Heinemann: Oxford, UK, 1991.
14. Mandenius, C.-F.; Titchener-Hooker, N.J. Measurement, Monitoring, Modelling and Control of Bioprocesses. In *Advances in Biochemical Engineering/Biotechnology*; Springer: Berlin, Germany, 2013.
15. Meyer, H.-P.; Schmidhalter, D. *Industrial Scale Suspension Culture of Living Cells*; Wiley: Hoboken, NJ, USA, 2014.
16. Shuler, M.L.; Kargi, F. *Bioprocess Engineering*, 2nd ed.; Prentice Hall: Upper Saddle River, NJ, USA, 2002.
17. Prokop, A. Implications of Cell Biology in Animal Cell Biotechnology. In *Animal Cell Bioreactors*; Ho, C.S., Wang, D.I.C., Eds.; Butterworth-Heinemann: Oxford, UK, 1991.
18. Pigou, M.; Morchain, J. Investigating the interactions between physical and biological heterogeneities in bioreactors using compartment, population balance and metabolic models. *Chem. Eng. Sci.* **2015**, *126*, 267–282. [CrossRef]
19. Meshram, M.; Naderi, S.; McConkey, B.; Ingalls, B.; Scharer, J.; Budman, H. Modeling the coupled extracellular and intracellular environments in mammalian cell culture. *Metab. Eng.* **2013**, *19*, 57–68. [CrossRef] [PubMed]
20. Sidoli, F.R.; Asprey, S.P.; Mantalaris, A. A Coupled Single Cell-Population-Balance Model for Mammalian Cell Cultures. *Ind. Eng. Chem. Res.* **2006**, *45*, 5801–5811. [CrossRef]
21. Mantzaris, N.V.; Daoutidis, P. Cell population balance modeling and control in continuous bioreactors. *J. Process Control* **2004**, *14*, 775–784. [CrossRef]
22. Mantzaris, N.V. Stochastic and deterministic simulations of heterogeneous cell population dynamics. *J. Theor. Biol.* **2006**, *241*, 690–706. [CrossRef] [PubMed]
23. Dorka, P.; Fischer, C.; Budman, H.; Scharer, J.M. Metabolic flux-based modeling of mAb production during batch and fed-batch operations. *Bioprocess Biosyst. Eng.* **2009**, *32*, 183–196. [CrossRef] [PubMed]

24. Fadda, S.; Cincotti, A.; Cao, G. A novel population balance model to investigate the kinetics of in vitro cell proliferation: Part I. model development. *Biotechnol. Bioeng.* **2012**, *109*, 772–781. [CrossRef] [PubMed]
25. Jandt, U.; Platas Barradas, O.; Pörtner, R.; Zeng, A.P. Synchronized Mammalian Cell Culture: Part II—Population Ensemble Modeling and Analysis for Development of Reproducible Processes. *Biotechnol. Prog.* **2015**, *31*, 175–185. [CrossRef] [PubMed]
26. Craven, S.; Whelan, J.; Glennon, B. Glucose concentration control of a fed-batch mammalian cell bioprocess using a nonlinear model predictive controller. *J. Process Control* **2014**, *24*, 344–357. [CrossRef]
27. Sbarciog, M.; Coutinho, D.; Wouwer, A.V. A simple output-feedback strategy for the control of perfused mammalian cell cultures. *Control Eng. Pract.* **2014**, *32*, 123–135. [CrossRef]
28. Amribt, Z.; Niu, H.X.; Bogaerts, P. Macroscopic modelling of overflow metabolism and model based optimization of hybridoma cell fed-batch cultures. *Biochem. Eng. J.* **2013**, *70*, 196–209. [CrossRef]
29. Mantzaris, N.V.; Liou, J.J.; Daoutidis, P.; Srienc, F. Numerical solution of a mass structured cell population balance model in an environment of changing substrate concentration. *J. Biotechnol.* **1999**, *71*, 157–174. [CrossRef]
30. Farzan, P.; Mistry, B.; Ierapetritou, M.G. Review of the Important Challenges and Opportunities related to Modeling of Mammalian Cell Bioreactors. *AIChE J.* **2017**, *63*, 398–408. [CrossRef]
31. Rocha, I. *Model-Based Strategies for Computer-Aided Operation of Recombinant E. coli Fermentation*; Universidade do Minho: Braga, Portugal, 2003.
32. Lopez-Meza, J.; Araíz-Hernández, D.; Carrillo-Cocom, L.M.; López-Pacheco, F.; del Refugio Rocha-Pizaña, M.; Alvarez, M.M. Using simple models to describe the kinetics of growth, glucose consumption, and monoclonal antibody formation in naive and infliximab producer CHO cells. *Cytotechnology* **2016**, *68*, 1287–1300. [CrossRef] [PubMed]
33. Cachaza, E.M.; Díaz, M.E.; Montes, F.J.; Galán, M.A. Simultaneous Computational Fluid Dynamics (CFD) simulation of the hydrodynamics and mass transfer in a partially aerated bubble column. *Ind. Eng. Chem. Res.* **2009**, *48*, 8685–8696. [CrossRef]
34. Wang, H.N.; Jia, X.; Wang, X.; Zhou, Z.; Wen, J.; Zhang, J. CFD modeling of hydrodynamic characteristics of a gas-liquid two-phase stirred tank. *Appl. Math. Model.* **2014**, *38*, 63–92. [CrossRef]
35. Azargoshasb, H.; Mousavi, S.M.; Amani, T.; Jafari, A.; Nosrati, M. Three-phase CFD simulation coupled with population balance equations of anaerobic syntrophic acidogenesis and methanogenesis reactions in a continouos stirred bioreactor. *J. Ind. Eng. Chem.* **2015**, *27*, 207–217. [CrossRef]
36. Kerdouss, F.; Bannari, A.; Proulx, P.; Bannari, R.; Skrga, M.; Labrecque, Y. Two-phase mass transfer coefficient prediction in stirred vessel with a CFD model. *Comput. Chem. Eng.* **2008**, *32*, 1943–1955. [CrossRef]
37. Micale, G.; Montante, G.; Grisafi, F.; Brucato, A.; Godfrey, J. CFD simulation of particle distribution in stirred vessels. *Chem. Eng. Res. Des.* **2000**, *78*, 435–444. [CrossRef]
38. Farzan, P.; Ierapetritou, M.G. Integrated Modeling to Capture the Interaction of Physiology and Fluid Dynamics in Biopharmaceutical Bioreactors. *Comput. Chem. Eng.* **2017**, *97*, 271–282. [CrossRef]
39. Bezzo, F.; Macchietto, S.; Pantelides, C.C. A general methodology for hybrid multizonal/CFD models: Part I. Theoretical framework. *Comput. Chem. Eng.* **2004**, *28*, 501–511. [CrossRef]
40. Delafosse, A.; Collignon, M.L.; Calvo, S.; Delvigne, F.; Crine, M.; Thonart, P.; Toye, D. CFD-based compartment model for description of mixing in bioreactors. *Chem. Eng. Sci.* **2014**, *106*, 76–85. [CrossRef]
41. Kagoshima, M.; Mann, R. Development of a networks-of-zones fluid mixing model for an unbaffled stirred vessel used for precipitation. *Chem. Eng. Sci.* **2006**, *61*, 2852–2863. [CrossRef]
42. Vrabel, P.; Van der Lans, R.G.J.M.; Cui, Y.Q.; Luyben, K.C.A. Compartment model approach: Mixing in large scale aerated reactors with multiple impellers. *Chem. Eng. Res. Des.* **1999**, *77*, 291–302. [CrossRef]
43. Bashiri, H.; Heniche, M.; Bertrand, F.; Chaouki, J. Compartmental modelling of turbulent fluid flow for the scale-up of stirred tanks. *Can. J. Chem. Eng.* **2014**, *92*, 1070–1081. [CrossRef]
44. Vrabel, P.; van der Lans, R.G.; Luyben, K.C.A.; Boon, L.; Nienow, A.W. Mixing in large-scale vessels stirred with multiple radial or radial and axial up-pumping impellers: Modelling and measurements. *Chem. Eng. Sci.* **2000**, *55*, 5881–5896. [CrossRef]
45. Ishii, M.; Hibiki, T. *Thermo-Fluid Dynamics of Two-Phase Flow*; Springer: Berlin, Germany, 2011.
46. ANSYS Inc. *ANSYS Fluent Theory Guide*, Release 15.0 ed.; ANSYS Inc.: Canonsburg, PA, USA, 2013.
47. Schmalzriedt, S.; Jenne, M.; Mauch, K.; Reuss, M. Integration of physiology and fluid dynamics. In *Process Integration in Biochemical Engineering*; Springer: Berlin, Germany, 2003; pp. 19–68.

48. ANSYS Inc. *ANSYS Fluent User's Guide*, Release 15.0 ed.; ANSYS Inc.: Canonsburg, PA, USA, 2013.
49. Adams, R.L.P. *Cell Culture for Biochemists*; Elsevier: New York, NY, USA, 1990.
50. Kaiser, S.C.; Löffelholz, C.; Werner, S.; Eibl, D. CFD for Characterizing Standard and Single-use Stirred Cell Culture Bioreactors. In *Computational Fluid Dynamics Technologies and Applications*; Minin, I.V., Minin, O.V., Eds.; InTech: Vienna, Austria, 2011; pp. 97–122.
51. Chalmers, J. Animal cell culture, effects of agitation and aeration on cell adaption. In *Encyclopedia of Cell Technology*; Spier, R.E., Ed.; Wiley-Interscience: Hoboken, NJ, USA, 2000.
52. Sarkar, J.; Shekhawat, L.K.; Loomba, V.; Rathore, A.S. CFD of mixing of multi-phase flow in a bioreactor using population balance model. *Biotechnol. Prog.* **2016**, *32*, 613–628. [CrossRef] [PubMed]
53. Alves, S.S.; Maia, C.I.; Vasconcelos, J.M.T.; Serralheiro, A.J. Bubble size in aerated stirred tanks. *Chem. Eng. J.* **2002**, *89*, 109–117. [CrossRef]
54. Chatterjee, A. An introduction to the proper orthogonal decomposition. *Curr. Sci.* **2000**, *78*, 808–817.
55. Chen, H.; Reuss, D.L.; Sick, V. On the use and interpretation of proper orthogonal decomposition of in-cylinder engine flows. *Meas. Sci. Technol.* **2012**, *23*, 085302. [CrossRef]
56. Xiu, Z.-L.; Deckwer, W.-D.; Zeng, A.-P. Estimation of rates of oxygen uptake and carbon dioxide evolution of animal cell culture using material and energy balances. *Cytotechnology* **1999**, *29*, 159–166. [CrossRef] [PubMed]
57. Mostafa, S.S.; Gu, X.J. Strategies for improved dCO(2) removal in large-scale fed-batch cultures. *Biotechnol. Prog.* **2003**, *19*, 45–51. [CrossRef] [PubMed]
58. Kolev, N.I. Solubility of O_2, N_2, H_2 and CO_2 in water. In *Multiphase Flow Dynamics 4 Turbulence, Gas Adsorption and Release, Diesel Fuel Properties*; Kolev, N.I., Ed.; Springer: Berlin, Germany, 2012; pp. 209–239.
59. Ozturk, S.S.; Palsson, B.O. Growth, metabolic, and antibody-production kinetics of hybridoma cell-culture: 2. Effects of serum concentration, dissolved-oxygen concentration, and medium PH in a batch reactor. *Biotechnol. Prog.* **1991**, *7*, 481–494. [CrossRef] [PubMed]
60. Xing, Z.Z.; Bishop, N.; Leister, K.; Li, Z.J. Modeling Kinetics of a Large-Scale Fed-Batch CHO Cell Culture by Markov Chain Monte Carlo Method. *Biotechnol. Prog.* **2010**, *26*, 208–219. [CrossRef] [PubMed]
61. Biegler, L.T.; Lang, Y.D.; Lin, W.J. Multi-scale optimization for process systems engineering. *Comput. Chem. Eng.* **2014**, *60*, 17–30. [CrossRef]
62. Bryson, J.A.E.; Ho, Y.-C. *Applied Optimal Control: Optimization, Estimation and Control*; CRC Press: Boca Raton, FL, USA, 1975.
63. Flores-Tlacuahuac, A.; Moreno, S.T.; Biegler, L.T. Global optimization of highly nonlinear dynamic systems. *Ind. Eng. Chem. Res.* **2008**, *47*, 2643–2655. [CrossRef]
64. Mahadevan, R.; Doyle, F.J. On-line optimization of recombinant product in a fed-batch bioreactor. *Biotechnol. Prog.* **2003**, *19*, 639–646. [CrossRef] [PubMed]
65. Banga, J.R.; Balsa-Canto, E.; Moles, C.G.; Alonso, A.A. Dynamic Optimization of Bioreactors: A Review. *Proc. Ind. Natl. Sci. Acad.* **2003**, *69*, 257–265.
66. Cuthrell, J.E.; Biegler, L.T. Simultaneous-Optimization and Solution Methods for Batch Reactor Control Profiles. *Comput. Chem. Eng.* **1989**, *13*, 49–62. [CrossRef]
67. Hedengren, J.D.; Shishavan, R.A.; Powell, K.M.; Edgar, T.F. Nonlinear modeling, estimation and predictive control in APMonitor. *Comput. Chem. Eng.* **2014**, *70*, 133–148. [CrossRef]
68. Constantinides, A.; Mostoufi, N. *Numerical Methods for Chemical Engineers with MATLAB Applications*; Prentice Hall: Upper Saddle River, NJ, USA, 2000.
69. Byrd, R.H.; Gilbert, J.C.; Nocedal, J. A trust region method based on interior point techniques for nonlinear programming. *Math. Program.* **2000**, *89*, 149–185. [CrossRef]
70. Farzan, P. A framework for development of integrated and computationally feasible models of large-scale mammalian cell bioreactors. In *Chemical and Biochemical Engineering*; Rutgers, The State University of New Jersey: New Brunswick, NJ, USA, 2018.
71. Zhao, Y.; Amemiya, Y.; Hung, Y. Efficient Gaussian Process Modeling using Experimental Design-Based Subagging. In Proceedings of the Conference on Experimental Design and Analysis (CEDA), Taipei, Taiwan, 15–17 December 2016; Institute of Statistical Science, Academia Sinica: Taipei, Taiwan, 2016.

![processes logo] *processes*

MDPI

Article

ADAR Mediated RNA Editing Modulates MicroRNA Targeting in Human Breast Cancer

Justin T. Roberts [1], Dillon G. Patterson [1], Valeria M. King [1], Shivam V. Amin [1],
Caroline J. Polska [1], Dominika Houserova [2], Aline Crucello [1], Emmaline C. Barnhill [1],
Molly M. Miller [1], Timothy D. Sherman [1] and Glen M. Borchert [1,2,*]

[1] Department of Biology, University of South Alabama, Mobile, AL 36688-0002, USA;
justin.roberts@ucdenver.edu (J.T.R.); dillon.patterson@emory.edu (D.G.P.);
vmk902@jagmail.southalabama.edu (V.M.K.); sva1002@jagmail.southalabama.edu (S.V.A.);
carolinepolska@gmail.com (C.J.P.); alinecrucello@gmail.com (A.C.);
ecb1621@jagmail.southalabama.edu (E.C.B.); mmm314@jagmail.southalabama.edu (M.M.M.);
tsherman@southalabama.edu (T.D.S.)
[2] Department of Pharmacology, USA College of Medicine, Mobile, AL 36688-0002, USA;
dh1001@jagmail.southalabama.edu
* Correspondence: borchert@southalabama.edu; Tel.: +1-251-460-7310

check for updates

Received: 6 April 2018; Accepted: 21 April 2018; Published: 25 April 2018

Abstract: RNA editing by RNA specific adenosine deaminase acting on RNA (ADAR) is increasingly being found to alter microRNA (miRNA) regulation. Editing of miRNA transcripts can affect their processing, as well as which messenger RNAs (mRNAs) they target. Further, editing of target mRNAs can also affect their complementarity to miRNAs. Notably, ADAR editing is often increased in malignancy with the effect of these RNA changes being largely unclear. In addition, numerous reports have now identified an array of miRNAs that directly contribute to various malignancies although the majority of their targets remain largely undefined. Here we propose that modulating the targets of miRNAs via mRNA editing is a frequent occurrence in cancer and an underappreciated participant in pathology. In order to more accurately characterize the relationship between these two regulatory processes, this study examined RNA editing events within mRNA sequences of two breast cancer cell lines (MCF-7 and MDA-MB-231) and determined whether or not these edits could modulate miRNA associations. Computational analyses of RNA-Seq data from these two cell lines identified over 50,000 recurrent editing sites within human mRNAs, and many of these were located in 3′ untranslated regions (UTRs). When these locations were screened against the list of currently-annotated miRNAs we discovered that editing caused a subset (~9%) to have significant alterations to mRNA complementarity. One miRNA in particular, miR-140-3p, is known to be misexpressed in many breast cancers, and we found that mRNA editing allowed this miRNA to directly target the apoptosis inducing gene *DFFA* in MCF-7, but not in MDA-MB-231 cells. As these two cell lines are known to have distinct characteristics in terms of morphology, invasiveness and physiological responses, we hypothesized that the differential RNA editing of *DFFA* in these two cell lines could contribute to their phenotypic differences. Indeed, we confirmed through western blotting that inhibiting miR-140-3p increases expression of the *DFFA* protein product in MCF-7, but not MDA-MB-231, and further that inhibition of miR-140-3p also increases cellular growth in MCF-7, but not MDA-MB-231. Broadly, these results suggest that the creation of miRNA targets may be an underappreciated function of ADAR and may help further elucidate the role of RNA editing in tumor pathogenicity.

Keywords: ADAR; breast; cancer; inosine; microRNA; microRNA targeting; RNA editing

1. Introduction

Transcript variation at the single nucleotide level is increasingly being found to have widespread occurrences within the transcriptome with fundamental roles in numerous biological processes including development and disease. Specifically, several independent studies have reported that there are hundreds of thousands of RNA editing sites catalyzed by the enzyme ADAR (adenosine deaminase acting on RNA) within human mRNAs [1–6]. Editing via ADAR is characterized by the conversion of the nucleic acid adenosine to inosine via deamination at the C6 position [7] (Figure 1). Since inosines have been shown to preferentially bind to cytosines, functionally the ADAR-catalyzed editing changes an 'A' to a 'G' in the transcript sequence [7]. Interestingly, the vast majority (>99%) of editing sites occur in the UTRs of primate-specific Alu elements [8–10], likely due to the common occurrence of two oppositely oriented Alus located in the same pre-mRNA pairing together to produce the long and stable double-stranded RNA structure that is required for ADAR to bind. As the ability to convert nucleotides adds a great deal of functionality to the transcriptome, it is not surprising that it has fundamental roles in many cellular activities. Editing events within mRNA coding for various neuroreceptors, such as serotonin and glutamate, have been intensively detailed in an array of organisms from flatworms to primates and found to be critical for routine neural activity [11]. With regards to gene regulation, sequence editing of RNA has widespread implications, including splice site alteration, localization and nuclear retention, and modification to the RNA secondary structure itself [12]. Further, given these abundant roles for ADAR editing in routine cellular function, it should also not be surprising that dysfunction of this important mechanism can have detrimental effects and, indeed, an increasing number of reports indicate a strong correlation between altered ADAR activity and a variety of pathologies. Specifically, because ADAR has been shown to be such an integral player in apoptotic regulation and cellular differentiation, cancer is of especially heightened interest and, in fact, more and more evidence is pointing to dysregulation of the editing process being a major factor in tumorigenesis [13–18].

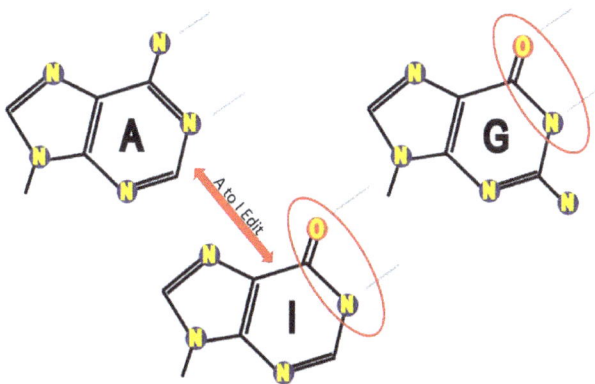

Figure 1. ADARs deaminate adenosine to inosine, potentially altering miRNA complementarities. A cartoon depicting adenosine (**left**), deaminated adenosine (inosine, in **center**), and guanine (**right**).

Editing events can also have widespread effects on the gene regulatory ability of noncoding RNAs, such as miRNAs [17,19]. MiRNAs are small regulatory RNA molecules roughly 20 to 23 nucleotides in length that regulate cell processes by binding to their target mRNAs and inhibiting translation [20]. MiRNAs are initially transcribed as primary miRNAs (pri-miRNAs) consisting of several thousand nucleotides in length that are then processed by into mature miRNAs by the enzymes Dicer and Drosha before entering the RNAi gene silencing complex where they regulate gene expression by binding to the 3' UTR of their mRNA targets via complimentary base pairing and silencing the gene by either repressing translation of the mRNA or triggering its degradation [21]. MiRNAs have been found to play a role in numerous cellular processes, from cell cycle control and apoptosis regulation to hormone production and immune response [22]. Importantly, misexpression of miRNAs has been implicated in a number of different disease states ranging from cardiovascular [23] and neurological disorders [24] to many types of cancer [25]. Having been associated with such a wide array of processes and pathologies, these molecules have garnered increased attention recently as investigators begin to evaluate their potential utility as biomarkers and therapies.

While these two mechanisms are fairly well understood independently, only recently have reports profiled the functional connection between RNA editing and miRNAs. For example, it has been shown that ADAR1 forms a complex with Dicer through direct protein interactions and enhances global miRNA processing [26]. Further, ADAR deamination of pri-miRNA transcripts can cause alterations to their structural conformations and subsequent maturation and processing by Drosha and Dicer [27,28]. In addition, while any editing of miRNA transcripts can have functional implications, arguably the most critical changes are to the seed regions of the miRNAs as this can drastically alter the set of genes able to be regulated [29]. This is especially true in cancer where altered miRNA regulation of oncogenes and tumor suppressors can lead to tumor formation [17]. Importantly, it should be noted that, in addition to editing the miRNA transcript, ADAR can also edit the 3′ UTR of target mRNAs. This modification dramatically increases the interplay between miRNAs and their targets by allowing a different set of miRNAs to regulate a given mRNA depending on if the transcript has been edited or not. Unfortunately, while the effects of editing the miRNA transcripts themselves have been well documented, this opposite effect of editing mRNAs in regions complementary to miRNA seeds is less understood. However, a number of reports suggesting this plays a significantly underappreciated role in miRNA targeting have surfaced within the last year [30–33]. To examine this, our study identified edit sites within two breast cancer cell lines (MCF-7 and MDA-MB-231) and analyzed the effect these edits had on subsequent regulation by miRNAs.

2. Materials and Methods

2.1. NGS Sequencing of MCF-7 and MDA-MB-231

Two breast cancer cell lines (MCF-7 and MDA-MB-231) grown under standard procedures were obtained from colleagues at the Mitchell Cancer Institute (Mobile, AL, USA). RNA was isolated and suspended in Trizol per standard manufacture protocol before being shipped to Otogenetics (Otogenetics Corporation, Atlanta, GA, USA) for commercial next-generation sequencing on an Illumina HiSeq2000 sequencer. Two RNA-Seq protocols were requested: (1) a total polyA selected RNA-Seq to provide mRNA transcripts, and (2) a small size selected RNA-Seq to provide small RNAs ranging from 17 to 35 nt in length. Raw paired-end reads were received totaling around 6 billion base pairs per cell line. Reads were uploaded to the NCBI Sequence Read Archive (SRA) and assigned the project number SRP101635.

2.2. Identification of A-to-G Edits in Breast Cancer Cells

Reads from the polyA selected RNA-Seq were filtered for low-quality reads and adapter contamination using Trimmomatic [34] and then aligned to the GRCh38 human reference genome using TopHat [35] (one mismatch allowed per alignment, only unique mappings reported). Edit sites were identified using the 'mpileup' command of SAMtools [36] which generates a VCF file containing location information for observed variations between the reads and the reference. All identified variations other than A-to-G and T-to-C were removed, and the remaining locations were cross-referenced with the dbSNP database to exclude variations that are known SNPs. To be considered a probable edit, at least 10% of transcriptome reads were required to differ from the reference genome at the edit position (with a minimum of 30 total reads).

2.3. Computational Identification of MiRNAs Biased towards Editing

The list of remaining putative edit sites (plus the sites identified in a previous study [37]) were used to generate a dataset consisting of two files each containing 201 bp sequences (edit site plus/minus 100 bp flanking sequences from the human reference genome). One file contained an 'unedited' version of the transcript with the edit site corresponding to the reference genome, and the other file contained an 'edited' version where the central site was edited. An in-house program written in Java was used to compare the reverse complement of the 7 nt seed sequences from all 2588 known human miRNAs in miRbase [38] to each possible 7-mer sequence within the generated dataset using a sliding window approach that counted perfect seed matches and recorded the position of each match in an Excel file (illustrated in Figure 2). Both the edited and unedited set of transcripts were analyzed for comparison, and after statistical analysis those miRNAs whose total number of seed matches increased or decreased significantly (10-fold or higher) in one set or the other were said to be biased towards editing.

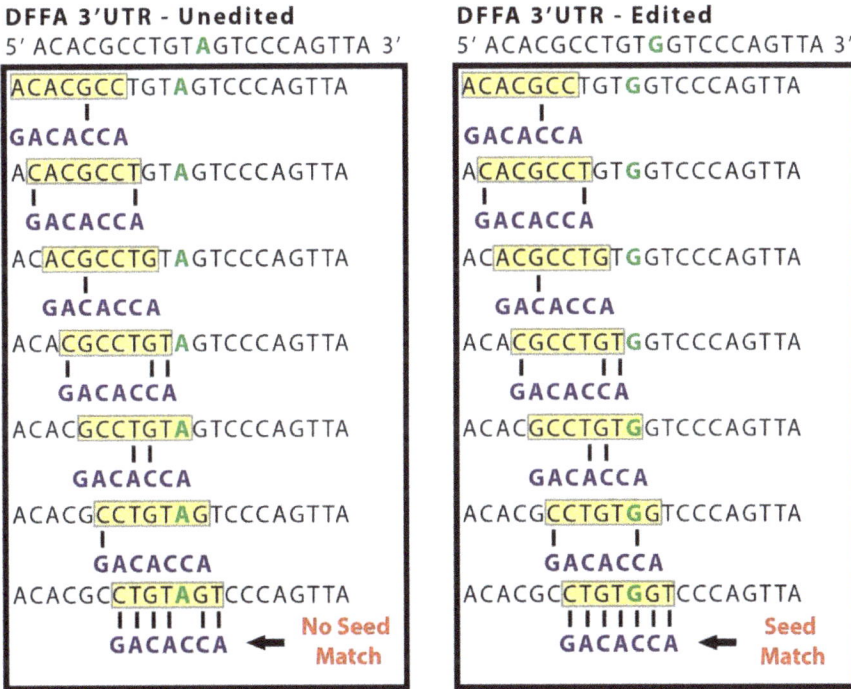

Figure 2. Effect of RNA editing on DFFA. A representative deamination site (green) occurring in the 3′ UTR of DNA fragmentation factor α (DFFA) is shown in both the unedited (**left**) and edited (**right**) state. The seed of miR-140-3p (blue) was screened using a sliding windows approach (depicted with a yellow box) against all possible seed matches within the DFFA sequence. Complimentary base pairing is indicated by the black lines.

2.4. Small RNA-Seq Analysis

To generate miRNA expressions data, reads from the small RNA-Seq experiment were aligned to known miRNA transcripts using the BLAST+ [39] sequence aligner. In order to be reported as valid the alignment was required to be over 99% similar with no more than one mismatch over 36 base pairs.

2.5. Cell Growth Assay

MDA-MB-231 cells were first transfected with either 100 nmol/L of miR-140 antagomir (Anti-140) (Cat # C-301055-01-0005, GE Healthcare Dharmacon, Chicago, IL, USA) or scrambled negative control (Ctrl-140) (catalog number CN-001000-01-05, Dharmacon) using Lipofectamine (Life Technologies, Carlsbad, CA, USA) according to the manufacturers protocol. Cell number was determined by trypan blue staining and manual counting at 24, 36, and 48 h post-transfection. Growth was determined as the relative cell number compared with vehicle-treated (0.1% DMSO) controls.

2.6. Western Blot Analysis

Following transfection of cells with anti-miRNAs, at 36 h existing media was replaced with lysis buffer containing protease inhibitors, incubated for 15 min at 4 °C, and then transferred to tubes. The cell proteins were electrophoresed through an 8% SDS–polyacrylamide gel and transferred to polyvinylidene fluoride membranes for the immobilization of the proteins. The membranes were blocked for 1 h in 2% non-fat milk in phosphate-buffered saline containing 0.05% Tween-20 surfactant and then washed and incubated with primary DFFA (ICAD) antibody (LF-PA0058, Thermo Scientific, Rockford, IL, USA) overnight at 4 °C. Following subsequent washing and incubation with goat anti-rabbit peroxidase-conjugated secondary antibody the immunoreactive bands were visualized and quantified using a Flurochem densitometer for the reporting of the protein levels.

3. Results

In order to characterize transcriptional differences between MCF-7 and MDA-MB-231 cells, RNA was isolated from each and split into "mRNA" (>200 nt RNAs) and "small RNA" (<200 nt RNAs including mature miRNAs) fractions. These samples were commercially sequenced resulting in over 2 billion nucleotides of small RNA reads and roughly 6 billion nucleotides of the longer mRNA reads.

3.1. Identification of RNA Edit Sites

Identification of putative RNA edit sites within each of the two cell lines was performed by mapping RNA-Seq reads to the GRCh38 human reference genome. As read alignments are reported with respect to the leading strand of the reference genome, a putative edit site would appear as an A-to-G mutation if the read arose from the forward strand, or a T-to-C mutation in the case of the reverse strand (Figure 3). In all, 19,462 unique edit sites were identified in MCF-7 and 35,090 sites were found in MDA-MB-231 (Table S1). That said, we found reads containing edits differed from the reference genome at the edit position 51.8% of the time in MCF-7s on average and 49.8% of the time in MDA-MB-231s.

```
REFERENCE_GENOME                        TAACTGGGACTACAGGCGTGT 21
HWI-ST1129_585_HA9B1ADXX_1_120          TAACTGGGACTACAGGCGTGT 21
HWI-ST1129_585_HA9B1ADXX_2_221          TAACTGGGACTACAGGCGTGT 21
HWI-ST1129_585_HA9B1ADXX_1_211          TAACTGGGACTTCAGGCGTGT 21
HWI-ST1129_585_HA9B1ADXX_2_210          TAACTGGGACTACAGGCGTGT 21
HWI-ST1129_585_HA9B1ADXX_2_120          CAACTGGGACTACAGGCGTGT 21
HWI-ST1129_585_HA9B1ADXX_2_120          TAACTGGGACTACAGGCGTGT 21
HWI-ST1129_585_HA9B1ADXX_2_210          TAACTGGGACTACAGGCGTGT 21
HWI-ST1129_585_HA9B1ADXX_1_111          TAACTGGGACTACAGGCGTGT 21
HWI-ST1129_585_HA9B1ADXX_2_220          TAACTGGGACTACAGGCGTGT 21
HWI-ST1129_585_HA9B1ADXX_2_221          CAACTGGGACTACAGGCGTGT 21
HWI-ST1129_585_HA9B1ADXX_1_220          TAACTGGGACTACAGGCGTGT 21
HWI-ST1129_585_HA9B1ADXX_2_220          TAACTGGGACTACAGGCGTGT 21
HWI-ST1129_585_HA9B1ADXX_1_111          TAACTGGGACTACAGGCGTGT 21
HWI-ST1129_585_HA9B1ADXX_2_110          TAACTGGGACTACAGGCGTGT 21
HWI-ST1129_585_HA9B1ADXX_1_111          TAACTGGGACTACAGGCGTGT 21
HWI-ST1129_585_HA9B1ADXX_2_211          TAACTGGGACTACAGGCGTGT 21
HWI-ST1129_585_HA9B1ADXX_2_211          TAACTGGGACTACAGGCGTGT 21
HWI-ST1129_585_HA9B1ADXX_2_220          TAACTGGGACTACAGGCGTGT 21
HWI-ST1129_585_HA9B1ADXX_2_210          TAACTGGGACTACAGGCGTGT 21
HWI-ST1129_585_HA9B1ADXX_2_220          TAACTGGGACTACAGGCGTGT 21
HWI-ST1129_585_HA9B1ADXX_2_121          TAACTGGGACTACAGGCGTGT 21
HWI-ST1129_585_HA9B1ADXX_2_120          TAACTGGGACTACAGGCGTGT 21
HWI-ST1129_585_HA9B1ADXX_1_221          TAACTGGGACCACAGGCGTGT 21
HWI-ST1129_585_HA9B1ADXX_2_110          TAACCGGGACCACAGGCGTGT 21
HWI-ST1129_585_HA9B1ADXX_1_120          TAACTGGGACCACAGGCGTGT 21
HWI-ST1129_585_HA9B1ADXX_2_121          TAACCGGGACCACAGGCGTGT 21
                                        *** ***** *********
```

Figure 3. Alignment of RNA-Seq reads to the human genome. Poly(A) selected RNA from two breast cancer cell lines (MCF-7 and MDA-MB-231) were sequenced with an Illumina Hi-Seq to provide high coverage mRNA transcripts. These transcripts were then compared to reference genome (top in red), with mismatches indicating a possible site of editing activity. Here one such site is shown within the red box, with mismatched reads outlined in green. Alignment was generated using ClustalW (http://www.genome.jp/tools-bin/clustalw) [40].

3.2. MiRNAs Biased towards Editing

Subsequent identification of miRNAs whose set of predicted target mRNAs were significantly affected due to our identified mRNA deaminations was achieved by screening the 'seed' regions from all 2588 currently-annotated human miRNAs in miRBase [38] against our full set of putative edit sites and an independently-generated publicly-available set of >12,000 A-to-I human edit sites [37] (Figure 2). Cataloging all of a miRNA's seed matches in both edited and unedited transcripts identified a subset whose mRNA target sets were significantly altered due to RNA editing (Tables 1, S2 and S3). In total, 206 miRNAs were shown to have altered target sites caused by ADAR-mediated single nucleotide mutations. Interestingly, we found that 86 of these miRNAs appeared to specifically target edited sequences and participate in regulations nonexistent prior to editing (Table S2) and, conversely, that the targets sites of the other 120 miRNAs were instead ablated upon ADAR editing due to a loss of sequence complementarity to their predicted mRNA targets (Table S3). As such, in order to ascertain whether any of these miRNAs were being actively expressed in our two cell lines, we next performed an expression analysis using our small RNA-Seq reads via BLAST+ [39]. Reads were aligned to known miRNAs, limited to only the highest scoring alignment per read, and required to be 100% identical to annotated miRNAs. Using these criteria, we identified 20 miRNAs for further evaluation based on their relative high expressions (>50 reads per million) in both MCF-7 and MDA-MB-231 (Table 1).

Table 1. List of top 10 miRs where ADAR editing of mRNAs alters complementarity to miR seed regions and either (**A**) creates novel target sites for regulation or (**B**) destroys predicted target sites. In addition to altered edit complementarity, microRNAs included were also required to be present at >50 reads per million in MCF-7 and MDA-MB-231 small RNA-Seq datasets.

miR	miRBase ID	Seed (RC)	Targets (Edited)	Targets (Unedited)	Expected
A					
hsa-miR-513a-5p	MIMAT0002877	CCTGTGA	258	0	0.63
hsa-miR-450b-3p	MIMAT0004910	GATCCCA	252	4	0.79
hsa-miR-769-3p	MIMAT0003887	GATCCCA	252	4	0.79
hsa-miR-6089	MIMAT0023714	CGGCCTC	219	0	3.83
hsa-miR-4691-3p	MIMAT0019782	GTGGCTG	181	0	1.16
hsa-miR-3189-3p	MIMAT0015071	CCCAAGG	140	5	0.48
hsa-miR-140-3p	MIMAT0004597	CTGTGGT	139	0	1.11
hsa-miR-3065-3p	MIMAT0015378	GGTGCTG	118	0	0.5
hsa-miR-3940-3p	MIMAT0018356	CCGGGCT	111	0	0.72
hsa-miR-3680-3p	MIMAT0018107	ATGCAAA	108	2	0.82
B					
hsa-miR-5089-5p	MIMAT0021081	AATCCCA	0	644	21.39
hsa-miR-6504-3p	MIMAT0025465	CTGTAAT	58	587	19.93
hsa-miR-6506-5p	MIMAT0025468	ATCCCAG	18	377	21.57
hsa-miR-619-5p	MIMAT0026622	ATCCCAG	18	377	21.57
hsa-miR-4775	MIMAT0019931	AAAATTA	0	351	19.37
hsa-miR-4735-5p	MIMAT0019860	AAATTAG	6	305	17.31
hsa-miR-6514-3p	MIMAT0025485	ACAGGCA	10	216	9.59
hsa-miR-4794	MIMAT0019967	TAGCCAG	10	173	8.05
hsa-miR-664a-5p	MIMAT0005948	TAGCCAG	10	173	8.05
hsa-miR-1273e	MIMAT0018079	TCAAGCA	2	169	5.22

3.3. MiR-140 Is Able to Target DFFA in MCF-7 but not MDA-MB-231

Next, after a thorough examination of the subset of miRNAs whose set of predicted target mRNAs were significantly affected by deamination in our cell lines, we selected miR-140-3p for a detailed experimental examination. Importantly, we found miR-140-3p was highly expressed in both cell lines and, notably, its set of target mRNAs was found to be significantly altered by RNA editing in MCF-7 cells, but not in the MDA_MB-231 cells. Importantly, we found A-to-G mutations caused dramatic changes to miR-140′s set of predicted mRNA target sites in MCF-7s, with deamination events leading to the creation of 91 new putative target sites in 34 mRNAs. Of note, through utilizing strategies we previously employed to successfully identify sites created in a publicly-available set of >12,000 A-to-I human edit sites [37,41] (Figure 4), we identified a particularly interesting target site created for miR-140-3p in MCF-7 cells—DNA fragmentation factor alpha (DFFA), also known as inhibitor of caspase-activated DNase (ICAD) (Figure 5). As the principle function of DFFA is to trigger DNA fragmentation during apoptosis, we hypothesized that the miRNA-mediated downregulation of this gene specifically in MCF-7 cells might directly contribute to their characteristically lower rate of cellular proliferation as compared to MDA-MB-231s.

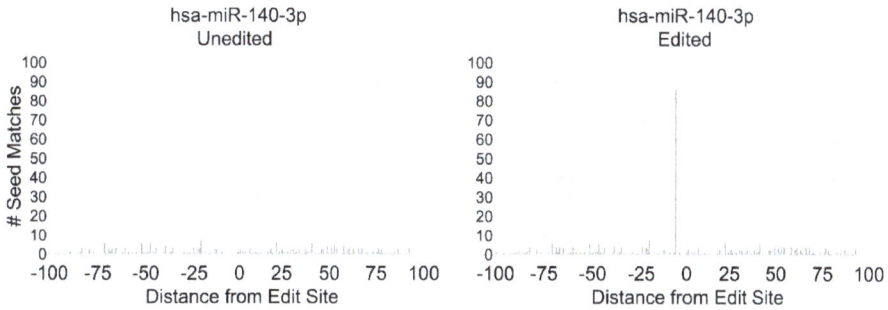

Figure 4. A-to-I edits create novel target sites for miR-140-3p. mRNA sequences from the edit sites previously identified [37] (each consisting of a central A-to-I deamination and 100 nt flanks) were screened for complementarity to human miRNAs. The graphs represent all miR-140-3p seed matches occurring at each possible position within both the unedited (**left**) and edited (**right**) states.

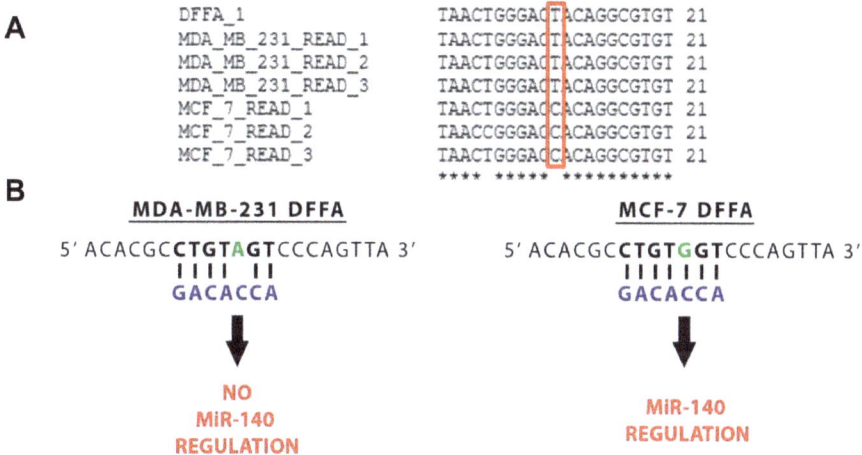

Figure 5. MiR-140 can regulate *DFFA* in MCF-7, but not MDA-MB-231. (**A**) Alignment of 21 nt segments of six RNA-Seq reads (three from each cell line) to a portion of the apoptosis inducing gene DFFA. Our edit identification algorithm identified an A-to-G edit site at basepair 10,460,668 on Chromosome 1, and corresponding reads mapping to that location were extracted and trimmed to 21 bp (edit site plus/minus 10 bp flanking regions). Edit location is outlined in red. The alignment was generated via ClustalW [40]. (**B**) Illustration showing complimentary base pairing between the miR-140 seed (blue) and the DFFA gene in both cell lines. The edit site is indicated in green.

3.4. Inhibiting miR-140-3p Increases DFFA Expression in MCF-7

In order to determine if miR-140-3p directly regulates the endogenous expression of DFFA, we performed DFFA Western blots (Figure 6A) to examine the effects of introducing a specific miR-140-3p antagomir as compared to a non-specific control. Excitingly, although we found a marked increase of DFFA levels following miR-140-3p inhibition in MCF-7s (where a target site is created by ADAR deamination), we found no appreciable effect of inhibiting miR-140-3p in MDA-MB-231s (in which DFFA does not undergo deamination). Furthermore, qPCR analysis of DFFA expression found no effect on DFFA mRNA levels following miR-140-3p inhibition in either cell line (data not shown) confirming miR-140-3p regulates DFFA post transcriptionally.

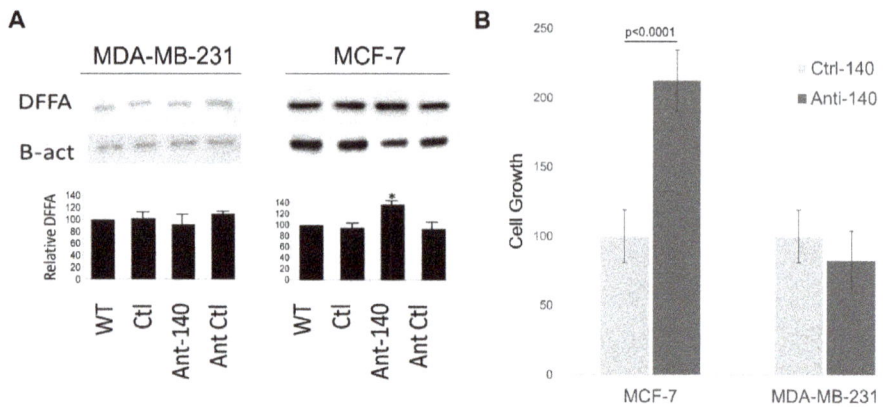

Figure 6. Depletion of DFFA protein expression and the effect of miR-140-3p on cellular growth. (**A**) Representative blots for DFFA and β-actin (loading control) are shown (n = 3). The miRNA is able to bind and regulate the *DFFA* gene in MCF-7, but not in MDA-MB-231 due the presence of an A-to-I edit. WT, wild type; Ctl, empty lipo transfection; Ant-140, miR-140 antagomir; Ant-Ctl, random antagomir. (**B**) Cell growth assay examining effects of transfecting a miR-140 inhibitor in both cell lines. Five microscopic fields randomly chosen from each assay were counted individually, and the statistical significance between treatment and control determined by *t*-test.

3.5. Inhibiting miR-140-3p Increases MCF-7 Cellular Proliferation

We next examined the effects of inhibiting miR-140-3p on cellular growth and similarly found cellular growth was largely unaffected by decreased miR-140-3p levels in MDA-MB-231, whereas we found there was over a 110% increase in MCF-7 cellular growth following miR-140-3p depletion at 24 h post transfection (Figure 6B). Importantly, these results strongly agree with our examination of *DFFA* regulations and further support the idea that miR-140-3p mediated downregulation of DFFA specifically in MCF-7 cells directly contributes to the characterized differences of these two cell lines in cellular growth.

4. Discussion

ADAR-mediated RNA editing is well characterized as having dramatic effects on a multitude of cellular processes [11,18,42,43]. However, the molecular mechanisms through which ADAR editing confers these effects remain largely undefined. That said, ADAR editing of miRNA transcripts has now been shown to affect their regulatory ability, in some cases leaving them unable to bind to their target transcripts and in others leading to unintended inhibition of new targets altogether [17,19,44]. To add to the relationship between A-to-I editing and miRNAs, we have now successfully shown that mRNA editing can also affect miRNA targeting by changing the complementarity between a 3' UTR binding site and the seed region of a miRNA. Results from our analysis strongly suggest that A-to-I editing is routinely employed to modify mRNA complementarities to a specific subset of 233 human microRNAs currently annotated in miRBase [38]. Interestingly, for 86 of these miRNAs ADAR editing leads to the generation of new regulatory targets, whereas A-to-I editing conversely results in a significant loss of complementarity to mRNAs and, therefore, a loss of putative targets for the other 120 miRNAs. We find these two subsets of ADAR editing-related miRNAs to be completely distinct—86 specifically targeting edited mRNAs and 120 specifically targeting unedited mRNAs (or whose regulation is blocked by editing). This latter observation is notable as the ability of ADAR to destroy mRNA targets has not been previously reported and is in direct contrast to previous work that suggested ADAR editing could likely only create targets for miRNAs [41].

Based on these results, we believe that the generation of novel miRNA regulatory networks is a critical function of ADAR editing, and, notably, that dysregulated editing may create susceptibilities that allow tumorigenesis and tumor progression to occur. Corroborating this idea, several studies have already established a clear precedent for ADAR activity being implicated in cancer biology. Recently, Chen et al. [15] described direct involvement of ADAR editing in human hepatocellular carcinoma (HCC), showing how the transcripts of an oncoprotein degrader and confirmed contributor to HCC pathology, antizyme inhibitor 1 (*AZIN1*), are modified at specific sites by ADAR1, and that ADAR1 is commonly upregulated in HCC patient tumors resulting in even higher *AZIN1* editing frequency and poorer prognosis. In addition, the authors were able to successfully demonstrate that higher levels of edited *AZIN1* promoted an increased incidence of tumor formation and invasive ability. Over-editing of *AZIN1* has also been implicated in other cancers, such as esophageal squamous cell carcinoma [13]. Other recent studies suggest that ADAR1 might also play a pathogenic role in chronic myeloid leukemia (CML). Jiang et al. [14] have recently shown that overexpression of ADAR1 in cultured blood progenitor cells can promote reprogramming of myeloid progenitor cells resulting in heightened hematopoietic differentiation toward the myeloid lineage. Increased ADAR1 levels were repeatedly found in CML patient samples leading the authors to speculate that ADAR played a causal role. In fact, a related study recently found CML could not be induced in mice following a bone marrow transplant of marrow cells carrying an ADAR deletion suggesting ADAR1 may be essential for leukemia cell survival [14].

In contrast to the previous examples linking hyper-editing to malignancy, the opposite scenario, hypo-editing, has also been implicated as contributing to various cancers, specifically in relation to miRNAs. For instance, it has been shown by Choudhury et al. [17] that reduced editing of miR-376a promotes glioblastoma cell invasion in orthotopic glioma. Normally-edited miR-376a targets and suppresses the receptor for the autocrine motility factor (AMF) that stimulates tumor motility via base pair complementarity with the 3' UTR of the AMF receptor mRNA; however, when unedited, the miRNA loses this ability. It was also demonstrated that unedited miR-376a binds to the 3' UTR of the *RAP2A* mRNA transcript (coding for a protein known to suppress glioblastoma cell invasion), causing the RAP2A protein's function to be inhibited. This report does an excellent job of demonstrating how ADAR-induced single base pair changes in miRNAs can alter their target specificity and ultimately lead to pathologically significant ramifications. Further, while it is clear that RNA editing can be fundamentally linked to cancer via sequence alteration and the expression/repression of oncogenes, there is also evidence of involvement in other tumorigenic pathways. For instance, a correlation has been shown between reduced editing of Alu elements and multiple tumors, including brain, prostate, lung, and kidneys [14,18]. Additionally, chronic inflammation related to viral infection has been previously implicated in tumorigenesis and this may be due, in part, to overexpression of ADAR1 mediated by inflammation [45]. Of note, in this work we identify 19,462 unique edit sites in MCF-7 cells versus 35,090 unique sites in MDA-MB-231s suggesting generally higher ADAR1 activity in this more aggressive breast cancer cell line.

Importantly, the work presented here represents the most comprehensive of only a handful of analyses of the effects of mRNA A-to-I editing on miRNA targeting published to date [30–32], and represents only the second ever experimental evidence indicating that the modulation of miRNA targeting through ADAR editing may directly contribute to breast cancer pathology [33]. When taken together, this report along with recently published studies suggesting mRNA editing can alter microRNA regulations [30–33] (all published within the last few months) strongly suggest that the participation of A-to-I editing in directing microRNA targeting is currently significantly underappreciated.

That said, our analysis of the RNA editing data from two breast cancer cell lines demonstrate that miR-140-3p is able to regulate the apoptosis inducing gene *DFFA* in MCF-7 but not in MDA-MB-231. DFFA is the larger of two protein subunits that comprise caspase-activated DNase (CAD) and, when bound to CAD, DFFA inhibits its ability to degrade DNA and condense chromatin, but during apoptosis caspase-3 cleaves DFFA resulting in DNA fragmentation [46]. As misexpression of an apoptotic contributor can have significant ramifications in terms of tumor development, the differential regulation of DFFA by miR-140 between our two cell lines is highly intriguing, especially as numerous reports have previously implicated a role for miR-140 in breast malignancy [47–49]. That said, the two cell lines involved in this study, MCF-7 and MDA-MB-231, have very distinct characteristics in terms of morphology, invasiveness, and physiological responses. While they are both adenocarcinomas (cancers of the breast epithelium tissue that originated in the mammary gland), the MCF-7 line was derived from an in situ carcinoma where the cancerous cells had not yet invaded surrounding tissues. These cells are weakly invasive, luminal epithelial-like, and are hormone responsive, requiring noticeably less aggressive therapies [50]. In contrast, the highly-invasive, fibroblast-like MDA-MB-231 line was derived from a metastatic carcinoma and is a triple-negative breast cancer making it highly chemoresistant and, thus, significantly more difficult to treat [51]. When taken in conjunction with reports of elevated ADAR activity in many breast cancers, it is feasible to assume that RNA editing could contribute to some of the characteristic phenotypic differences observed between these two cell lines. Excitingly, we suggest the work presented here strongly supports this as we find ADAR editing directly mediates the regulation of *DFFA* in MCF-7s whereas the absence of *DFFA* editing in MDA-MB-231 conversely disallows *DFFA* regulation by miR-140-3p in these cells. Simply put, we find miR-140 is able to bind and regulate *DFFA* due to editing in MCF-7s, so inhibition of the miRNA increases growth. As it is unable to bind in MDA-MB-231, no effect is seen. As such, it is tempting to speculate that the differential regulation of *DFFA* by miR-140-3p between these two breast cancer lines directly contributes to their observed differences in cellular proliferation and cellular survival (Figure 6). That said, miR-140-3p undoubtedly regulates multiple mRNAs and the observed effects on cellular growth may be mediated through more than DFFA restriction alone. Of note, Salem et al. [52] recently demonstrated that transfecting several breast cancer cell lines with miR-140-3p isoform mimics commonly resulted in a decrease in breast cancer cell viability (nicely complementing the increased cellular growth we observe in MCF-7s following transfection of miR-140-3p inhibitor). Additionally, and also in agreement with our findings, this group similarly observed no change in MDA-MB-231 viability following manipulation of miR-140-3p levels via transfection of a miR-140-3p mimic.

While this work represents the first direct indication of a contributory role for A-to-I editing in modulating miRNA targeting in malignancy, we suggest the repeated observation of a correlation between altered ADAR activity and various pathologies suggests altered miRNA regulations due to alterations in A-to-I profiles may represent a significant currently underappreciated contributor to an array of pathologies. Perhaps of broader importance. However, our findings lead us to believe that many miRNA targets can only be identified by analyzing expressed sequences, and that accurate miRNA target prediction may ultimately require analyzing transcriptomes and not genomes.

Supplementary Materials: The following are available online at http://www.mdpi.com/2227-9717/6/5/42/s1, Table S1: Comprehensive list of unique edit sites identified. To be considered a probable edit, at least 10% of transcriptome reads were required to differ from the reference genome at the edit position (with a minimum of 30 total reads). In all, 19,462 unique edit sites were identified in MCF-7 and 35,090 sites were found in MDA-MB-231. Chr, chromosome; Edit Location, bp position; Reference Base, expected nucleotide; Alternate Base, unexpected nucleotide; Quality Score, as in SAMtools(36); Info, as in SAMtools(36); Freq, Alternate Base/Reference Base %. Table S2: List of 86 miRNAs representing 72 unique seeds where ADAR editing of mRNA transcripts alters complementarity to their seed region and creates novel target sites for regulation. MiRNA information was obtained from miRBase, 'RC' indicates the seed is reverse complemented, 'Edited Targets' is the total number of seed matches when the transcripts are edited whereas 'Unedited Targets" indicates the number of seed matches in the absence of editing activity. 'Expected Targets' is the average number of seed matches found within the windows flanking the edit site. Table S3: List of 120 miRs representing 93 unique seeds where ADAR editing of mRNA transcripts destroys complementarity to miR seed regions and effectively inhibits regulation. MiRNA information was obtained from miRBase, 'RC' indicates the seed is reverse complemented, 'Edited Targets' is the total number of seed matches when the transcripts are edited whereas 'Unedited Targets" indicates the number of seed matches in the absence of editing activity. 'Expected Targets' is the average number of seed matches found within the windows flanking the edit site.

Author Contributions: J.T.R. and G.M.B. conceived and designed the experiments; J.T.R., D.G.P., V.M.K., S.V.A., C.J.P., D.H., A.C., E.C.B., M.M.M. performed the experiments; J.T.R., V.M.K., and G.M.B. analyzed the data; T.D.S. and G.M.B. contributed reagents/materials/analysis tools; J.T.R., D.G.P., V.M.K., S.V.A., C.J.P., D.H., A.C., E.C.B., M.M., T.D.S. and G.M.B. wrote/edited the paper.

Funding: Funding was provided in part by NSF CAREER grant 1350064 (G.M.B.) awarded by Division of Molecular and Cellular Biosciences (with co-funding from NSF EPSCoR) and in part by the Abraham A. Mitchell Cancer Research Fund (G.M.B.). Graduate funding was provided in part by Alabama Commission on Higher Education ALEPSCoR grants 150380 (J.T.R.), 160330 (V.M.K.), 170232 (V.M.K.) and 170233 (D.H.).

Acknowledgments: We thank the University of South Alabama College of Arts and Sciences, College of Medicine and the Mitchell Cancer Institute for ongoing support.

Conflicts of Interest: The authors declare no conflict of interest.

References

1. Bazak, L.; Haviv, A.; Barak, M.; Jacob-Hirsch, J.; Deng, P.; Zhang, R.; Levanon, E.Y. A-to-I RNA editing occurs at over a hundred million genomic sites, located in a majority of human genes. *Genome Res.* **2014**, *24*, 365–376. [CrossRef] [PubMed]

2. Peng, Z.; Cheng, Y.; Tan, B.C.; Kang, L.; Tian, Z.; Zhu, Y.; Guo, J. Comprehensive analysis of RNA-Seq data reveals extensive RNA editing in a human transcriptome. *Nat. Biotechnol.* **2012**, *30*, 253–260. [CrossRef] [PubMed]

3. Li, J.B. Genome-wide identification of human RNA editing sites by parallel DNA capturing and sequencing. *Science* **2009**, *324*, 1210–1213. [CrossRef] [PubMed]

4. Park, E.; Williams, B.; Wold, B.J.; Mortazavi, A. RNA editing in the human ENCODE RNA-seq data. *Genome Res.* **2012**, *22*, 1626–1633. [CrossRef] [PubMed]

5. Bahn, J.H. Accurate identification of A-to-I RNA editing in human by transcriptome sequencing. *Genome Res.* **2012**, *22*, 142–150. [CrossRef] [PubMed]

6. Maas, S. Genome-wide evaluation and discovery of vertebrate A-to-I RNA editing sites. *Biochem. Biophys. Res. Commun.* **2011**, *412*, 407–412. [CrossRef] [PubMed]

7. Nishikura, K. Functions and regulation of RNA editing by ADAR deaminases. *Annu. Rev. Biochem.* **2010**, *79*, 321–349. [CrossRef] [PubMed]

8. Chen, L.L.; Carmichael, G.G. Gene regulation by SINES and inosines: Biological consequences of A-to-I editing of Alu element inverted repeats. *Cell Cycle* **2008**, *7*, 3294–3301. [CrossRef] [PubMed]

9. Athanasiadis, A.; Rich, A.; Maas, S. Widespread A-to-I RNA editing of Alu-containing mRNAs in the human transcriptome. *PLoS Biol.* **2004**, *2*, e391. [CrossRef] [PubMed]

10. Kim, D.D.; Kim, T.T.; Walsh, T.; Kobayashi, Y.; Matise, T.C.; Buyske, S.; Gabriel, A. Widespread RNA editing of embedded alu elements in the human transcriptome. *Genome Res.* **2004**, *14*, 1719–1725. [CrossRef] [PubMed]

11. Seeburg, P.H. A-to-I editing: New and old sites, functions and speculations. *Neuron* **2002**, *35*, 17–20. [CrossRef]

12. Rieder, L.E.; Reenan, R.A. The intricate relationship between RNA structure, editing, and splicing. *Semin. Cell Dev. Biol.* **2012**, *23*, 281–288. [CrossRef] [PubMed]

13. Qin, Y.R.; Qiao, J.J.; Chan, T.H.; Zhu, Y.H.; Li, F.F.; Liu, H.; Chen, L. Adenosine-to-inosine RNA editing mediated by ADARs in esophageal squamous cell carcinoma. *Cancer Res.* **2014**, *74*, 840–851. [CrossRef] [PubMed]

14. Jiang, Q.; Crews, L.A.; Barrett, C.L.; Chun, H.J.; Court, A.C.; Isquith, J.M.; Zipeto, M.A.; Dao, K.H.T. ADAR1 promotes malignant progenitor reprogramming in chronic myeloid leukemia. *Proc. Natl. Acad. Sci. USA* **2013**, *110*, 1041–1046. [CrossRef] [PubMed]

15. Chen, L.; Li, Y.; Lin, C.H.; Chan, T.H.; Chow, R.K.; Song, Y.; Qi, L. Recoding RNA editing of AZIN1 predisposes to hepatocellular carcinoma. *Nat. Med.* **2013**, *19*, 209–216. [CrossRef] [PubMed]

16. Steinman, R.A. Deletion of the RNA-editing enzyme ADAR1 causes regression of established chronic myelogenous leukemia in mice. *Int. J. Cancer* **2013**, *132*, 1741–1750. [CrossRef] [PubMed]

17. Choudhury, Y. Attenuated adenosine-to-inosine editing of microRNA-376a* promotes invasiveness of glioblastoma cells. *J. Clin. Investig.* **2012**, *122*, 4059–4076. [CrossRef] [PubMed]

18. Paz, N. Altered adenosine-to-inosine RNA editing in human cancer. *Genome Res.* **2007**, *17*, 1586–1595. [CrossRef] [PubMed]

19. Blow, M.J.; Grocock, R.J.; van Dongen, S.; Enright, A.J.; Dicks, E.; Futreal, P.A.; Stratton, M.R. RNA editing of human microRNAs. *Genome Biol.* **2006**, *7*, R27. [CrossRef] [PubMed]

20. Kim, V.N. MicroRNA biogenesis: Coordinated cropping and dicing. *Nat. Rev. Mol. Cell Biol.* **2005**, *6*, 376–385. [CrossRef] [PubMed]

21. Zamore, P.D. A microRNA in a multiple-turnover RNAi enzyme complex. *Science* **2002**, *297*, 2056–2060.

22. He, L.; Hannon, G.J. MicroRNAs: Small RNAs with a big role in gene regulation. *Nat. Rev. Genet.* **2004**, *5*, 522–531. [CrossRef] [PubMed]

23. Maegdefessel, L. The emerging role of microRNAs in cardiovascular disease. *J. Intern. Med.* **2014**, *276*, 633–644. [CrossRef] [PubMed]

24. Sun, E.; Shi, Y. MicroRNAs: Small molecules with big roles in neurodevelopment and diseases. *Exp. Neurol.* **2014**, *268*, 46–53. [CrossRef] [PubMed]

25. Jansson, M.D.; Lund, A.H. MicroRNA and cancer. *Mol. Oncol.* **2012**, *6*, 590–610. [CrossRef] [PubMed]

26. Ota, H. ADAR1 forms a complex with Dicer to promote microRNA processing and RNA-induced gene silencing. *Cell* **2013**, *153*, 575–589. [CrossRef] [PubMed]

27. Yang, W. Modulation of microRNA processing and expression through RNA editing by ADAR deaminases. *Nat. Struct. Mol. Biol.* **2006**, *13*, 13–21. [CrossRef] [PubMed]

28. Kawahara, Y.; Zinshteyn, B.; Chendrimada, T.P.; Shiekhattar, R.; Nishikura, K. RNA editing of the microRNA-151 precursor blocks cleavage by the Dicer-TRBP complex. *EMBO Rep.* **2007**, *8*, 763–769. [CrossRef] [PubMed]

29. Kawahara, Y.; Zinshteyn, B.; Sethupathy, P.; Iizasa, H.; Hatzigeorgiou, A.G.; Nishikura, K. Redirection of silencing targets by adenosine-to-inosine editing of miRNAs. *Science* **2007**, *315*, 1137–1140. [CrossRef] [PubMed]

30. Nakano, M.; Fukami, T.; Gotoh, S.; Takamiya, M.; Aoki, Y.; Nakajima, M. RNA Editing Modulates Human Hepatic Aryl Hydrocarbon Receptor Expression by Creating MicroRNA Recognition Sequence. *J. Biol. Chem.* **2016**, *291*, 894–903. [CrossRef] [PubMed]

31. Zhang, L.; Yang, C.S.; Varelas, X.; Monti, S. Altered RNA editing in 3′ UTR perturbs microRNA-mediated regulation of oncogenes and tumor-suppressors. *Sci. Rep.* **2016**, *6*, 23226. [CrossRef] [PubMed]

32. Soundararajan, R.; Stearns, T.M.; Griswold, A.L.; Mehta, A.; Czachor, A.; Fukumoto, J.; Kolliputi, N. Detection of canonical A-to-G editing events at 3′ UTRs and microRNA target sites in human lungs using next-generation sequencing. *Oncotarget* **2015**, *6*, 35726–35736. [CrossRef] [PubMed]

33. Nakano, M.; Fukami, T.; Gotoh, S.; Nakajima, M. A-to-I RNA Editing Up-regulates Human Dihydrofolate Reductase in Breast Cancer. *J. Biol. Chem.* **2017**, *292*, 4873–4884. [CrossRef] [PubMed]

34. Bolger, A.M.; Lohse, M.; Usadel, B. Trimmomatic: A flexible trimmer for Illumina sequence data. *Bioinformatics* **2014**, *30*, 2114–2120. [CrossRef] [PubMed]

35. Kim, D.; Pertea, G.; Trapnell, C.; Pimentel, H.; Kelley, R.; Salzberg, S.L. TopHat2: Accurate alignment of transcriptomes in the presence of insertions, deletions and gene fusions. *Genome Biol.* **2013**, *14*, R36. [CrossRef] [PubMed]

36. Li, H. The Sequence Alignment/Map format and SAMtools. *Bioinformatics* **2009**, *25*, 2078–2079. [CrossRef] [PubMed]

37. Levanon, E.Y. Systematic identification of abundant A-to-I editing sites in the human transcriptome. *Nat. Biotechnol.* **2004**, *22*, 1001–1005. [CrossRef] [PubMed]

38. Kozomara, A.; Griffiths-Jones, S. miRBase: Annotating high confidence microRNAs using deep sequencing data. *Nucleic Acids Res.* **2014**, *42*, D68–D73. [CrossRef] [PubMed]

39. Camacho, C.; Coulouris, G.; Avagyan, V.; Ma, N.; Papadopoulos, J.; Bealer, K.; Madden, T.L. BLAST+: Architecture and applications. *BMC Bioinform.* **2009**, *10*, 421. [CrossRef] [PubMed]

40. Thompson, J.D.; Gibson, T.J.; Higgins, D.G. Multiple Sequence Alignment Using ClustalW and ClustalX. *Curr. Protocals Bioinform.* **2002**. [CrossRef] [PubMed]

41. Borchert, G.M. Adenosine deamination in human transcripts generates novel microRNA binding sites. *Hum. Mol. Genet.* **2009**, *18*, 4801–4807. [CrossRef] [PubMed]

42. Rosenthal, J.J.; Seeburg, P.H. A-to-I RNA editing: Effects on proteins key to neural excitability. *Neuron* **2012**, *74*, 432–439. [CrossRef] [PubMed]

43. Maldonado, C.; Alicea, D.; Gonzalez, M.; Bykhovskaia, M.; Marie, B. Adar is essential for optimal presynaptic function. *Mol. Cell. Neurosci.* **2013**, *52*, 173–180. [CrossRef] [PubMed]
44. Garcia-Lopez, J.; de Jde, D.; Mazo, J. Hourca Del Reprogramming of microRNAs by adenosine-to-inosine editing and the selective elimination of edited microRNA precursors in mouse oocytes and preimplantation embryos. *Nucleic Acids Res.* **2013**, *41*, 5483–5493. [CrossRef] [PubMed]
45. Yang, J.H. Widespread inosine-containing mRNA in lymphocytes regulated by ADAR1 in response to inflammation. *Immunology* **2003**, *109*, 15–23. [CrossRef] [PubMed]
46. Liu, X.; Zou, H.; Slaughter, C.; Wang, X. DFF, a heterodimeric protein that functions downstream of caspase-3 to trigger DNA fragmentation during apoptosis. *Cell* **1997**, *89*, 175–184. [CrossRef]
47. Song, B. Mechanism of chemoresistance mediated by miR-140 in human osteosarcoma and colon cancer cells. *Oncogene* **2009**, *28*, 4065–4074. [CrossRef] [PubMed]
48. Wolfson, B.; Eades, G.; Zhou, Q. Roles of microRNA-140 in stem cell-associated early stage breast cancer. *World J. Stem Cells* **2014**, *6*, 591–597. [CrossRef] [PubMed]
49. Li, Q. Downregulation of miR-140 promotes cancer stem cell formation in basal-like early stage breast cancer. *Oncogene* **2013**, *33*, 2589–2600. [CrossRef] [PubMed]
50. Sorlie, T. Introducing molecular subtyping of breast cancer into the clinic. *J. Clin. Oncol.* **2009**, *27*, 1153–1154. [CrossRef] [PubMed]
51. Pozo-Guisado, E.; Alvarez-Barrientos, A.; Mulero-Navarro, S.; Santiago-Josefat, B.; Fernandez-Salguero, P.M. The antiproliferative activity of resveratrol results in apoptosis in MCF-7 but not in MDA-MB-231 human breast cancer cells: Cell-specific alteration of the cell cycle. *Biochem. Pharmacol.* **2002**, *64*, 1375–1386. [CrossRef]
52. Salem, O.; Erdem, N.; Jung, J.; Munstermann, E.; Worner, A.; Wilhelm, H.; Körner, C. The highly expressed 5′isomiR of hsa-miR-140-3p contributes to the tumor-suppressive effects of miR-140 by reducing breast cancer proliferation and migration. *BMC Genom.* **2016**, *17*, 566. [CrossRef] [PubMed]

processes

Article

FluxVisualizer, a Software to Visualize Fluxes through Metabolic Networks

Tim Daniel Rose [1,2,3] and Jean-Pierre Mazat [2,3,*]

1   Institute for Quantitative and Theoretical Biology, Heinrich-Heine-University, 40225 Duesseldorf, Germany; tim@rose-4-you.de
2   IBGC-CNRS UMR 5095, 1 rue Camille Saint Saens, CS 61390, 33077 Bordeaux-cedex, France
3   University of Bordeaux France, 33076 Bordeaux-cedex, France
*   Correspondence: jean-pierre.mazat@u-bordeaux.fr; Tel.: +33-556-999-041

Received: 2 March 2018; Accepted: 17 April 2018; Published: 24 April 2018

Abstract: FluxVisualizer (Version 1.0, 2017, freely available at https://fluxvisualizer.ibgc.cnrs.fr) is a software to visualize fluxes values on a scalable vector graphic (SVG) representation of a metabolic network by colouring or increasing the width of reaction arrows of the SVG file. FluxVisualizer does not aim to draw metabolic networks but to use a customer's SVG file allowing him to exploit his representation standards with a minimum of constraints. FluxVisualizer is especially suitable for small to medium size metabolic networks, where a visual representation of the fluxes makes sense. The flux distribution can either be an elementary flux mode (EFM), a flux balance analysis (FBA) result or any other flux distribution. It allows the automatic visualization of a series of pathways of the same network as is needed for a set of EFMs. The software is coded in python3 and provides a graphical user interface (GUI) and an application programming interface (API). All functionalities of the program can be used from the API and the GUI and allows advanced users to add their own functionalities. The software is able to work with various formats of flux distributions (Metatool, CellNetAnalyzer, COPASI and FAME export files) as well as with Excel files. This simple software can save a lot of time when evaluating fluxes simulations on a metabolic network.

Keywords: metabolic network visualization; metabolic modelling; elementary flux modes visualization; flux balance analysis

1. Introduction

The study of genome-scale metabolic models has grown strongly in recent years. This has stimulated the development of visualization software of large models of metabolism [1–3] for reviews. At the same time, new methods of studying metabolic fluxes have emerged which lead to the enumeration of EFMs, Flux Balance Analysis (FBA) [4,5] for reviews alongside more traditional methods such as dynamical systems and Metabolic Control Analysis [6–8] using the rate functions of the metabolic steps [9,10]. However only FBA can be applied to the greatest genome-scale models. As a matter of fact, the number of EFMs highly increases and it is not possible to calculate them. Furthermore, the rate equations are not entirely known at the level of genome scale model, particularly the number of enzymes. Consequently, it is not possible to derive a pertinent dynamical system describing the behaviour of a genome-scale model in a physiological context. For these reasons, reduced metabolic models, or core models, are often derived to study particular problems or for a manoeuvrable approach to metabolism [11–13]. In this type of approach drawing a reduced metabolic network plays an essential role. It usually summarizes the results or hypotheses of the authors in the form of pathways of different colours or of different sizes according to the flux values. Many software already exist for automatically generating flux maps for a metabolic network (at the genome-scale level or not): Omix (Omix Visualization GmbH & Co. KG, Lennestadt, Germany, 2018) [14], MetDraw (freely available at http://www.metdraw.com) [15], MetExploreViz (available at: http://metexplore.toulouse.inra.fr/metexploreViz/doc/) [16], BiGG (freely available for academic use at http://bigg.ucsd.edu) [3], Fluxviz (Cytoscape pen-source plug-in available at http://apps.cytoscape.org/apps/fluxviz) [17], VisANT (VisANT 5.0 freely available at: http://visant.bu.edu) [18], among many others. However, 'visualization relies mainly on human perceptual and cognitive capabilities for extracting information' [19], so many scientists prefer to use their own network representations of their models which are perfectly fitted to their needs because they are accustomed to recognizing these classic metabolic pathways and metabolites at a glance. Furthermore, even on core models one can be confronted with a great (huge in some instances) number of flux data so that hand-drawing is time consuming. FluxVisualizer is a software that does not seek to compete with the software mentioned above in drawing metabolic networks from, for instance, an Systems Biology Markup Language (SBML) file, which most software already does very well. The first aim of FluxVisualizer is to use a customer's SVG representation of a metabolic network to simultaneously visualize reactions and flux values, that is, to automatically draw from the customer's network what the biochemist usually draws by hand. The second aim of FluxVisualizer is to automatically generate a series of pathways on a metabolic network. This is often necessary when dealing with a list of elementary flux modes (EFMs) or a series of results obtained in Flux Balances Analysis (FBA) particularly in varying the constraints (Flux Variability Analysis) or with time (dynamic FBA). In these cases, researchers are faced with tedious time-consuming series of drawings that have to be automated.

FluxVisualizer is an open source software, which offers a simple way to represent fluxes by colour and/or width on a Scalable Vector Graphics (SVG) image. We chose the SVG format, because it is widely used and can be built and edited by a variety of programs. The XML structure of the SVG format makes it easy to access, edit and save in any necessary quality. Furthermore, already existing software [3,14–18] can output an SVG file of a metabolic network (inputted as XML file for instance). FluxVisualizer can handle different classical formats of flux distributions (Metatool output files [20], CellNetAnalyzer export files [21], COPASI export files [22] and FAME export files [23]) and more generally any CSV or TSV files, so that an Excel file of flux values can be directly represented on the customer's SVG image. FluxVisualizer can automatically describe a series of pathways of the same metabolic network, for instance a series of EFMs, resulting in a set of different SVG files of the same basic metabolic map. The program provides a graphical user interface (GUI) and an application programming interface (API) for python3. All functionalities of the program can be used from the API and the GUI, whereas the API has more possibilities to adapt the output and allows advanced users to add their own functionalities to the program.

2. Overview of FluxVisualizer

Figure 1 illustrates the idea of the algorithm. Starting from a SVG image of the metabolic network (Figure 1a) an "Example Flux" is plotted on Figure 1b with the option "Auto width" that automatically adjusts the width of the arrows to the flux values between two chosen extrema. The pathway with the flux values is written in place of the "place-here" label in Figure 1a.

place_here

(a)

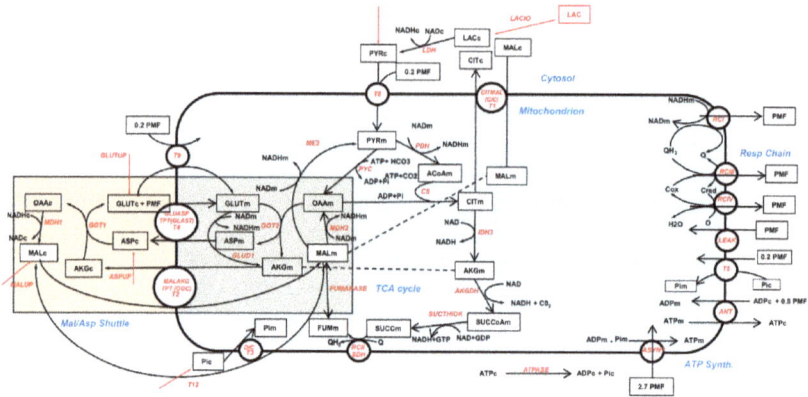

EXAMPLE FLUX

(b)

0.22 T6   0.22 PDH   0.0 NAD_balance_mito 0.22 CS   0.22 IDH3   0.22 AKGDH 0.22 SUCTHIOK
0.22 SDH   0.22 FUMARASE 0.22 MDH2   0.88 RCI   0.08 Leak   1.1 RCIII   1.1 RCIV   2.66 ASYNT
2.88 ANT   2.88 T5   2.88 ATPASE

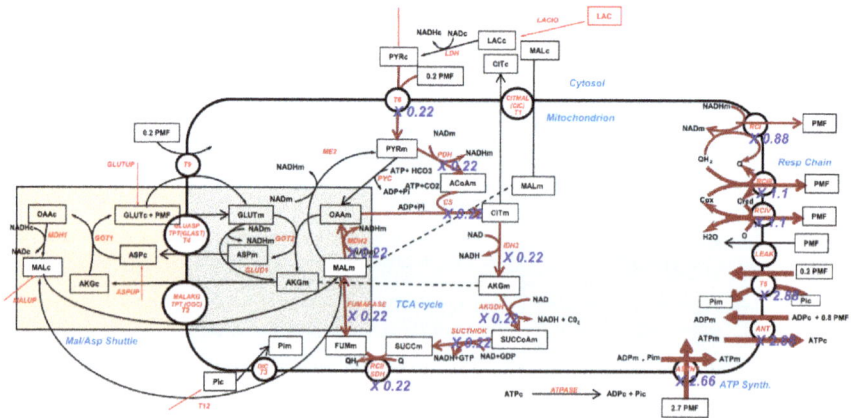

Figure 1. An example of customer's metabolic network as an SVG file (**a**) and (**b**) after the visualization of the pathway 0.22 T6 . . . . on this SVG image with the "Auto width" option. The width of the fluxes is proportional to the coefficient of the step between the min. width 1.5 and the max. width: 3.0. Note that the flux values are written in blue italic along the reactions arrows and that the complete pathway is written in the place of the "place_here" label in (**a**). The light blue background represents mitochondrion and the light brown background depicts the Malate Aspartate Shuttle (MAS) connecting the redox status of cytosol and mitochondria.

2.1. Main Window (Figure 2)

After starting FluxVisualizer the main window appears (Figure 2). The user can change the ID format and adapt it to the format used in their SVG file. The algorithm will replace the word "REACTION" with the actual reaction name and the word "COUNTER" with a number. To indicate Reversibility the word "REV" has to be added to the ID format (see the manual). All other letters and characters will remain the same for every actual ID in the SVG file. Below the ID format the user has the choice between three ways to define the width factor with which the original width of the arrows is multiplied when a reaction is part of the flux distribution. It is important to mention that all flux constraints and width factors are always considered as absolute flux values (A flux of −5 will have the same width as a flux of 5). This can either be a constant width factor, an automatically fitted width or a variable width. If a constant width is chosen, the arrow widths of the non-zero fluxes are multiplied with the value in the text field "Width factor." If the "Auto width" is selected, the reaction arrow with the minimum flux (absolute) will be multiplied with the "min. width" value and the maximum value (absolute) will be multiplied with the "max. width" value. The width of all fluxes in between will be obtained by a linear intrapolation in between the minimum and maximum. This option is used to draw the fluxes in Figure 1b showing a broader arrow in reactions RCI (Respiratory Complex I), RCIII (Respiratory Complex III) and so forth (see the right part of the figure) illustrating the high NADH (Reduced Nicotinamide adenine dinucleotide) production by the Krebs cycle giving a higher Oxidative Phosphorylation flux than the Krebs cycle flux. One can also notice a slightly higher flux in RCIII than in RCI due to the entry of succinate in the respiratory chain and the maximum flux for T5, ANT (Adenine Nucleotide Translocator) and ATPSYNT (ATP synthase) evidencing the nearly 3 ATP (Adenosine Triphosphate) synthesized per NADH molecule. This option immediately gives a visual idea of the various fluxes in the network. If the check box "Variable width" is selected, three flux boundaries can be inserted separating four width factors chosen by the user. The program will then visualize the flux with these different widths according to the boundaries in the text boxes. Before proceeding it is necessary to open a SVG image of the network under study. If the image is not readable by the program, opening the file will set up a warning.

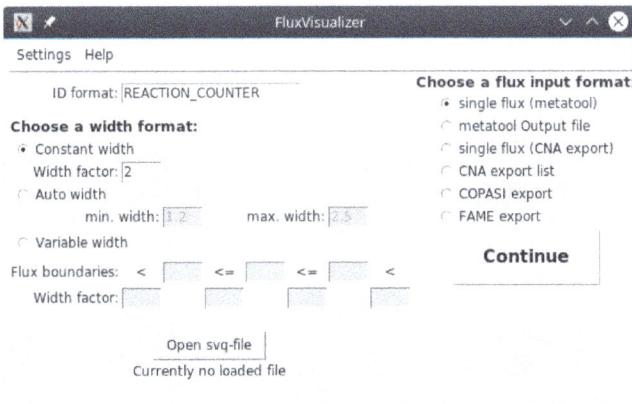

Figure 2. Main window of FluxVisualizer showing the various width formats on the left and the different input formats on the right.

2.2. Secondary Windows (Figure 3)

On the right side of the main window it is possible to decide which input format of the flux distribution will be used. The user can choose between single pathway representations (Figure 3a with different formats: single flux (metatool), single flux (CNA export), COPASI export and FAME export.

Conversely metatool and CNA can produce a file containing a series of pathways (EFMs) which can be automatically represented on the same SVG image previously selected in the main window (Figure 2). In this later case, after indicating the input file and the output folder, the series of corresponding SVG images is saved in the output folder (Figure 3b).

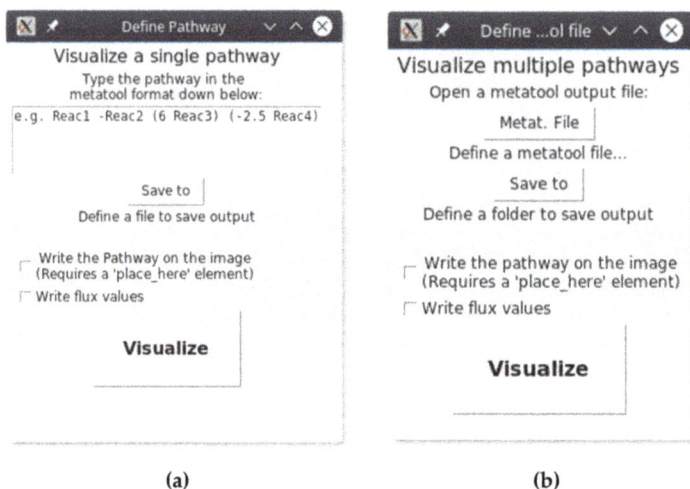

(a)   (b)

Figure 3. Windows appearing after the user has pressed the "continue" button in the main window. By closing the windows in the upper right corner, the user goes back to the main window (Figure 2). The (**a**) figure concerns the single pathway representation on a single SVG image. The (**b**) figure concerns the case of a list of pathways (metatool or CNA format) corresponding to a series of SVG images automatically created and saved in a folder.

2.3. Additional Options

Additional options are proposed on the secondary windows (Figure 3). The first one allows writing the complete pathway with the flux values provided that the "place_here" label exists on the SVG image (see Figure 1). The second option allows writing the flux value on the SVG file below the name of the reaction (see Figure 1). The colour of the arrows of the non-zero fluxes, the size and the number of decimals of the flux values can be defined in the "settings" menu in the main window (Figure 2).

3. Implementation and Requirements

FluxVisualizer is written in Python 3.5.2 (https://www.python.org/) and requires the following modules: tkinter (8.6) lxml (3.5.0) and re (2.2.1). The program, with a manual is freely available at https://fluxvisualizer.ibgc.cnrs.fr.

SVG file Requirements: Reactions of the metabolic network are drawn with arrows. Flux through a reaction is visualized by increasing the width of an arrow and/or colouring it. To visualize fluxes, FluxVisualizer needs to recognize the image elements to be changed, essentially the reaction arrow. To this aim, these elements have an annotation ID with the exact name of the reaction as it appears in the pathway entered in the second windows (Figure 3). If it is not the case, (SVG output of another program), ID can be changed easily by any SVG editing tools (Note that MetDraw output [15] can be straightforward used by FluxVisualizer). An example of IDs for reactions is given in Figure 4. The default ID format is REACTION_COUNTER where REACTION will be replaced by the actual reaction name and COUNTER will be replaced by a number (In case that a reaction consists of several arrows (Figure 4b)). Another important parameter is the word REV. It indicates reversibility of reactions. It is part of the ID and if a reaction has a positive flux, it will be removed; if a reaction has a negative flux, it will be replaced by a "-". An example of reversibility (ID format: REACTION_REVCOUNTER) is given in Figure 4 on the left side and its use on Figure 5. A COUNTER is not required but if a counter is used, it must be used for every reaction ID, even if the reaction consists only of one arrow.

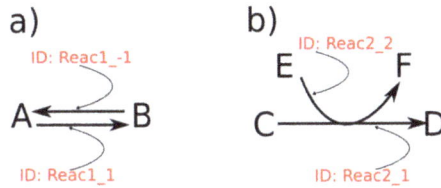

Figure 4. (**a**) A simple example of IDs of reaction arrows with reversibility (ID format: REACTION_REVCOUNTER) and (**b**) with several arrows for one reaction (ID format: REACTION_COUNTER).

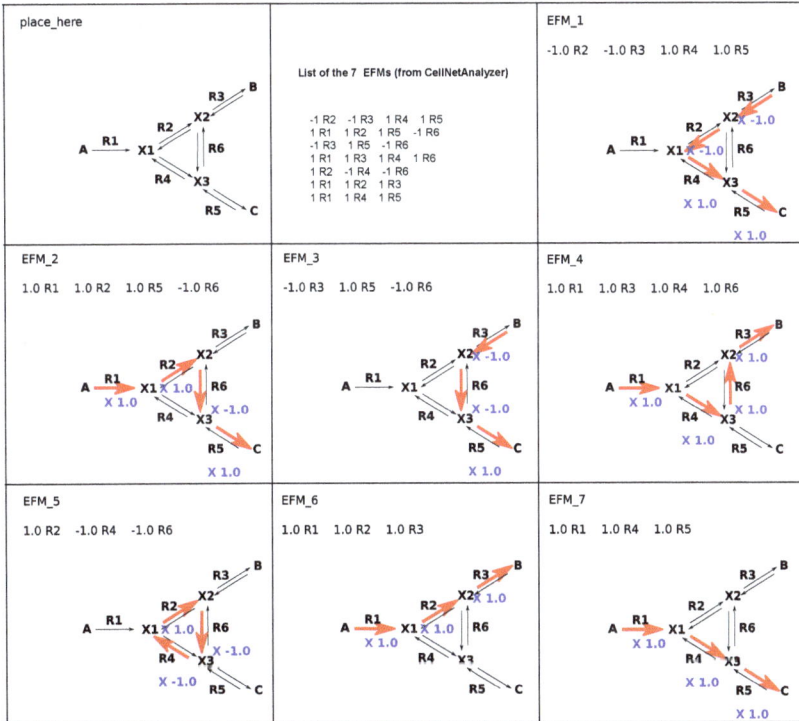

Figure 5. A simple example of the generation of Elementary Mode of Flux (EFMs) representation. The customer's SVG (Scalable Vector Graphics) file is given in the top left of the figure (labelled 'place_here'). The list of EFMs from CellNetAnalyzer in this form is in middle top images. With the option 'cna (CellNetAnalyzer) export list' (Figure 2) this list is entered as CNA file (Figure 3a) and generates the successive EFMs, EFM_1 until EFM_7.

Reaction Names (optional): The reaction names on the SVG representation are necessary if you want to write down the flux values on the SVG image. The ID must be identical to the reaction name (without the counter and additional characters). It must be in a SVG tspan element of a text element. The SVG tspan element of the reaction name is not allowed to carry any other text than the reaction name.

Placing the Flux text (optional): It is also possible to write down the flux distribution as text on the image. To enable this the network file requires a text element saying: "place_here" (Figure 1a). This text element fulfils the same constraints as the reaction names mentioned above. At the position of this text element, the program will write down the flux distribution (Figure 1b, the text "place_here" in (a) is replaced by "Example Flux: 0.22 T6 ..." in (b)).

4. Discussion and Conclusions

As mentioned in the introduction, many software already exists to automatically generate flux maps in metabolic network (at the genome-scale level or not). Usually they are able to draw the entire metabolic network and the pathways of interest with possibly several layouts and zoom functions.

However, there is, among biochemists, a long tradition of metabolic network representation with simple arrows as exemplified in biochemistry textbooks (and in Figures 1 and 5) with particular disposition in space. A simple vertical line with a circle underneath irreversibly evokes glycolysis and the Krebs cycle for a biochemist. It is important to keep this implicit knowledge to facilitate the interpretation of metabolic results or hypotheses. It is undoubtedly the reason why experimental biochemists often draw their own metabolic network representation by hand and report flows also by hand with values and/or colours. This process is tiresome and time-consuming and was automatized with FluxVisualizer. This is why FluxVisualizer does not draw a representation of metabolic networks but starts from the drawings of the biochemist himself with a minimum of constraints.

It is thus difficult to compare FluxVisualizer with other software mainly dealing (among other functionalities) with network representation. These software can be used in synergy with FluxVisualizer in providing an SVG representation FluxVisualizer will exploit. This is done on Figure 6 using MetExplore, which is rather easy to use and can export an SVG image, in this case the same metabolic network as in Figure 1a. It must be noted however, that the presence of the nodes, rectangles and circles representing the reactions and metabolites unnecessarily clutters the larger schemes and makes them less readable. Furthermore, it is necessary to re-arrange spatially this diagram to reproduce the biochemist's layout of Figure 1a. It necessitates a long and tedious work. Undeniably, SVG representation is not (yet) a standard among experimental biologists and this requirement may turn off biologists from FluxVisualizer use, contradicting the purpose of FluxVisualizer to be tailored to biochemists. This is a difficulty for a biochemist to a friendly use of FluxVisualizer. There is another way to take this deterrent step forward. Very often, biologists use LibreOffice (https://fr.libreoffice.org) (or Microsoft Office (https://products.office.com)) suite to draw their diagrams of metabolic networks. It is easy to save them in the pdf format which can be read by Inkscape, for instance, and be converted to the SVG format. This was done in the case of Figure 1 which, initially, was a PowerPoint file. It is then necessary to check the arrow's ID in order for them to match exactly the name of the reaction. The last solution is to build directly the diagram with Inkscape, for instance, which is not so difficult to manage. This was done for Figure 5.

Another difficulty in visualizing the FluxVisualizer results are the overlaps of the flux values with the rest of the initial image (see Figure 1b). These unescapable overlaps, are not too troublesome for small metabolic networks for which the attribution of a flux value to an enzymatic step remains clear but it is a limit in the size of the network. These overlaps cannot be avoided in an automatic process but their effects can be diminished in playing with the different options of positioning the flux values offered in FluxVisualizer.

Although there is, in principle, no limit in the size of the metabolic network that FluxVisualizer can deal with (genome-scale network should be, in principle, handled) the increasing number of overlaps will decrease the readability of the results on the metabolic network as the number of steps increases. It is difficult to define a precise limit in size and data. Typically, the field of use of this software consists of metabolic networks between a few steps (around 10) to less than 200 reactions, which is probably the limit for a discriminating visualization. This is also the range of the size of customer's network in the literature (mainly between 10 and 50 which can already generate a very great number of EFMs and possible solutions).

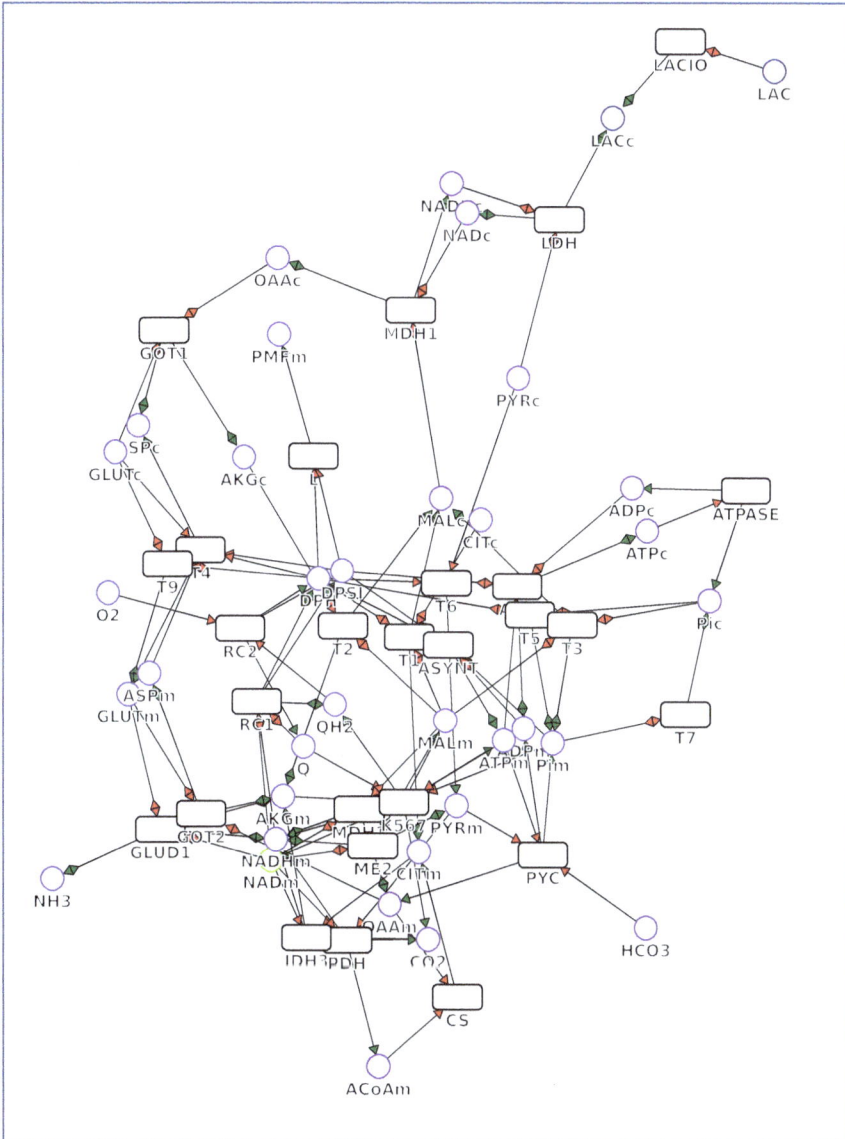

Figure 6. SVG representation of the network of Figure 1a with MetExplore, which is representative of most of the software cited in the text. The red arrows correspond to substrates and the green ones to products of the reactions in rectangles. The metabolites are indicated with circles.

To sum it up, we think that FluxVisualizer fulfils the two aims stated in the introduction: (i) use the biochemist's own diagram and to replace a tedious manual operation of colouring or/and resizing steps by hand with a fast automatic process and (ii) automatically deliver a series of pathway representations of the same metabolic map from a text list of these pathways (EFMs for instance). This is a great time saver that justifies the time spent presenting the metabolic network in SVG format. We believe that this simple software, freely available at https://fluxvisualizer.ibgc.cnrs.fr, has its place in the theoretical toolbox of the experimental biochemist.

Acknowledgments: Supported by the Plan cancer 2014–2019 No BIO 2014 06 and the French Association against Myopathies. We thank Axel Cattouillart for his technical assistance, Sylvain Prigent for his help in manipulating other tools, Anne Devin for her criticisms and English review and Oliver Ebenhoeh as TDR's teacher.

Author Contributions: Tim Daniel Rose developed FluxVisualizer, wrote the manual and participated to the redaction of the paper. Jean-Pierre Mazat conceived the project and wrote the paper.

Conflicts of Interest: The authors declare no conflict of interest.

References

1. Gehlenborg, N.; O'Donoghue, S.; Baliga, N.S.; Goesmann, A.; Hibbs, M.A.; Kitano, H.; Kohlbacher, O.; Neuweger, H.; Schneider, R.; Tenenbaum, D.; et al. Visualization of omics data for systems biology. *Nat. Methods* **2010**, *7*, S56–S68. [CrossRef] [PubMed]
2. O'Donoghue, S.I.; Baldi, B.F.; Maier-Hein, L.; Stenau, E.; Hogan, J.M.; Humphrey, S.; Kaur, S.; McCarthy, D.J.; Moore, W.J.; Procter, J.B.; et al. Visualization of Biomedical Data. *Annu. Rev. Biomed. Data Sci.* **2018**, in press.
3. Schellenberger, J.; Park, J.O.; Conrad, T.M.; Palsson, B.Ø. BiGG: A Biochemical Genetic and Genomic knowledgebase of large scale metabolic reconstructions. *BMC Bioinform.* **2010**, *11*, 213. [CrossRef] [PubMed]
4. Maarleveld, T.R.; Khandelwal, R.A.; Olivier, B.G.; Teusink, B.; Bruggeman, F.J. Basic concepts and principles of stoichiometric modeling of metabolic networks. *Biotechnol. J.* **2013**, *8*, 997–1008. [CrossRef] [PubMed]
5. Antoniewicz, M.R. Methods and advances in metabolic flux analysis: A mini-review. *J. Ind. Microb. Biotechnol.* **2015**, *42*, 317–325. [CrossRef] [PubMed]
6. Kacser, H.; Burns, J.A. The control of flux. *Symp. Soc. Exp. Biol.* **1973**, *32*, 65–104. [CrossRef]
7. Heinrich, R.; Rapoport, T.A. A linear steady-state treatment of enzymatic chains; its application for the analysis of the crossover theorem and of the glycolysis of human erythrocytes. *Acta Biol. Med. Ger.* **1973**, *31*, 479–494. [PubMed]
8. Reder, C. Metabolic control theory: A structural approach. *J. Theor. Biol.* **1988**, *135*, 175–201. [CrossRef]
9. Sauro, H. *Systems Biology: Introduction to Pathway Modeling*; Ambrosius Publishing: Bangalore, India, 2014; ISBN 978-0-9824773-7-3.
10. Klipp, E.; Liebermeister, W.; Wierling, C.; Kowald, A. *Systems Biology: A Textbook*, 2nd ed.; Wiley-Blackwell: Berlin, Germany, 2016; ISBN 978-3-527-33636-4.
11. Ataman, M.; Hernandez Gardiol, D.F.; Fengos, G.; Hatzimanikatis, V. redGEM: Systematic reduction and analysis of genome-scale metabolic reconstructions for development of consistent core metabolic models. *PLoS Comput. Biol.* **2017**, *13*, e1005444. [CrossRef] [PubMed]
12. Smith, A.C.; Eyassu, F.; Mazat, J.-P.; Robinson, A.J. MitoCore: A curated constraint-based model for simulating human central metabolism. *BMC Syst. Biol.* **2017**, *11*, 114. [CrossRef] [PubMed]
13. Orth, J.; Fleming, R.; Palsson, B. Reconstruction and Use of Microbial Metabolic Networks: The Core Escherichia coli Metabolic Model as an Educational Guide. *EcoSal Plus* **2010**. [CrossRef] [PubMed]
14. Omix Visualization | Welcome. Available online: https://www.omix-visualization.com/ (accessed on 21 April 2018).
15. Jensen, P.A.; Papin, J.A. MetDraw: Automated visualization of genome-scale metabolic network reconstructions and high-throughput data. *Bioinformatics* **2014**, *30*, 1327–1328. [CrossRef] [PubMed]
16. Chazalviel, M.; Frainay, C.; Poupin, N.; Vinson, F.; Merlet, B.; Gloaguen, Y.; Cottret, L.; Jourdan, F. MetExploreViz: Web component for interactive metabolic network visualization. *Bioinformatics* **2017**, *34*, 312–313. [CrossRef] [PubMed]
17. König, M.; Holzhütter, H.G. Fluxviz—Cytoscape plug-in for visualization of flux distributions in networks. *Genome Inform.* **2010**, *24*, 96–103. [PubMed]

18. Granger, B.R.; Chang, Y.C.; Wang, Y.; DeLisi, C.; Segrè, D.; Hu, Z. Visualization of metabolic interaction networks in microbial communities using VisANT 5.0. *PLoS Comput. Boil.* **2016**, *12*, e1004875. [CrossRef] [PubMed]
19. Turkay, C.; Jeanquartier, F.; Holzinger, A.; Hauser, H. On computationally-enhanced visual analysis of heterogeneous data and its application in biomedical informatics. In *Interactive Knowledge Discovery and Data Mining in Biomedical Informatics*; Holzinger, A., Jurisica, I., Eds.; Springer: Berlin/Heidelberg, Germany, 2014; pp. 117–140.
20. Pfeiffer, T.; Sánchez-Valdenebro, I.; Nuño, J.C.; Montero, F.; Schuster, S. METATOOL: For studying metabolic networks. *Bioinformatics* **1999**, *15*, 251–257. [CrossRef] [PubMed]
21. Klamt, S.; Saez-Rodriguez, J.; Gilles, E.D. Structural and functional analysis of cellular networks with CellNetAnalyzer. *BMC Syst. Biol.* **2007**, *1*, 2. [CrossRef] [PubMed]
22. Hoops, S.; Sahle, S.; Gauges, R.; Lee, C.; Pahle, J.; Simus, N.; Singhal, M.; Xu, L.; Mendes, P.; Kummer, U. COPASI—A COmplex PAthway SImulator. *Bioinformatics* **2006**, *22*, 3067–3074. [CrossRef] [PubMed]
23. Boele, J.; Olivier, B.G.; Teusink, B. FAME, the Flux Analysis and Modeling Environment. *BMC Syst. Biol.* **2012**, *6*, 8. [CrossRef] [PubMed]

![processes logo] *processes*

MDPI

Article

Measuring Cellular Biomass Composition for Computational Biology Applications

Ashley E. Beck [1], Kristopher A. Hunt [2] and Ross P. Carlson [3,]*

[1] Microbiology and Immunology, Center for Biofilm Engineering, Montana State University, Bozeman, MT 59717, USA; ashley.beck@montana.edu

[2] Civil and Environmental Engineering, University of Washington, Seattle, WA 98195, USA; hunt0362@uw.edu

[3] Chemical and Biological Engineering, Center for Biofilm Engineering, Montana State University, Bozeman, MT 59717, USA

[*] Correspondence: rossc@montana.edu; Tel.: +1-406-994-3631

Received: 27 January 2018; Accepted: 17 April 2018; Published: 24 April 2018

check for updates

Abstract: Computational representations of metabolism are increasingly common in medical, environmental, and bioprocess applications. Cellular growth is often an important output of computational biology analyses, and therefore, accurate measurement of biomass constituents is critical for relevant model predictions. There is a distinct lack of detailed macromolecular measurement protocols, including comparisons to alternative assays and methodologies, as well as tools to convert the experimental data into biochemical reactions for computational biology applications. Herein is compiled a concise literature review regarding methods for five major cellular macromolecules (carbohydrate, DNA, lipid, protein, and RNA) with a step-by-step protocol for a select method provided for each macromolecule. Additionally, each method was tested on three different bacterial species, and recommendations for troubleshooting and testing new species are given. The macromolecular composition measurements were used to construct biomass synthesis reactions with appropriate quality control metrics such as elemental balancing for common computational biology methods, including flux balance analysis and elementary flux mode analysis. Finally, it was demonstrated that biomass composition can substantially affect fundamental model predictions. The effects of biomass composition on in silico predictions were quantified here for biomass yield on electron donor, biomass yield on electron acceptor, biomass yield on nitrogen, and biomass degree of reduction, as well as the calculation of growth associated maintenance energy; these parameters varied up to 7%, 70%, 35%, 12%, and 40%, respectively, between the reference biomass composition and ten test biomass compositions. The current work furthers the computational biology community by reviewing literature regarding a variety of common analytical measurements, developing detailed procedures, testing the methods in the laboratory, and applying the results to metabolic models, all in one publicly available resource.

Keywords: biomass reaction; computational biology; macromolecular composition; metabolic model; methods

1. Introduction

The in silico study of metabolism has largely transitioned from a specialty discipline to a mainstream biological approach due to improvements in software usability, increases in computational power, and the accumulation of omics databases. Cellular growth is an essential component of many of these computational biology studies [1–3]. Understanding the foundation of growth from the level of mass and energy fluxes remains critical for interpretation and integration of in silico metabolic models and omics datasets. The macromolecular composition of a cell is one such area of basic knowledge. Macromolecular composition of both prokaryotic and eukaryotic cells is governed by allocation of resources and can shift depending on cell cycle, specific growth rate, and diel cycle (e.g., cyanobacteria and green algae) [4–6].

Stoichiometric modeling approaches analyze steady state fluxes based on metabolic reactions identified from an organism's genomic potential, enzyme-coding genes identified in the genome sequence [7]. These methods can be applied to microbial communities as well as individual species [8,9]. Optimal metabolic pathways are often assessed in terms of growth: constraint-based approaches, such as flux balance analysis [10], typically use production of biomass as an objective function, and macromolecular composition dictates the metabolic precursors necessary for growth. Different weightings of macromolecular components in the biomass synthesis reaction can influence results by shifting requirements for precursors [11]. However, the proportions of biomass components are not specified by the genome sequence [12]. While technologies for automatic model construction are rapidly increasing, stoichiometric coefficients for the biomass reaction are still necessary [13]. Often, coefficients for this essential reaction are borrowed from literature reported for *Escherichia coli* or an organism similar in physiology or phylogeny to the organism being modeled (e.g., [14,15]). However, these values may not be representative of the organism under study. In biotechnology applications, a specific macromolecular component may be targeted, such as lipids extracted for biofuels [16] or starch compounds for biochemical production. Accurate quantification of these components is important for comparison of production potential under different conditions. Additionally, ratios of macromolecule pools, such as protein, DNA, or RNA, from a microbial population can be correlated to important culture properties, including specific growth rate [17].

A variety of methods for quantification of any given macromolecule can be found in the literature (e.g., [18]). Many of these methods date back several decades, and numerous adaptations have arisen over the years. Selecting and implementing a method with an assurance of valid and accurate results relevant to computational biology applications can present a significant challenge, particularly when testing new organisms. Additionally, not all reported methods have been developed for or tested on prokaryotes, and different organisms may respond differently to treatment conditions. For example, cell wall type may influence the efficacy of reagents or procedures, resulting in a method with varying degrees of efficiency for different types of microorganisms. External factors, such as materials used, can also affect the outcome of an analysis, and specific procedural details not included in publications can hinder reproducibility. Recently, methods for determining multiple biomass components with a single technique, e.g., gas chromatography-mass spectrometry, have been developed [19] but still rely on adequate cell lysis techniques and standard compounds for quantification. A concise collection of information about the variety of existing methods for each macromolecule, including advantages and disadvantages of methods, specific procedural details, and points for potential pitfalls, is a useful resource that is lacking from the published literature.

The current work fills this gap with objectives: (1) to review and compare existing literature regarding methods to measure five major macromolecules (carbohydrate, DNA, lipid, protein, and RNA); (2) to develop a select step-by-step protocol for each macromolecule and test the efficacy on different types of bacterial samples; and (3) to demonstrate the application to computational biology by generating biomass synthesis reactions. Three bacterial species were used as test cases in the current work: *E. coli* (Gram-negative, mesophilic model laboratory organism), *Synechococcus* sp. PCC 7002 (Gram-negative, mesophilic cyanobacterium; *Synechococcus* 7002 hereafter), and *Alicyclobacillus acidocaldarius* (Gram-positive, thermophilic acidophile). These microorganisms encompass a range of physiological capabilities and characteristics, including photosynthesis and alicyclic fatty acids. The impact of biomass composition on model predictions was demonstrated using essential parameters, including biomass yield on electron donor, biomass yield on electron acceptor, biomass yield on nitrogen, biomass degree of reduction, and growth associated maintenance energy. The results highlight the importance of appropriate methods for the accurate determination of macromolecule composition. Compiling a literature review in conjunction with laboratory-tested protocols with demonstrated application to metabolic models, all within a single source, serves as a useful resource for the computational biology community that should facilitate model building transparency and reproducibility.

2. Culturing Methods

2.1. Strains and Media

E. coli str. K-12 substr. MG1655 was grown in standard M9 minimal salts medium (6 g/L Na_2HPO_4, 3 g/L KH_2PO_4, 1 g/L NH_4Cl, 0.5 g/L NaCl), supplemented with 1 mL/L 1 M $MgSO_4 \cdot 7H_2O$ and 10 mL/L trace metals (0.55 g/L $CaCl_2$, 0.10 g/L $MnCl_2 \cdot 4H_2O$, 0.17 g/L $ZnCl_2$, 0.043 g/L $CuCl_2 \cdot 2H_2O$, 0.06 g/L $CoCl_2 \cdot 6H_2O$, 0.06 g/L $Na_2MoO_4 \cdot 2H_2O$, 0.06 g/L $Fe(NH_4)_2(SO_4)_2 \cdot 6H_2O$, 0.20 g/L $FeCl_3 \cdot 6H_2O$) [20,21].

Synechococcus 7002 was grown in A+ synthetic seawater medium (18 g/L NaCl, 0.6 g/L KCl, 1 g/L $NaNO_3$, 5 g/L $MgSO_4 \cdot 7H_2O$, 0.05 g/L KH_2PO_4, 0.27 g/L $CaCl_2$, 0.03 g/L Na_2 ethylenediaminetetraacetic acid, 0.004 g/L $FeCl_3 \cdot 6H_2O$, 1 g/L Tris pH 8.2), supplemented with 4 mg/L vitamin B12 and 1 mL/L P1 trace metal mix (34.26 g/L H_3BO_3, 4.32 g/L $MnCl_2 \cdot 4H_2O$, 0.315 g/L $ZnCl_2$, 0.03 g/L MoO_3, 0.003 g/L $CuSO_4 \cdot 5H_2O$, 0.0122 g/L $CoCl_2 \cdot 6H_2O$) [22,23].

A. acidocaldarius str. acidocaldarius DSM446 was grown in *Bacillus acidocaldarius* medium (BAM) (1 g/L KH_2PO_4, 1.5 g/L $(NH_4)_2SO_4$), supplemented with 0.2 g/L $MgSO_4 \cdot 7H_2O$, 0.1 g/L $CaCl_2 \cdot 2H_2O$, and 1 mL/L trace metal mix (10 g/L $FeSO_4 \cdot 7H_2O$, 0.1 g/L H_3BO_3, 0.15 g/L $MnSO_4 \cdot H_2O$, 0.18 g/L $ZnSO_4 \cdot 7H_2O$, 0.2 g/L $CuSO_4 \cdot 5H_2O$, 0.3 g/L $Na_2MoO_4 \cdot 2H_2O$, 0.18 g/L $CoCl_2 \cdot 6H_2O$) (modified from Farrand et al. [24]).

2.2. Culture Conditions

E. coli cultures were grown at 37 °C shaking at 150 rpm. Inoculum cultures were prepared in 8 mL of M9 + 10 g/L glucose in disposable culture tubes, inoculated with multiple isolated colonies from an agar plate streaked from a 20% glycerol freezer stock, and grown to OD_{600} < 0.6 (exponential phase). Cells were then centrifuged at 4000 rpm for 10 min and re-suspended to OD_{600} ~0.05 in 50 mL of fresh M9 + 1 g/L glucose in 250-mL baffled shake flasks. Cultures were grown to OD_{600} ~0.6 (exponential phase) and then harvested for analysis (collected in chilled 50-mL polypropylene centrifuge tubes on ice followed by centrifugation).

Synechococcus 7002 cultures were grown at 38 °C without shaking under continuous light. Inoculum cultures were prepared in 25 mL of A+ media in 250-mL non-baffled shake flasks, inoculated with multiple isolated colonies from an agar plate (transferred monthly for propagation), and grown to OD_{730} < 0.5. Cells were then centrifuged at 4000 rpm for 10 min and re-suspended in 25 or 50 mL of fresh A+ media to an OD_{730} ~0.1. Cultures were grown to OD_{730} 0.4–0.5 and harvested for analysis.

A. acidocaldarius cultures were grown at 60 °C shaking at 200 rpm. Inoculum cultures were prepared in 50 mL of BAM + 5 g/L glucose in 250-mL baffled shake flasks, inoculated with multiple isolated colonies from an agar plate streaked from a 20% glycerol freezer stock, and grown to OD_{600} < 0.6. Cells were then centrifuged at 4000 rpm for 10 min and re-suspended to OD_{600} ~ 0.1 in 50 mL of fresh BAM + 5 g/L glucose. Cultures were grown to OD_{600} ~ 0.6 and then harvested for analysis.

2.3. Dry Weight Determination

Optical density was correlated to biomass for each organism to determine amount of dry weight used for macromolecular analyses. Because optical density can fluctuate at high cell concentrations due to shading effects, samples were diluted to an optical density reading below 0.3 to remain within the linearity of the spectrophotometer. Biomass to OD_{600} correlation for *E. coli* of 0.5 g/L cell dry weight per unit OD_{600} was obtained from Folsom, Parker, and Carlson [25], which used the same strain and was performed in the same laboratory using the methods below.

Biomass to OD_{730} correlation for *Synechococcus* 7002 was determined from biomass combined from 50-mL shake flask cultures. Cells were harvested by centrifugation (4000 rpm, 20 min, 4 °C), re-suspended in A+ media and centrifuged again, and a series of dilutions was prepared. Three milliliters of each dilution were aliquoted into pre-dried, pre-weighed aluminum pans, dried at 80 °C for 24 h, and weighed on a microbalance with accuracy to 0.001 mg (Mettler Toledo MT5). Pans were dried and weighed again to confirm stability. The correlation curve is provided in Appendix A (Figure A1a).

Biomass to OD_{600} correlation for *A. acidocaldarius* was determined from biomass grown in a batch fermentor aerated at 1 vessel volume per minute and agitated at 600 rpm. Cells were harvested by centrifugation (6000 rpm, 5 min, 4 °C), re-suspended in water and centrifuged again, and a series of dilutions was prepared in pre-weighed 50-mL polypropylene centrifuge tubes, which had been dried at 100 °C for one week before pre-weighing. Tubes were dried at 100 °C for one week and weighed on an analytical balance with accuracy to 0.1 mg. Tubes were dried and weighed again to confirm stability. The correlation curve is provided in Appendix A (Figure A1b).

3. Modeling Methods

A metabolic network model for *A. acidocaldarius* was constructed in CellNetAnalyzer [26,27] from the annotated genome [28] with the aid of MetaCyc, KEGG, BRENDA, and NCBI [29–31] databases. Reversible exchange reactions were defined for protons and water. Irreversible exchange reactions were defined to permit ammonium, sulfate, oxygen, and glucose or xylose uptake and carbon dioxide evolution, as well as secretion of possible byproducts, including acetate, lactate, ethanol, and formate. Macromolecular synthesis reactions were defined for nucleic acids, glycogen, lipid, and protein. Synthesis reactions utilized two phosphate bonds per nucleic acid monomer, one phosphate bond per glycogen monomer, and four phosphate bonds per protein monomer [32]. Nucleotide distributions were set based on percent GC content of the genome for DNA and nucleotide sequence of the rRNA genes for RNA. Fatty acid distribution was assigned based on literature values [33,34]. The amino acid distribution was set using the experimentally measured values in the current study. All reactions were balanced for elements, charge, and electrons. Thermodynamic considerations were built into the model via reaction reversibilities based on data from BRENDA [31]. Model simulations were performed with elementary flux mode analysis. Flux vectors v satisfying the stoichiometric matrix S at steady state (Sv = 0) subject to conservation of mass, specified irreversibilities, and indecomposability constraints were computed, resulting in the collection of minimal pathways through the network, called elementary flux modes (EFMs) [35]. EFMs were enumerated using EFMtool [36]. Analysis of resulting EFMs (e.g., biomass yield) was performed with MATLAB. Maintenance energy was fit to experimental glucose and oxygen yield data for *A. acidocaldarius* obtained from [37]. Both growth associated (dominant in fast-growing environmental conditions) and non-growth associated (dominant in slow-growing environmental conditions) maintenance terms were determined. The metabolic model with supporting details, CellNetAnalyzer metabolite and reaction input, an SBML file, and maintenance calculations can be found in the Supplementary Materials (Files S1, S2, and S3).

4. Carbohydrate

4.1. Literature Review

Carbohydrates are common cellular energy storage molecules and constituents of cell walls. HPLC methods can be used to separate and quantify specific sugars [38,39]; however, methods for quantifying total carbohydrates were the focus of the current work. Chaplin [40] reviewed many colorimetric methods for carbohydrate quantification and detailed the advantages and disadvantages of each. The phenol sulfuric acid method [41,42] is widely used, and the L-cysteine and anthrone methods [40,43] are also frequently found in the literature. An issue with many methods is interference from other cellular constituents. For example, protein interferes with measured absorbance in the phenol sulfuric acid assay [40]. In the L-cysteine assay, pentose, heptose, and deoxy sugars contribute to absorbance, and absorbance stability varies among different carbohydrates. Pentoses also contribute to signal in the anthrone assay, but the absorbance fades rapidly and presents minimal interference. Different hexoses may also produce differential responses in the anthrone assay; for example, mannose produces 55% percent of the measured absorbance intensity of glucose [43]. Minimizing interference from pentoses is a key consideration when selecting assays to avoid measuring nucleotide bases twice in both nucleic acid and carbohydrate assays.

Glycogen is the most common form of carbohydrate storage for bacteria [44]. Glycogen content can indicate cellular responses to changing nutrient conditions; for instance, *E. coli* and *Synechococcus* 7002 have both been found to increase glycogen storage during nitrogen limitation [45,46]. Glycogen can be precipitated from cells with KOH, but alkalinity causes some degradation of glycogen. An alternative method using sodium sulfate to adsorb and co-precipitate glycogen has been developed for mosquitoes [47] and adapted to bacterial samples [48] and was selected for the current study. The anthrone assay was selected for quantification of hexoses due to minimal pentose interference. The method employs sulfuric acid to hydrolyze polysaccharides to glucose monomers. In the presence of anthrone, glucose monomers are converted to hydroxyaldehydes and dehydrated to hydroxymethylfurfurals [49], which form blue-green colored complexes with anthrone. The current study tested the hexose quantification assay on cell pellets, glycogen extracts, and the residue remaining after the glycogen extraction process. The sum of the glycogen extract and residue measurements was compared with the total cell pellet measurement to verify recovery of all cellular carbohydrates. Differentiation between glycogen and other cellular carbohydrates, such as cell wall sugars, can provide useful parameters for metabolic models.

4.2. Procedure (After Del Don et al., 1994)

4.2.1. Reagents

- Cell pellet (0.5–1 mg dry biomass, fresh or frozen, washed with carbon-free media).
- Anthrone reagent: (per reaction) mix 10 mg anthrone and 250 µL fresh absolute ethanol (anthrone will partially dissolve), add 75% sulfuric acid to a final volume of 5 mL, and stir until anthrone is completely dissolved [18]. Prepare fresh daily (within 24 h of use) and store at 4 °C.
- 2% sodium sulfate (*w/v*).
- Methanol.
- Glucose standards (prepare from fresh 1 mg/mL glucose solution). A linear response was observed using 10–250 µg/mL standards (e.g., 10, 50, 90, 130, 170, 210, 250 µg/mL). The limit of detection with anthrone has been previously reported as 5 µg/mL [48].

4.2.2. Quantification of Glycogen

(1) Re-suspend cell pellet in 200 µL 2% sodium sulfate in 2-mL Eppendorf tube.
(2) Seal tube with parafilm to prevent cap from popping open and heat for 10 min at 70 °C (VWR analog heat block).
(3) Add 1 mL methanol, and vortex in two 10-s rounds to co-precipitate sodium sulfate and glycogen.
(4) Centrifuge for 15 s at 10,000 rpm to pellet the precipitate (Eppendorf 5415D microcentrifuge) and decant the supernatant.
(5) Wash the precipitate with 1 mL methanol (add methanol, vortex, centrifuge, and decant).
(6) Re-suspend the pellet in 1 mL water, transfer to a clean glass test tube, and place on ice to chill.
(7) Add 5 mL ice-cold anthrone reagent (mixing is unnecessary).
(8) Chill on ice for 5 min, vortex briefly to homogenize the solution, and incubate in a boiling water bath for 10 min.

(9) Place on ice for 5–10 min until cool, vortex briefly, and measure absorbance at 625 nm (Genesys 6 spectrophotometer).

Notes: Use a neutral reaction (containing no glucose) as the blank. Perform a standard curve with each assay, and treat standards identically to samples with anthrone reagent. Different sources report slightly varying absorbance wavelengths and water bath incubation times; the most widely supported parameters were implemented in the current work.

4.2.3. Quantification of Hexoses Excluding Glycogen

Collect the methanol decanted after the precipitation and wash steps (Steps 4–5) in an aluminum pan and evaporate in a fume hood. The methanol contains non-glycogen hexoses, which did not adsorb to and precipitate with sodium sulfate. Re-suspend in 1 mL water, transfer to a clean glass test tube, and place on ice to chill (Step 6); then perform the anthrone reaction as in Steps 7–9.

4.2.4. Quantification of Total Carbohydrate

Skip the glycogen precipitation and wash steps (Steps 1–5). Re-suspend the cell pellet in 1 mL water, transfer to a clean glass test tube, and place on ice to chill (Step 6). Then, perform the anthrone reaction as in Steps 7–9.

4.3. Test Results

Assay linearity was observed within 10–250 µg/mL glucose (Figure 1a). The sum of the glycogen extract and residue measurements was equivalent to the total carbohydrate measurement for *E. coli*, *Synechococcus* 7002, and *A. acidocaldarius* (Figure 1b, $p > 0.05$), indicating that glycogen can be accurately distinguished from other cellular carbohydrates. The glycogen mass percentage obtained for *E. coli* is similar to previously published values measured under carbon limitation (3.6%) [45] and in balanced growth (2.5%) [32]. Previous measurements for *Synechococcus* 7002 estimated 10–12% of dry biomass was carbohydrates under carbon- and light-limited chemostat conditions and 61% of dry biomass was carbohydrates under nitrogen-limited conditions [46]. The 17% of dry biomass value measured here falls close to the carbon- or light-limited conditions. No literature comparison was available for *A. acidocaldarius*.

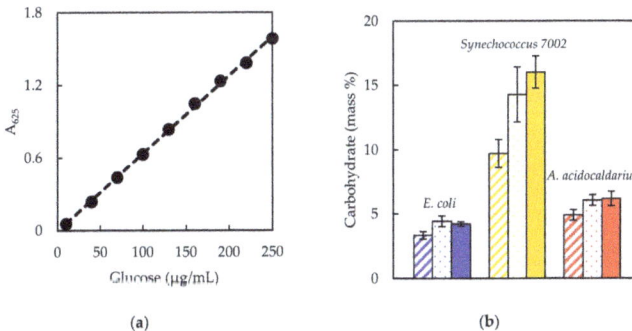

Figure 1. (**a**) Representative glucose standard curve for anthrone assay. The trendline is described by the equation y = 0.0064x − 0.0082 with R^2 value 0.999. (**b**) Carbohydrate mass percentages of dry biomass measured for *E. coli*, *Synechococcus* 7002, and *A. acidocaldarius*. Striped columns indicate glycogen extract measurements, dotted columns indicate the sum of glycogen extract and residue measurements, and solid columns indicate total carbohydrate measurements. Each column is the average of biological triplicate samples with error bars representing standard deviation.

When testing the anthrone assay on glycogen extract, residue, and total carbohydrate samples, it was found that not all three measurements were captured within the standard curve for the entire set of samples; the residue represented a relatively small proportion of total cellular carbohydrates. Thus, different amounts of biomass were tested for each species to identify a quantity that would place all measurements within the standard curve. One milligram dry weight was found appropriate for *E. coli* and *A. acidocaldarius*, and 0.5 mg dry weight was used for *Synechococcus* 7002 (i.e., organisms with higher carbohydrate content require less biomass for the assay).

Additionally, the procedure outlined in Del Don et al. [48] prescribes washing the glycogen pellet with methanol until the pellet is white. However, in the current work, it was observed that glycogen pellets from cyanobacterial samples remained slightly blue after three successive methanol washes, most likely due to photosynthetic pigments. The anthrone assay was tested on glycogen extracts from *Synechococcus* 7002 samples after one, two, or three washes. The carbohydrate content was not significantly different among the three treatments ($p > 0.05$ for all pair-wise T-tests), indicating that a single wash is sufficient (data not shown).

5. DNA

5.1. Literature Review

DNA represents a small but important component of cellular biomass, and its content changes with specific growth rate. For example, slower-growing cells contain more DNA on a cell mass basis than fast-growing cells [4,17]. De Mey et al. [50] provided a summary and comparison of methods for quantifying DNA and RNA. De Mey et al. tested different absorbance, colorimetric, and fluorescence-based assays on purified DNA solutions and reported accuracy and sensitivity for bacteria of differing GC contents [50]. UV absorbance is precise but requires a pure sample (for example, from kit extraction) to minimize interference from RNA and protein. Orcinol can be used to quantify DNA colorimetrically, but has differing sensitivities for different nucleic acids and is not as precise for mixtures of DNA and RNA. The diphenylamine assay [51] is a commonly used method but has lower sensitivity for low GC content and is not as precise as other mentioned methods [50]. The diphenylamine assay also seems to be sensitive to the purity and preparation of the reagents [18]. Fluorescence methods for DNA detection are becoming popular [50]. Hoechst 33258 is a DNA-intercalating dye and is reported to be biased toward AT content [50]. However, cell lysate can be used due to low affinity of the dye for protein and RNA. Thiazole orange is another dye with good precision, but it requires a pure sample and is biased toward GC content [50]. Additional fluorescent dyes that require pure sample for good quantification include PicoGreen and RiboGreen [52].

Considerations when selecting a DNA quantification method include the purity of the sample, interfering substances, and bias toward nucleotide content. Based on these considerations, the Hoechst fluorescent assay was selected for the current study. It is more quantitative than extraction kits and is safer and more precise than the classic diphenylamine method. It is recognized that AT nucleotide bias and the DNA standard used will influence the resulting estimation. Downs and Wilfinger [53] developed and validated an alkali lysis procedure with subsequent Hoechst quantification of DNA using rat pituitary cells. Downs and Wilfinger showed equivalent accuracy but greater precision than the diphenylamine assay [53].

5.2. Procedure (After Downs and Wilfinger, 1983)

5.2.1. Reagents

- Cell pellet (0.4–1 mg dry biomass, fresh or frozen).
- Alkali extraction solution: 1 N NH_4OH, 0.2% Triton X-100.
- Assay buffer: 100 mM NaCl, 10 mM EDTA, 10 mM Tris, pH 7.0 with HCl.
- Standard buffer: 100 mM NaCl, 10 mM EDTA, 10 mM Tris, pH 7.0 with HCl, 0.025 N NH_4OH, 0.005% Triton X-100.

- DNA standards: ~300 μg/mL stock solution calf thymus DNA (Sigma D1501, Merck KGaA, St. Louis, MO, USA), stored at 4 °C. (Concentration was measured with a NanoDrop 1000 spectrophotometer (Wilmington, DE, USA) and was verified after several days and again after several weeks to ensure a stable concentration.) Prepare a 100 μg/mL working stock solution with standard buffer. Dilute the working stock into a standard series with standard buffer.
- Hoechst reagent: Prepare a 200 μg/mL intermediate Hoechst stock from 10 mg/mL stock solution (Biotium 40044, Fremont, CA, USA). Prepare a 1 μg/mL Hoechst working stock fresh daily from the intermediate stock with assay buffer. Store solutions at 4 °C wrapped in aluminum foil to protect from light.

Note: all solutions were prepared using nuclease-free water.

5.2.2. Assay

(1) Re-suspend cell pellet to 50 μL total volume in nuclease-free water in a 2-mL Eppendorf tube.
(2) Add 50 μL of alkali extraction solution.
(3) Incubate at 37 °C for 3 h (VWR analog heat block).
(4) Dilute to 2 mL total volume by adding 1.9 mL assay buffer.
(5) Transfer to a 15-mL polypropylene centrifuge tube and centrifuge (3400 rpm, 30 min, 4 °C).
(6) Aliquot 295 μL of Hoechst working reagent in a clear-bottom black 96-well plate (Corning 3603, Corning, NY, USA).
(7) Add 50 μL of sample to the well (manual mixing via pipette is unnecessary as the plate reader mixes by shaking).
(8) Use a fluorescent plate reader (Synergy HT, Gen5 software, BioTek, Winooski, VT, USA) to read the wells according to the settings in Table 1.

Notes: Perform a standard curve with each assay. Perform three reaction wells of each sample or standard for technical replicates.

Table 1. Fluorescent plate reader settings for Hoechst DNA assay. Data from Beck et al. [54].

Setting	Options
Plate type	96 well plate
Set temperature	Setpoint 30 °C, preheat before moving to next step
Shake	Double orbital 30 s, frequency 180 cpm
Read	Fluorescence endpoint, 352 nm excitation, 461 nm emission, bottom optics, gain 100, Xenon flash light source, high lamp energy, normal read speed, 100 ms delay, 10 measurements/data point

5.3. Test Results

Downs and Wilfinger [53] reported using 0.1 μg/mL Hoechst. However, saturation of calf thymus standard DNA was observed with 0.1 μg/mL Hoechst in the current work (Figure 2a). More recent protocols [55] have suggested that 1 μg/mL Hoechst dye may be used to detect higher quantities of DNA (up to 10 μg) but may not be as sensitive for lower DNA quantities. Based on standard curves using 0.1 μg/mL and 1 μg/mL Hoechst, 1 μg/mL was selected for the current work due to its improved detection range (Figure 2a). The lowest standard concentration used in the assay was 0.25 μg/mL. Hoechst fluorescent response was determined to be linear up to 40 μg/mL DNA; however, a standard curve up to 10 μg/mL was sufficient to capture sample measurements. Additionally, calf thymus DNA standards were subjected to the lysis procedure to ensure that lysis does not cause loss of DNA. Standard curves showed equivalent fluorescent response regardless of whether the lysis procedure was performed, indicating that the lysis step did not influence DNA recovery (Figure A2a, Appendix A).

After initial testing, standards were not subjected to the lysis steps along with samples but were subjected only to the Hoechst treatment.

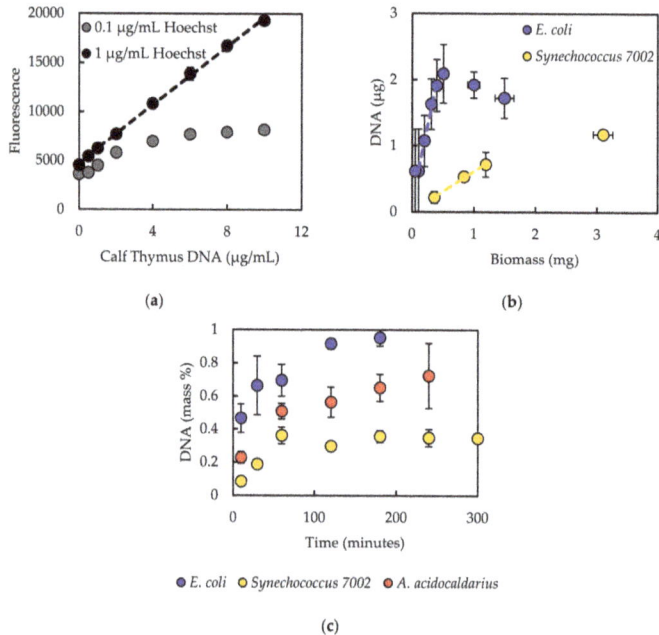

(a)

(b)

(c)

Figure 2. (**a**) Standard curve with calf thymus DNA using 0.1 µg/mL or 1 µg/mL Hoechst reagent. Error bars on the curve with 1 µg/mL Hoechst reagent represent standard deviation of duplicates. The trendline fitting the curve with 1 µg/mL Hoechst reagent is described by the equation $y = 1477x + 4803$ with R^2 value 0.999. (**b**) Relationship between dry biomass sample amount and resulting DNA quantity measured for *E. coli* and *Synechococcus* 7002. Dashed lines designate the linear regions. (**c**) DNA recovery depends on length of lysis step. Error bars represent standard deviation of 2–8 biological replicates. An incubation period of 180 min was selected for the recommended protocol.

Different quantities of biomass were also tested to ensure that DNA recovery was in the linear range of the assay (Figure 2b). DNA recovery from *E. coli* plateaued between 0.5 and 1 mg biomass, thus 0.4 mg was selected as an appropriate amount of starting material for *E. coli* and *A. acidocaldarius*. DNA recovery from *Synechococcus* 7002 plateaued around 1.2 mg, and 0.8–1 mg biomass was used for starting material. Organisms with higher DNA content (*E. coli*, *A. acidocaldarius*) required less biomass for the assay. Incubation time in the lysis solution also influenced DNA recovery, with longer incubation times resulting in increased DNA recovery (Figure 2c). Downs and Wilfinger [53] developed their assay on mammalian cells, which are more easily lysed in contrast to bacterial cells. Incubation of *E. coli* samples over a time series of 10, 30, 60, 120, and 180 min resulted in about twice the amount of DNA recovered. A similar result was observed for *A. acidocaldarius*. DNA recovery for *Synechococcus* 7002 samples increased nearly four-fold with longer incubation times. These results highlight the differing sensitivities of different cell types to assay conditions: *E. coli* and *Synechococcus* 7002 are both Gram-negative bacteria, but cyanobacteria are known to have thicker cell walls with more peptidoglycan [56]. *A. acidocaldarius* is a Gram-positive bacterium but is adapted to acidophilic environments and may be more sensitive to alkaline conditions. A lysis period of 180 min was selected as an adequate incubation time.

Impacts of sample treatment were also investigated with *E. coli*, including freezing of the sample and washing of the cell pellet prior to treatment. Fresh and frozen samples from the same culture were assayed and not found to be significantly different (Figure A2b). Downs and Wilfinger [53] reported washing samples with a cell wash solution (150 mM NaCl, 15 mM citrate, 3 mM EDTA, pH 7.0 with HCl) before performing the lysis step. In the current work, washing the sample with cell wash solution resulted in lower DNA recovery as compared to not washing (Figure A2b) and may indicate cell loss or lysis during the washing process. Extracellular DNA was not expected to comprise a significant proportion of total DNA in the planktonic, exponentially growing samples; however, this may not be the case for all samples, such as biofilm or natural environmental samples.

The DNA percentage of dry biomass obtained for *E. coli* is lower than the value of 3.1% reported in Neidhardt et al. [32], which could be due to differences in methods used or *E. coli* strains (K-12 vs. B/r), while the percentage obtained for *Synechococcus* 7002 is similar to the results reported in Vu et al. [46] measured with the diphenylamine method. Differences in growth rate or growth phase may contribute to differences in measured percentages. No literature comparison was available for *A. acidocaldarius*.

6. Lipid

6.1. Literature Review

Lipids are essential to cellular membranes and are carbon and energy storage molecules. Measurement of total lipid is commonly performed gravimetrically, with an absorbance-based sulfo-phospho-vanillin assay [57], or using gas chromatography. A common gravimetric method is the Bligh and Dyer chloroform–methanol extraction [58]. Other solvents have been used to mitigate the hazards of chloroform, and a variety of modifications to the original Bligh and Dyer procedure exist [16]. There is debate regarding the performance of these different methods, and methods may vary depending on downstream applications. Much research has been done on lipid extraction from biofuel-producing organisms such as algae, and recent testing and comparison of methods have shown microwave extraction with GC analysis to provide optimal results [16]. However, cyanobacteria synthesize predominantly diacylglycerol lipids as opposed to the triacylglycerol lipids of algae [59]. Cyanobacterial lipids are also located throughout the cytoplasm in the thylakoid membranes rather than in granular pockets as in algae. Many lipids are also associated with protein and photosynthetic components through hydrogen bonding. Different methods have been tested for the cyanobacterium *Synechocystis* sp. PCC 6803, and the traditional Bligh and Dyer and Folch methods were found to produce optimal results [59]. Different cell disruption methods have also been tested for *Synechocystis* sp. PCC 6803, and microwave extraction and autoclaving were found to be the most efficient disruption methods [60]. The traditional Bligh and Dyer method was selected for the current study for analysis of total cell lipid [58].

6.2. Procedure (After Bligh and Dyer, 1959)

6.2.1. Reagents

- Cell pellet (10 mg dry biomass, fresh or frozen, washed with carbon-free media).
- Chloroform.
- Methanol.
- Water.

6.2.2. Assay

(1) Re-suspend cell pellet to 0.6 mL total volume with water in a 15-mL polypropylene centrifuge tube.
(2) Sequentially add chloroform (0.75 mL) and methanol (1.5 mL) (vortexing between additions is not necessary).
(3) Vortex 15 min.

(4) Sequentially add chloroform (0.75 mL) and water (0.75 mL), vortexing 10–15 s after each addition.
(5) Centrifuge (4000 rpm, 15 min, 20 °C).
(6) Transfer the lower chloroform phase, which contains the lipids, via micropipette to a pre-weighed aluminum pan.
(7) Evaporate the chloroform in a fume hood and weigh after 12, 24, and 36 h to confirm complete evaporation.

Notes: Weights were measured with a Mettler Toledo MT5 microbalance with accuracy to 0.001 mg and were recorded as an average of three measurements. A blank reaction containing 0.6 mL water was also performed as a control.

6.3. Test Results

Typical protocols for this method recommend a minimum of 30 mg biomass [16]. However, 30 mg biomass requires a large culture volume. Smaller biomass quantities were tested, and the assay was observed to produce a linear response within 10–35 mg biomass (Figure 3a). Thus, 10 mg starting material was used in the current work.

Additional concerns for photosynthetic organisms when selecting an appropriate method for lipid quantification include interference from chlorophyll, which is also extracted by the solvents. Previous work [61] suggested that DMSO will remove chlorophyll prior to lipid extraction. DMSO was tested on cyanobacterial samples in the current work by vortexing the cell pellet in 10 mL DMSO, subsequently washing with water (re-suspending, centrifuging, decanting) until the supernatant was colorless, and then following the chloroform–methanol extraction procedure. However, DMSO treatment appeared to remove all lipid signal, resulting in no mass recovered (data not shown); thus, it was recognized that results of this method for cyanobacterial samples will encompass chlorophyll and photosynthetic pigments as well as lipid. Autoclaving samples was also tested as an alternative method of cell disruption for all three species but was not found to significantly improve lipid recovery (Figure A3, Appendix A).

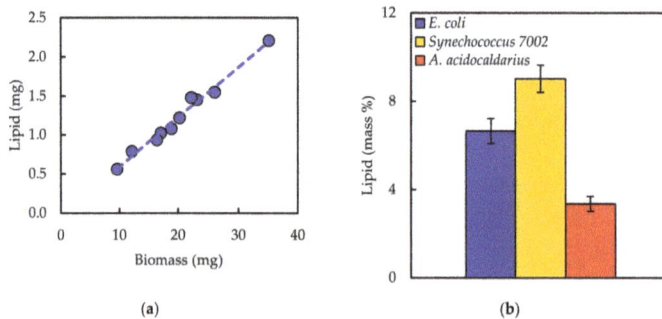

(a) (b)

Figure 3. (a) Lipid recovery is linear for biomass samples within 10–35 mg dry weight, measured with *E. coli*. (b) Lipid mass percentages of dry biomass measured for *E. coli*, *Synechococcus* 7002, and *A. acidocaldarius*. Error bars represent standard deviations from 5–9 biological replicates.

Lipid percentages of dry biomass for all three species are shown in Figure 3b. The lipid percentage obtained for *E. coli* (6.7%) is lower than the value of 9.1% reported by Neidhardt et al. [32], which may be due to differences in methods or strains, while the measured percentages for *Synechococcus* 7002 (9.0%) are also comparable with previously measured lipid and chlorophyll values by Vu et al. [46], i.e., 8.8%, 5.6%, and 3.8% for carbon-, light-, and nitrogen-limited conditions, respectively, who also used the Bligh and Dyer method. The percentage obtained for *A. acidocaldarius* (3.4%) is similar to a previously published report of 3.6% [62], which used a 2:1 chloroform/methanol extraction method.

Differences between the current measurements and previously reported values may reflect differences in culturing conditions or the influence of specific procedural details.

Polypropylene centrifuge tubes were used for safety during centrifugation rather than glass tubes, but it was noted that polypropylene is not completely chemically resistant to chloroform and may cause leaching of compounds from the polymer into chloroform. This error was accounted for by performing a blank reaction (0.6 mL water). The mass of the blank was then subtracted from the mass of the biological sample to obtain the mass of lipid. Additionally, removal of the lower chloroform phase can be difficult to perform reproducibly. Glass Pasteur pipettes were used initially; however, micropipettes with 200-μL tips provided more control over phase removal and yielded the most reproducible results.

7. Protein

7.1. Literature Review

Protein is typically the largest fraction of bacterial biomass. Many methods have been reported for determining protein quantity, including UV absorption spectroscopy and dye-based assays such as Bradford, Lowry, BCA, and others, for which Noble et al. [63] and Noble and Bailey [64] provided thorough discussions. UV absorption depends on tyrosine and tryptophan residues and the molar extinction coefficient of the protein under examination, and it requires a highly purified sample. Dye-based assays are influenced by different amino acid distributions and are subject to different interfering compounds as well as variability between proteins. Bovine serum albumin is a commonly used protein standard, but its amino acid sequence may not be representative of the total cellular protein. Amino acid analysis, or hydrolysis of cellular protein followed by derivatization and identification of individual amino acids via HPLC, is an alternative to these methods and is often considered the gold standard for protein analysis [63,64].

Amino acid analysis was selected for the current study due to improved accuracy and less bias as opposed to UV absorbance or dye-based methods. Some amino acids, such as cysteine and tryptophan, degrade during hydrolysis; special hydrolysis conditions may be used to retain them [65], or their proportions may be estimated based on genome codon distribution. Amino acid analysis provides experimental amino acid distribution in addition to total protein quantification, which serve as important parameters for metabolic modeling.

7.2. Procedure (After Henderson et al., 2000)

7.2.1. Reagents

- Cell pellet (1–3 mg dry biomass, fresh or frozen, washed with carbon-free media).
- 6 M HCl and 6 M NaOH.
- 0.1 M HCl, 0.22 μm filtered.
- Borate buffer: 0.4 N borate, titrate to pH 10.2 with NaOH, 0.22 μm filtered.
- OPA and FMOC derivatizing reagents (Agilent 5061–3335 and 5061–3337, Santa Clara, CA, USA).
- Amino acid standard: 1 nmol/mL (Agilent 5061–3330).
- Solvent A: 40 mM sodium phosphate buffer Na_2HPO_4, titrate to pH 7.8 with 10 M NaOH, 0.22 μm filtered. (A ratio of different sodium salts can be used to prepare a 40 mM phosphate solution with an initial pH closer to 7.8, e.g., 1:1 molar ratio of NaH_2PO_4 and Na_2HPO_4.)
- Solvent B: 45:45:10 acetonitrile:methanol:water ($v/v/v$), 0.22 μm filtered (nylon filter recommended for organic solvents).

7.2.2. Assay

(1) Transfer cell pellet to borosilicate HPLC vial with PTFE/silicone cap.
(2) Add 50 μL 6 M HCl per mg biomass.
(3) Tightly cap the vial and hydrolyze at 105 °C for 24 h (VWR analog heat block).

(4)   Neutralize to pH 7.0 with 6 M NaOH.

(5)   Filter with 0.22 μm PVDF centrifugal filters (Millipore UFC30GV00, Burlington, MA, USA); centrifuge for 5 min at 10,000 rpm.

(6)   Transfer filtrate to clean borosilicate HPLC vial, seal with parafilm, and poke a hole in the top (e.g., with a small pipette tip). Place at −80 °C to freeze before lyophilizing.

(7)   Lyophilize for 24 h (VirTis benchtop lyophilizer).

(8)   Store at −80 °C until HPLC analysis.

(9)   Re-suspend lyophilized material in 0.1 M HCl, and perform HPLC analysis according to the protocol validated by Agilent Technologies [66].

Notes: PVDF was selected as the filter membrane material due to its low protein binding capacity; materials with high protein binding such as nylon may affect amino acids. A fluorescence detector or diode array detector can be used for detection. The current work employed an Agilent 1100 HPLC system equipped with autosampler and fluorescence detector (for *A. acidocaldarius* samples) or diode array detector (for *E. coli* and *Synechococcus* 7002 samples). A diode array detector was less sensitive than a fluorescence detector (limit of detection ~100 μM vs. ~2 μM). o-phthalaldehyde (OPA) reagent with 3-mercaptopropionic acid as a stabilizing agent was used for detection of primary amino acids, and 9-fluorenylmethyl chloroformate (FMOC) reagent was used for secondary amino acids. OPA and FMOC reagents were replaced daily in amber vials and were used within 10 days upon opening an ampule. The flow rate was modified from 2 mL/min [66] to 1 mL/min to permit increased resolution of peaks (see gradient settings in Table 2). The injector program followed the steps described in Henderson et al. [66] but did not make use of the optional acetonitrile needle rinse. The integration parameters for collecting the data were set to a slope sensitivity of 1, peak width of 0.04, area reject of 1, and height reject of 0.4, with shoulders off. Amino acids were identified manually in the calibration table, and undesired peaks were discarded (derivatization byproduct peaks at the end of an injection).

Table 2. Amino acid analysis HPLC gradient settings. Data from Beck et al. [54].

Time (min)	% Solvent B
0	0
3.8	0
36.2	57
37.2	100
44.6	100
46.4	0
47	0

7.3. Test Results

Upon preparing for HPLC analysis, the lyophilized material was re-suspended in 100 μL 0.1 M HCl per mg biomass hydrolyzed. Different dilutions of the re-suspension were measured to ensure adequate detection of both more abundant and less abundant amino acids. Peak identity was confirmed for each amino acid by testing individual solutions of each amino acid. A representative chromatogram is shown in Figure 4. An internal standard, α-aminobutyric acid, was used in samples and standards alike for peak area normalization across injections. Standard curves were constructed, resulting in linear regressions with fits of 0.99 or greater. The experimental amino acid distribution and total protein quantification for the three bacterial species are shown in Table 3. Since cysteine and tryptophan were degraded during hydrolysis and methionine was present in low quantities with high variability (likely oxidized during hydrolysis), the distribution of these three amino acids was calculated according to the percentage found in the protein-coding genes of the genome. Reasonable correlations were observed between the experimentally measured and genome-predicted distributions (Figure A4, Appendix A). Interestingly, leucine content was observed to be consistently over-predicted in the genome and under-measured in the laboratory among the three species tested, although no explanation has been linked to this observation in the literature.

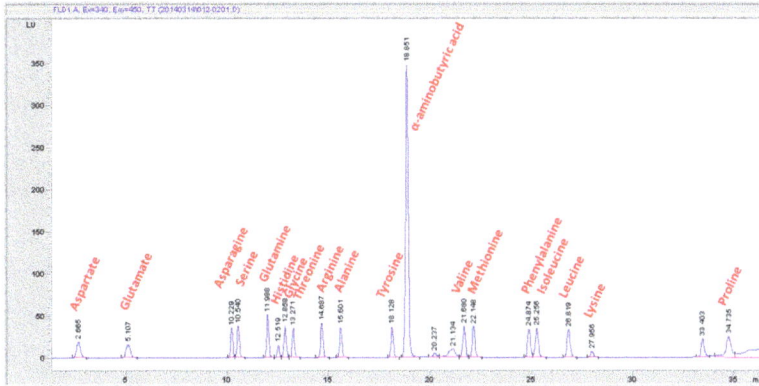

Figure 4. Representative chromatogram from fluorescence detector for amino acid standards. The background chromatogram shows the full length of the injection, including the derivatization byproducts eluting at the end. The outset chromatogram labels the individual peaks.

Table 3. Amino acid distributions and protein quantification for *E. coli*, *Synechococcus* 7002, and *A. acidocaldarius*. Amino acid mass and mole percentages of total protein are reported as averages of three biological replicates; percent relative standard deviations are in parentheses. HPLC quantification was obtained with a diode array detector for *E. coli* and *Synechococcus* 7002 and a fluorescence detector for *A. acidocaldarius*. Cysteine, methionine, and tryptophan were factored in according to genome-based codon proportions. Asparagine and glutamine were converted to aspartate and glutamate, respectively, during derivatization with OPA. The final row includes the total protein mass percent of cell dry weight, averaged from three biological replicates with standard deviation in parentheses.

Amino Acid	*E. coli*		*Synechococcus* 7002		*A. acidocaldarius*	
	Mass %	Mole %	Mass %	Mole %	Mass %	Mole %
Alanine	7.1 (3.5)	10.1 (3.2)	8.5 (1.4)	12.0 (1.4)	8.4 (1.0)	11.8 (0.9)
Arginine	6.9 (4.9)	5.0 (4.6)	8.3 (0.5)	6.0 (0.6)	7.7 (1.9)	5.6 (1.9)
Asparagine/Aspartate	9.9 (1.3)	9.5 (1.0)	10.2 (3.3)	9.7 (3.3)	10.0 (1.0)	9.5 (1.0)
Cysteine	1.1 (NA)	1.2 (NA)	0.9 (NA)	1.0 (NA)	0.9 (NA)	0.9 (NA)
Glutamine/Glutamate	14.0 (4.9)	12.1 (4.5)	14.6 (2.7)	12.5 (2.6)	13.2 (0.7)	11.3 (0.8)
Glycine	6.4 (4.0)	10.9 (3.6)	6.1 (3.0)	10.2 (2.9)	5.7 (0.3)	9.6 (0.3)
Histidine	2.0 (4.9)	1.6 (5.2)	1.8 (9.7)	1.4 (9.8)	2.1 (1.8)	1.7 (1.8)
Isoleucine	4.9 (6.4)	4.8 (6.7)	4.6 (2.8)	4.4 (3.0)	4.1 (0.4)	3.9 (0.5)
Leucine	7.4 (3.3)	7.2 (2.9)	9.0 (2.0)	8.6 (2.0)	7.6 (0.2)	7.3 (0.2)
Lysine	6.7 (3.4)	5.8 (3.1)	5.3 (0.2)	4.6 (0.1)	4.4 (1.1)	3.8 (1.1)
Methionine	3.2 (NA)	2.8 (NA)	2.2 (NA)	1.9 (NA)	3.0 (NA)	2.6 (NA)
Phenylalanine	5.1 (13.3)	3.9 (13.6)	5.1 (0.9)	3.9 (0.7)	3.6 (0.9)	2.8 (1.0)
Proline	3.6 (8.2)	4.0 (8.6)	4.1 (5.2)	4.5 (5.3)	4.3 (0.8)	4.8 (0.8)
Serine	4.0 (11.3)	4.9 (11.7)	4.1 (1.7)	4.9 (1.6)	4.7 (0.6)	5.6 (0.6)
Threonine	4.9 (2.2)	5.3 (2.0)	4.9 (0.9)	5.2 (0.7)	5.8 (0.4)	6.2 (0.5)
Tryptophan	2.4 (NA)	1.5 (NA)	2.4 (NA)	1.5 (NA)	2.6 (NA)	1.6 (NA)
Tyrosine	3.9 (19.7)	2.8 (20.0)	2.5 (10.4)	1.8 (10.6)	4.4 (1.0)	3.0 (1.1)
Valine	6.3 (5.9)	6.8 (5.5)	5.4 (0.9)	5.8 (1.0)	7.5 (0.5)	8.1 (0.5)
Total Mass % **Protein**	35.2 (1.3)		27.2 (1.5)		38.5 (2.2)	

8. RNA

8.1. Literature Review

RNA is a major macromolecule class which contributes to ribosome assembly and cellular information processing. Methods used for quantifying RNA include UV absorbance, orcinol colorimetric reaction, and thiazole orange fluorescent dye [50]. UV absorbance is precise but requires a pure sample and is not feasible for a mixture of DNA and RNA. Orcinol is not as precise for RNA or for a mixture of DNA and RNA, and carbohydrates may also interfere. Thiazole orange has good precision, but fluorescence is biased toward GC content, and it is less sensitive for RNA than for DNA [50]. Additionally, kits are available for RNA extraction but focus predominantly on downstream applications such as PCR and RNAseq, and thus remain questionable as quantitative methods. Fluorescent dyes such as RiboGreen and PicoGreen have also been reported for quantifying extracted RNA [52] but are usually used in combination with kit extractions.

Major concerns in selecting an RNA quantification method include sample purity, accuracy, and bias of nucleotide content. Many studies have used the colorimetric orcinol reaction to quantify RNA after hot perchloric acid extraction. However, Benthin et al. [67] developed an alternative method with the bacterium *Lactobacillus* that utilizes alkali (KOH) lysis in combination with cold perchloric acid extraction, followed by UV absorbance. The method provided similar accuracy to the orcinol reaction but showed improved precision [67]. Benthin's KOH-UV method was selected for the current study as a more reliable and safe method, using cold rather than hot perchloric acid, and to eliminate interference from carbohydrates, which occurs in the orcinol reaction. The results can be quantified with UV absorbance using average nucleotide molar extinction coefficients, which eliminates the need to prepare a standard from a different source. This method has been used in metabolic modeling studies to quantify RNA percentage [68–70].

8.2. Procedure (After Benthin et al., 1991)

8.2.1. Reagents

- Cell pellet (2–8 mg dry biomass, fresh or frozen, washed with carbon-free media).
- $HClO_4$ solutions: 0.5 M, 0.7 M, and 3 M.
- 0.3 M KOH solution.

8.2.2. Assay

(1) Wash cell pellet three times with 3 mL 0.7 M $HClO_4$ to degrade cell walls. Vortex to re-suspend in between washing. Centrifuge 4000 rpm for 10 min at 4 °C and decant between washes.
(2) Re-suspend pellet in 3 mL 0.3 M KOH to lyse cells.
(3) Incubate in a 37 °C water bath for 1 h, shaking at 15-min intervals.
(4) Cool and add 1 mL 3 M $HClO_4$.
(5) Centrifuge and decant supernatant into a new 50-mL polypropylene centrifuge tube.
(6) Wash pellet twice with 4 mL 0.5 M $HClO_4$ (re-suspend and mix), centrifuge, and decant supernatant into the 50-mL tube. The 0.5 M $HClO_4$ extracts the RNA, while DNA, which is stable even in strong alkali, and protein, which does not solubilize in the alkali, remain in the precipitate.
(7) Add 3 mL 0.5 M $HClO_4$ to the collection of extracts to obtain a total volume of 15 mL, and centrifuge once more to remove any non-visible precipitates of $KClO_4$.
(8) Measure absorbance at 260 nm against a 0.5 M $HClO_4$ blank.
(9) Calculate RNA quantity by assuming 1 unit of absorbance at 260 nm corresponds to 38 μg/mL RNA on average [71].

Notes: Quartz cuvettes are commonly used for measuring UV absorbance; however, disposable UV cuvettes can also be used (VWR 47727-024, rated to 220 nm and tested for chemical compatibility with concentrated hydrochloric acid). Linearity of the spectrophotometer within the range of sample absorbance should be confirmed by successively diluting the sample with 0.5 M HClO$_4$ and confirming a linear fit to the resulting absorbance measurements. Absorbance at 280 nm can also be measured, and the A$_{260}$/A$_{280}$ can be calculated to assess RNA purity.

8.3. Test Results

Benthin et al. [67] recommended using a quantity of biomass corresponding to ~0.4 mg of RNA. ~2 mg and ~8 mg biomass was used for *E. coli* and *Synechococcus* 7002, respectively, based on previous estimates of RNA content [32,46]. Correlation between biomass and RNA content was tested for *Synechococcus* 7002, and a linear response was observed within 2–8 mg biomass (Figure 5a). RNA mass percentages for all three species are shown in Figure 5b. The percentage obtained for *E. coli* is similar to the 20.5% value reported in Neidhardt et al. [32]. The percentage of dry biomass obtained for *Synechococcus* 7002 is higher than the 4.0% average value measured in Vu et al. [46] via the orcinol method but could reflect different growth states. No literature comparison was available for *A. acidocaldarius*.

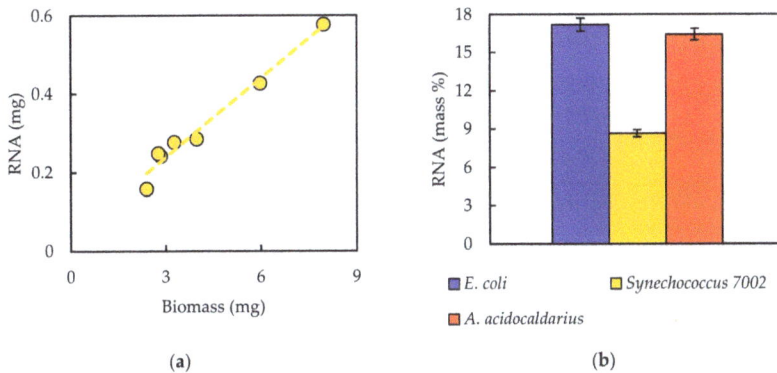

Figure 5. (**a**) RNA recovery is linear for biomass samples within 2–8 mg dry weight, measured for *Synechococcus* 7002; and (**b**) RNA mass percentages on a dry biomass basis measured for *E. coli*, *Synechococcus* 7002, and *A. acidocaldarius*. Error bars represent standard deviation from three biological replicates.

9. Model Biomass Reaction

Experimentally measured biomass composition provides a species-relevant basis for representing cellular growth in computational models. The results of the macromolecular assays for *E. coli*, *Synechococcus* 7002, and *A. acidocaldarius* are summarized in Table 4. The mass percentages for the five assays do not necessarily sum to 100% of cell dry weight. The reduced mass recovery may be due to loss of biomass during centrifugation and transfer of material while performing the assays. Some bacteria may also possess other storage compounds that are not accounted for in these analyses, such as polyhydroxyalkanoates or polyphosphates. Ash weight typically accounts for 5–10% of cell dry weight [72], or perhaps even more for some organisms (e.g., 20–30% ash content has been measured in phytoplankton [73]). To adjust for losses during sample processing, measurements can be normalized to the total mass recovered such that the sum of biomass recovered from all measurements is 100% (Table 4).

Table 4. Summary of average macromolecular composition, based on mass fraction of dry biomass, for *E. coli*, *Synechococcus* 7002, and *A. acidocaldarius*; standard deviations are included in parentheses. Column *N* contains the scaled biomass composition, normalized to the sum of the measurements.

Macromolecule	*E. coli*	N	*Synechococcus* 7002	N	*A. acidocaldarius*	N
Carbohydrate	4.2 (0.2)	6.5	16.9 (0.1)	27.2	6.2 (0.6)	9.5
DNA	1.0 (0.1)	1.6	0.4 (0.0)	0.6	0.7 (0.1)	1.1
Lipid	6.7 (0.6)	10.4	9.0 (0.6)	14.5	3.4 (0.3)	5.2
Protein	35.2 (1.3)	54.7	27.2 (1.5)	43.7	38.5 (2.2)	59.0
RNA	17.2 (0.5)	26.7	8.7 (0.6)	14.0	16.4 (0.5)	25.2
Total	**64.3**	**99.9**	**62.2**	**100**	**65.2**	**100**

An in silico cellular growth reaction is a collection of macromolecular synthesis reactions scaled to account for biomass composition. The macromolecular synthesis reactions are constructed by accounting for the appropriate ratios of the monomers, polymerization energy requirements, and reaction byproducts. Macromolecular monomer distributions are either measured directly, such as the amino acid composition measured here, or can be estimated from appropriate omics datasets or the literature. DNA composition is typically estimated from GC content, and RNA composition may be estimated from rRNA-encoding genes; rRNA accounts for approximately 81% of cellular RNA [32]. Polymer lengths for the macromolecular synthesis reactions can be scaled to a convenient number of monomers, such as 10 or 100, with the appropriate polymerization energy requirements and byproducts. The polymerization energy error introduced with these scaled molecules is assumed minor.

Once formulas for individual macromolecules are calculated, model reactions can be quality control checked for balance of elemental formulas and degree of reduction to ensure adherence to the mass balance constraint required for stoichiometric modeling. Identification of imbalanced reactions can then be further investigated; often the issue can be traced to balancing of redox pairs or hydrolysis products, free protons, and water. Table 5 demonstrates the construction of a DNA macromolecule synthesis reaction for *A. acidocaldarius*, including the definition of monomer composition, polymerization energy requirements, and byproducts. The elemental and electron balances are included and validate conservation relationships [32]. The Supplementary Materials contain a workbook for the major biomass macromolecules that can be modified for different biomass measurements (File S4).

Table 5. Example calculation of DNA macromolecular formula for *A. acidocaldarius* with 61.9% GC content, assuming a polymer length of 1 monomer. Polymerization byproducts (diphosphate) are subtracted from the sum of dNTP monomer constituents to obtain the formula for a DNA macromolecule. Overall DNA synthesis reaction is shown in the last row.

Monomer	Stoichiometry/Formula	C	H	O	N	P
dATP	$0.19/C_{10}H_{12}N_5O_{12}P_3$	1.91	2.29	2.29	0.95	0.57
dCTP	$0.31/C_9H_{12}N_3O_{13}P_3$	2.79	3.71	4.02	0.93	0.93
dGTP	$0.31/C_{10}H_{12}N_5O_{13}P_3$	3.10	3.71	4.02	1.55	0.93
dTTP	$0.19/C_{10}H_{13}N_2O_{14}P_3$	1.91	2.48	2.67	0.38	0.57
Diphosphate	$1/HO_7P_2$	0.00	1.00	7.00	0.00	2.00
DNA molecule	$1/C_{9.69}H_{11.19}N_{3.81}O_6P_1$	9.69	11.19	6.00	3.81	1.00
0.19 dATP + 0.31 dCTP + 0.31 dGTP + 0.19 dTTP = 1 DNA + 1 diphosphate						

The overall cell growth reaction has a form analogous to A carbohydrate + B DNA + C lipid + D protein + E RNA = 1 biomass, where A, B, C, D, and E are stoichiometric coefficients corresponding to the measured mass fraction. Some biomass reactions may also include additional constituents, such as chlorophyll, salts, and metabolite pools, including vitamins. The coefficients for the macromolecular constituents A–E are obtained by converting the experimental mass fraction measurements to molar coefficients, thereby yielding the appropriate stoichiometries. The following steps convert experimentally measured mass fractions of macromolecules to molar coefficients for use in the biomass reaction:

(1) Record mass fractions as g macromolecule per g cell dry weight (see Table 4).
(2) Tabulate the molar mass of each macromolecule representation. Multiply the macromolecular formula by the atomic mass of the respective elements, and sum over all elements to obtain g/mol macromolecule.
(3) Divide the mass fraction of the macromolecule by its molar mass to obtain mol macromolecule/g cell dry weight. The basis for cell dry weight normalization can be selected as desired; 1, 10, or 100 kg cell dry weight typically results in reasonably scaled coefficients for elementary flux mode and flux balance analyses. One kilogram cell dry weight often provides a convenient basis, as when inputs are scaled to a mM basis in FBA, the resulting output biomass scales to grams.
(4) Incorporate the molar coefficients into the biomass reaction. The stoichiometries can be multiplied by the macromolecular formulas and summed over all the macromolecules to obtain an overall formula for biomass, which allows model output to be analyzed in terms of carbon moles of biomass (Table 6).

The Supplementary Materials detail the macromolecule and biomass calculations for each species, as well as demonstrate a quality control check for balancing mass, charge, electrons, and elemental composition (File S4).

Table 6. Species-specific biomass reactions for *E. coli*, *Synechococcus* 7002, and *A. acidocaldarius*, without consideration of maintenance energy. Molar coefficients represent 100 kg dry biomass.

Species	Biomass Reaction
E. coli	5.05 DNA + 8.40 RNA + 5.02 Protein + 13.9 Lipid + 4.03 Glycogen = 1 Biomass
Synechococcus 7002	2.09 DNA + 4.40 RNA + 4.05 Protein + 18.5 Lipid + 16.8 Glycogen = 1 Biomass
A. acidocaldarius	3.49 DNA + 7.91 RNA + 5.46 Protein + 6.67 Lipid + 5.86 Glycogen = 1 Biomass

In addition to the macromolecular constituents that comprise a cell, metabolic models often account for maintenance energy requirements. Maintenance energy is an implicit energy consumption term accounting for a myriad of cellular processes, such as protein turnover and osmotic pressure maintenance. Maintenance energy is typically estimated by fitting the in silico model to experimental biomass-on-substrate yield data. For example, experiments correlating substrate consumption rate (for heterotrophs) or photon absorption rate (for photoautotrophs) with growth rate can be used to determine the yield [74,75]. For elementary flux mode analysis applications, a single maintenance energy term, set for a defined growth rate, can be added to the biomass reaction. For flux balance analysis applications, maintenance energy requirements can be broken down into growth and non-growth associated maintenance (GAM and NGAM) terms. The Supplementary Materials contain a genome-enabled model constructed for *A. acidocaldarius* (File S1). Calculations fitting maintenance energy to observed yield data for both glucose and oxygen consumption from Farrand et al. [37] for both EFMA and FBA application are provided in MATLAB and Excel formats (Files S1, S2, and S3). The specific growth rate-dependent (μ, h^{-1}) maintenance energy requirement (q_{ATP}) for *A. acidocaldarius* was calculated to be $q_{ATP} = 13.4\mu + 4.2$ mmol cellular energy per g biomass per hour, where GAM was 13.4 mmol cellular energy (phosphodiester bonds) per g biomass and NGAM was 4.2 mmol cellular energy per g biomass per hour. Using multiple datasets to fit the maintenance energy provides a metric of accuracy for the model, as they should provide similar results. The calculated maintenance terms for *A. acidocaldarius* were similar regardless of fitting with glucose or oxygen consumption data (Files S1, S2, and S3).

Finally, the *A. acidocaldarius* model was used to quantify potential pitfalls associated with inaccurate biomass compositions. Ten different biologically relevant variations of biomass composition were generated and tested in addition to the experimentally measured composition (see File S5). The optimal in silico biomass yield on electron donor (glucose) and associated biomass yield on electron acceptor (oxygen) was determined for each biomass composition. A sampling of the data is presented as a function of the biomass degree of reduction in Figure 6 and Table 7. The data point at degree of reduction 4.03 represents the experimentally measured composition for *A. acidocaldarius*; this point is used as a reference. The in silico biomass per glucose and biomass per oxygen yields change nonlinearly relative to degree of reduction. The biomass per oxygen yields change up to 70% from the reference composition, demonstrating the strong influence biomass composition can have on simulation results (Figure 6, Table 7). Common modeling practices for determining maintenance energy parameters fit model output to experimental yield data, which can mask the effects of inaccurate biomass composition. GAM values for each biomass composition were also calculated (Figure 6, Table 7). The GAM values changed up to 40% over the reference case. This represents a substantial 40% change in specific energy generation-associated fluxes, such as ATPase. Furthermore, the biomass yield on nitrogen was calculated for each biomass composition. The biomass per nitrogen yields varied up to 35% for the considered biomass compositions (see File S5). This variation in nitrogen content would have substantial impact on predictions for nitrogen-limited culturing conditions, such as those commonly used in bioprocesses to induce accumulation of bioplastics or lipids (e.g., [76,77]). This analysis highlights the importance of accurate species- and condition-specific measurements for biomass composition.

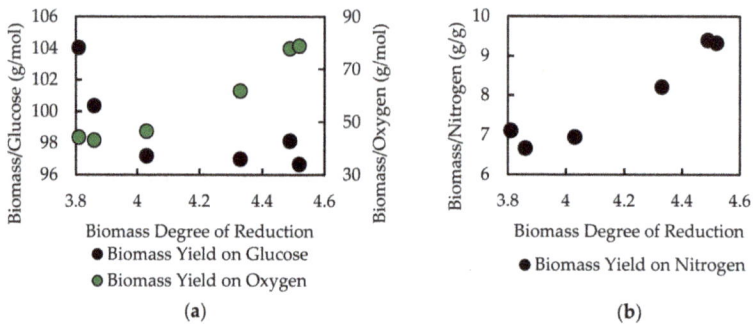

(a)

(b)

Figure 6. Biomass composition impacts stoichiometric model simulation results. (**a**) Mass percentages of the five macromolecules were varied (see Table 7), resulting in a range of biomass degrees of reduction. EFMA simulations of the *A. acidocaldarius* metabolic model with the different biomass compositions revealed substantial variation in biomass per electron donor yield (g biomass per mol glucose) and almost a doubling of oxygen required for biomass synthesis (g biomass per mol oxygen) as a function of biomass degree of reduction. Similar differences were also observed in growth associated maintenance (GAM, mmol cellular energy per g biomass) to fit the data for glucose consumption from Farrand et al. [37]. Fitted GAM values changed by as much as 40% between the experimentally determined biomass composition (degree of reduction 4.03) and the modulated biomass compositions (parenthetical percentages). (**b**) Biomass yield on nitrogen (calculated from the elemental composition) varied up to 35%, demonstrating the sensitivity of modeling results to biomass composition when considering nitrogen-limited conditions. Calculations, additional data, and further details are included in the Supplementary Materials (File S5).

Table 7. Sampling of biomass composition variations used to test the effect on model simulations in Figure 6. The experimentally determined biomass composition had degree of reduction 4.03. Calculations, additional data, and further details are included in File S5.

DNA	RNA	Protein	Lipid	Poly-Saccharide	Degree of Reduction	Elemental Composition	GAM (mmol/g)
0.05	0.40	0.40	0.02	0.13	3.81	$CH_{1.51}O_{0.44}N_{0.26}P_{0.02}S_{0.005}$	18.9 (140%)
0.05	0.35	0.50	0.02	0.08	3.86	$CH_{1.43}O_{0.53}N_{0.27}P_{0.04}S_{0.004}$	16.1 (119%)
0.01	0.25	0.59	0.05	0.10	4.03	$CH_{1.44}O_{0.49}N_{0.28}P_{0.03}S_{0.004}$	13.5 (100%)
0.05	0.15	0.50	0.20	0.10	4.33	$CH_{1.58}O_{0.40}N_{0.21}P_{0.02}S_{0.004}$	13.3 (99%)
0.05	0.15	0.40	0.30	0.10	4.49	$CH_{1.62}O_{0.39}N_{0.18}P_{0.02}S_{0.003}$	14.3 (106%)
0.05	0.10	0.45	0.30	0.10	4.52	$CH_{1.63}O_{0.37}N_{0.17}P_{0.02}S_{0.003}$	13.1 (97%)

10. Conclusions

Computational biology representations of metabolism often include cellular growth reactions necessitating knowledge of biomass composition for accurate predictions. The current work surveyed analytical methods for the five major macromolecules (carbohydrate, DNA, lipid, protein, and RNA), provided step-by-step procedures for a select method for each macromolecule, tested the methods on three different bacterial species, and demonstrated application of analytical measurements to a computational representation of cellular growth. The data include a quantitative analysis of potential pitfalls associated with inaccurate biomass representations. The literature survey included references to more in-depth reviews for each macromolecule for further exploration and also provided a rationale for the selected method. Table 8 provides a summary of the selected methods and their advantages and disadvantages. The three bacterial species used for testing (*E. coli*, *Synechococcus* 7002, and *A. acidocaldarius*) represent a range of physiological characteristics, including Gram-negative and Gram-positive, mesophilic and thermophilic, and neutrophilic and acidophilic, as well as chemoheterotrophic and photoautotrophic, which assessed the robustness of the methods. Testing of methods highlighted potential pitfalls and provided guidelines for troubleshooting when testing a new method or when applying a method to new organisms. Based on the current study, recommendations for verifying a new protocol or testing a new organism include ensuring that the test response is linear for both the amount of biomass used and the amount of reagent, testing the standard range, and confirming the effect of any sample pre-treatment steps on standards. It is also important to consider the organism being studied and the downstream application of the measurement (e.g., glycogen vs. total carbohydrate).

Table 8. Summary of selected methods with advantages and disadvantages for each class of macromolecule.

Macromolecule	Selected Method	Advantages/Disadvantages
Carbohydrate	Sodium sulfate co-precipitation, anthrone detection	Differentiate glycogen from total cellular carbohydrate/Colorimetric
DNA	Alkaline lysis, Hoechst 33258 fluorescence	Can use cell lysate/AT bias, DNA standard
Lipid	Chloroform–methanol extraction, gravimetric	No standard needed/Not specific for types of fatty acids
Protein	Hydrochloric acid hydrolysis, OPA and FMOC derivatization	Provides amino acid distribution/More involved than colorimetric assay
RNA	Alkaline lysis, perchloric acid extraction, UV absorbance	No standard needed/Use of perchloric acid

The presented methods of experimental measurement and conversion to computational biology reactions need to be integrated with the maturing quality standards for model construction [78,79]. The predicted elemental composition of the synthesized biomass is a relevant metric for the quality of the overall reaction. Average elemental compositions have been measured for several common microorganisms, providing a convenient check [80]. The elemental composition is linked to the biomass degree of reduction, which is an energetic measure of biomass and a critical parameter for computational biology analysis of consortia simulations. The degree of reduction of biomass for an average cell is approximately 4.2 or 4.8 on an NH_4^+ or N_2 basis, respectively [80]. These values can shift due to large quantities of cellular storage polymers, such as polysaccharides or polyhydroxyalkanoates. Additionally, biomass composition is known to shift with growth rate and culturing stress [45,81]; the provided approach can be used to create culturing condition-specific cellular growth reactions. Altogether, the current work serves as a useful resource for the broader computational biology community, which will enable more accurate representations of biomass synthesis and therefore more accurate metabolism simulations.

Supplementary Materials: The following are available online at http://www.mdpi.com/2227-9717/6/5/38/s1, File S1: *A. acidocaldarius* Model, File S2: *A. acidocaldarius* SBML, File S3: *A. acidocaldarius* Maintenance, File S4: Biomass Composition, File S5: Biomass Composition Variation.

Acknowledgments: This work is a contribution of the PNNL Foundational Scientific Focus Area (Principles of Microbial Community Design) subcontracted to Montana State University. Ashley E. Beck was supported by the Office of the Provost at Montana State University through the Molecular Biosciences Program and NSF (DMS-1361240). The authors would also like to thank James Folsom and Zackary Jay for helpful discussions.

Author Contributions: A.E.B. and R.P.C. conceived and designed the experiments; A.E.B. performed the experiments; A.E.B. and K.A.H. analyzed the data; and A.E.B., K.A.H., and R.P.C. wrote the paper.

Conflicts of Interest: The authors declare no conflict of interest. The funding sponsors had no role in the design of the study; in the collection, analyses, or interpretation of data; in the writing of the manuscript; or in the decision to publish the results.

Appendix A

Additional data supporting method testing are provided herein.

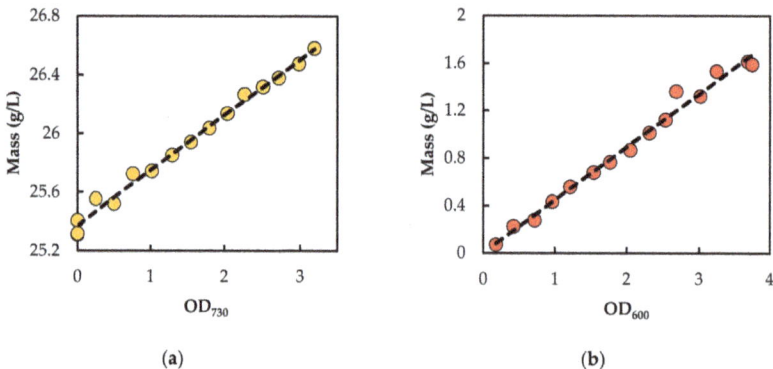

(a)   (b)

Figure A1. (a) Biomass–OD_{730} correlation curve determined for Synechococcus 7002, described by the equation y = 0.377x + 25.373 with R^2 value of 0.991. The average A+ media control (OD_{730} of 0) was 25.347 g/L and was subtracted from the biological samples to obtain biomass values. (b) Biomass–OD_{600} correlation curve determined for A. acidocaldarius, described by the equation y = 0.445x + 0.004 with R^2 value of 0.987.

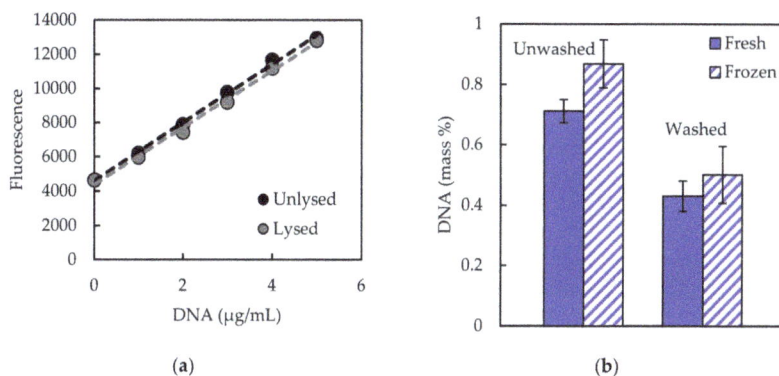

(a)

(b)

Figure A2. (a) Lysis of DNA standard did not significantly influence Hoechst fluorescent response. Error bars represent three technical replicates of standard solutions. Neither slopes nor intercepts of the two curves are significantly different ($p > 0.05$, T-tests). (b) Washing samples with cell wash solution resulted in significantly lower DNA recovery from *E. coli* samples ($p < 0.05$, T-test). Freezing samples did not have a significant impact on DNA recovery ($p > 0.05$, T-test).

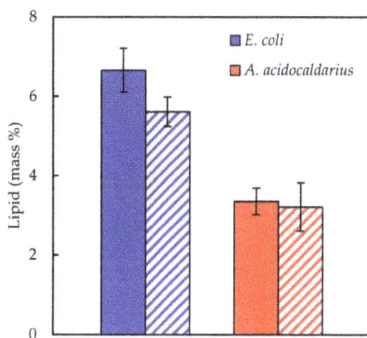

Figure A3. Autoclaving samples prior to chloroform–methanol lipid extraction did not significantly enhance lipid recovery. Solid columns represent unautoclaved samples and striped columns represent autoclaved samples. Lipid recovery was not significantly affected by autoclaving for *A. acidocaldarius* ($p > 0.05$, T-test), whereas lipid recovery was slightly lower for autoclaved *E. coli* samples ($0.05 < p < 0.01$, T-test). $n = 9$ and 3 for unautoclaved and autoclaved *E. coli* samples, respectively, and $n = 6$ and 3 for unautoclaved and autoclaved *A. acidocaldarius*, respectively.

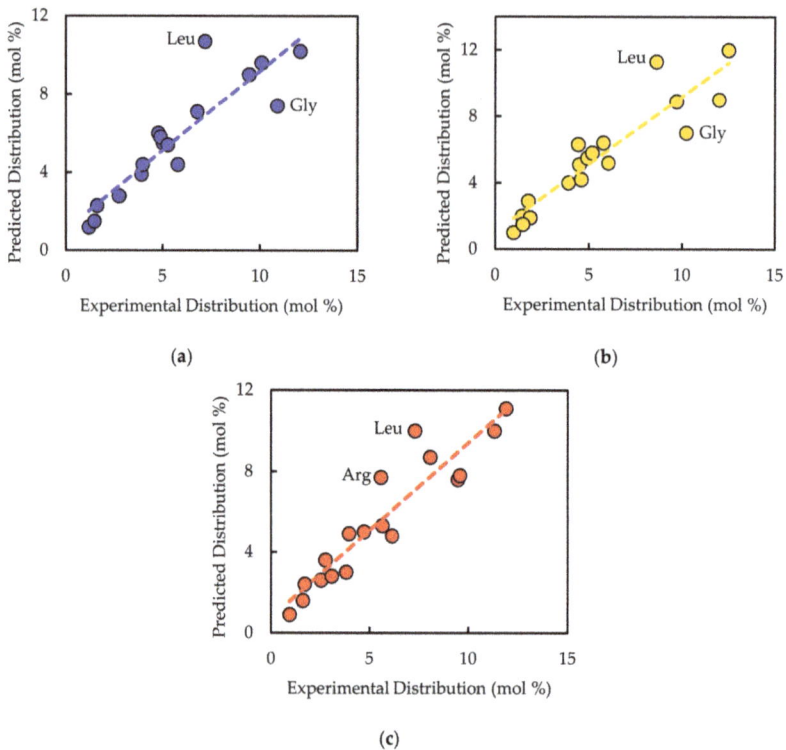

Figure A4. Correlations between experimentally measured and genome-based amino acid distributions for: (a) *E. coli* (NCBI reference sequence NC_000913.3; trendline y = 0.81x + 1.04 with R^2 value of 0.82); (b) *Synechococcus* 7002 (NCBI reference sequence NC_010475.1; trendline y = 0.81x + 1.07 with R^2 value of 0.85); and (c) *A. acidocaldarius* (NCBI reference sequence NC_013205.1; trendline y = 0.87x + 0.73 with R^2 value of 0.87). The two most outlying amino acids are labeled.

References

1. Feist, A.M.; Palsson, B.O. The biomass objective function. *Curr. Opin. Microbiol.* **2010**, *13*, 344–349. [CrossRef] [PubMed]
2. Hunt, K.A.; Jennings, R.D.; Inskeep, W.P.; Carlson, R.P. Stoichiometric modelling of assimilatory and dissimilatory biomass utilisation in a microbial community. *Environ. Microbiol.* **2016**, *18*, 4946–4960. [CrossRef] [PubMed]
3. Pramanik, J.; Keasling, J.D. Stoichiometric model of *Escherichia coli* metabolism: Incorporation of growth-rate dependent biomass composition and mechanistic energy requirements. *Biotechnol. Bioeng.* **1997**, *56*, 398–421. [CrossRef]
4. White, D.; Drummond, J.; Fuqua, C. *The Physiology and Biochemistry of Prokaryotes*, 4th ed.; Oxford University Press: New York, NY, USA, 2012.
5. Vrede, T.; Dobberfuhl, D.R.; Kooijman, S.; Elser, J.J. Fundamental connections among organism C: N: P stoichiometry, macromolecular composition, and growth. *Ecology* **2004**, *85*, 1217–1229. [CrossRef]
6. Zuniga, C.; Levering, J.; Antoniewicz, M.R.; Guarnieri, M.T.; Betenbaugh, M.J.; Zengler, K. Predicting dynamic metabolic demands in the photosynthetic eukaryote *Chlorella vulgaris*. *Plant Physiol.* **2017**. [CrossRef]
7. Maarleveld, T.R.; Khandelwal, R.A.; Olivier, B.G.; Teusink, B.; Bruggeman, F.J. Basic concepts and principles of stoichiometric modeling of metabolic networks. *Biotechnol. J.* **2013**, *8*, 997–1008. [CrossRef] [PubMed]

8. Zomorrodi, A.R.; Islam, M.M.; Maranas, C.D. d-OptCom: Dynamic multi-level and multi-objective metabolic modeling of microbial communities. *ACS Synth. Biol.* **2014**, *3*, 247–257. [CrossRef] [PubMed]

9. Zomorrodi, A.R.; Maranas, C.D. OptCom: A multi-level optimization framework for the metabolic modeling and analysis of microbial communities. *PLoS Comput. Biol.* **2012**, *8*, e1002363. [CrossRef] [PubMed]

10. Orth, J.D.; Thiele, I.; Palsson, B.O. What is flux balance analysis? *Nat. Biotechnol.* **2010**, *28*, 245–248. [CrossRef] [PubMed]

11. Cankorur-Cetinkaya, A.; Dikicioglu, D.; Oliver, S.G. Metabolic modeling to identify engineering targets for *Komagataella phaffii*: The effect of biomass composition on gene target identification. *Biotechnol. Bioeng.* **2017**, *114*, 2605–2615. [CrossRef] [PubMed]

12. Thiele, I.; Palsson, B.Ø. A protocol for generating a high-quality genome-scale metabolic reconstruction. *Nat. Protoc.* **2010**, *5*, 93–121. [CrossRef] [PubMed]

13. Henry, C.S.; DeJongh, M.; Best, A.A.; Frybarger, P.M.; Linsay, B.; Stevens, R.L. High-throughput generation, optimization and analysis of genome-scale metabolic models. *Nat. Biotechnol.* **2010**, *28*, 977–982. [CrossRef] [PubMed]

14. Oberhardt, M.A.; Puchałka, J.; Fryer, K.E.; Dos Santos, V.A.M.; Papin, J.A. Genome-scale metabolic network analysis of the opportunistic pathogen *Pseudomonas aeruginosa* PAO1. *J. Bacteriol.* **2008**, *190*, 2790–2803. [CrossRef] [PubMed]

15. Stolyar, S.; Van Dien, S.; Hillesland, K.L.; Pinel, N.; Lie, T.J.; Leigh, J.A.; Stahl, D.A. Metabolic modeling of a mutualistic microbial community. *Mol. Syst. Biol.* **2007**, *3*, 92. [CrossRef] [PubMed]

16. Lohman, E.J.; Gardner, R.D.; Halverson, L.; Macur, R.E.; Peyton, B.M.; Gerlach, R. An efficient and scalable extraction and quantification method for algal derived biofuel. *J. Microbiol. Methods* **2013**, *94*, 235–244. [CrossRef] [PubMed]

17. Bremer, H.; Dennis, P.P. Modulation of chemical composition and other parameters of the cell by growth rate. *Escherichia coli Salmonella Cell. Mol. Biol.* **1996**, *2*, 1553–1569.

18. Herbert, D.; Phipps, P.J.; Strange, R.E. Chemical analysis of microbial cells. In *Methods in Microbiology*; Norris, J.R., Ribbons, D.W., Eds.; Academic Press: New York, NY, USA, 1971; Vol. 5, pp. 209–344.

19. Long, C.P.; Antoniewicz, M.R. Quantifying biomass composition by gas chromatography/mass spectrometry. *Anal. Chem.* **2014**, *86*, 9423–9427. [CrossRef] [PubMed]

20. Ausubel, F.M.; Brent, R.; Kingston, R.E.; Moore, D.D.; Seidman, J.G.; Smith, J.A.; Struhl, K. *Short Protocols in Molecular Biology*, 2nd ed.; Greene Publishing Associates and John Wiley & Sons: New York, NY, USA, 1992; p. 834.

21. Bernstein, H.C.; Paulson, S.D.; Carlson, R.P. Synthetic *Escherichia coli* consortia engineered for syntrophy demonstrate enhanced biomass productivity. *J. Biotechnol.* **2012**, *157*, 159–166. [CrossRef] [PubMed]

22. Ludwig, M.; Bryant, D.A. Transcription profiling of the model cyanobacterium *Synechococcus* sp. strain PCC 7002 by next-gen (SOLiD (tm)) sequencing of cDNA. *Front. Microbiol.* **2011**, *2*, 41. [CrossRef] [PubMed]

23. Stevens, S.E.; Patterson, C.O.; Myers, J. Production of hydrogen peroxide by blue-green algae—Survey. *J. Phycol.* **1973**, *9*, 427–430. [CrossRef]

24. Farrand, S.G.; Linton, J.D.; Stephenson, R.J.; McCarthy, W.V. The use of response surface analysis to study the growth of *Bacillus acidocaldarius* throughout the growth range of temperature and pH. *Arch. Microbiol.* **1983**, *135*, 272–275. [CrossRef]

25. Folsom, J.P.; Parker, A.E.; Carlson, R.P. Physiological and proteomic analysis of *Escherichia coli* iron-limited chemostat growth. *J. Bacteriol.* **2014**, *196*, 2748–2761. [CrossRef] [PubMed]

26. Klamt, S.; Saez-Rodriguez, J.; Gilles, E.D. Structural and functional analysis of cellular networks with CellNetAnalyzer. *BMC Syst. Biol.* **2007**, *1*, 2. [CrossRef] [PubMed]

27. Klamt, S.; von Kamp, A. An application programming interface for CellNetAnalyzer. *Biosystems* **2011**, *105*, 162–168. [CrossRef] [PubMed]

28. Mavromatis, K.; Sikorski, J.; Lapidus, A.; Del Rio, T.G.; Copeland, A.; Tice, H.; Cheng, J.-F.; Lucas, S.; Chen, F.; Nolan, M. Complete genome sequence of *Alicyclobacillus acidocaldarius* type strain (104-IA). *Stand. Genom. Sci.* **2010**, *2*, 9. [CrossRef] [PubMed]

29. Caspi, R.; Billington, R.; Ferrer, L.; Foerster, H.; Fulcher, C.A.; Keseler, I.M.; Kothari, A.; Krummenacker, M.; Latendresse, M.; Mueller, L.A.; et al. The MetaCyc database of metabolic pathways and enzymes and the BioCyc collection of pathway/genome databases. *Nucleic Acids Res.* **2016**, *44*, D471–D480. [CrossRef] [PubMed]

30. Kanehisa, M.; Goto, S.; Sato, Y.; Furumichi, M.; Tanabe, M. KEGG for integration and interpretation of large-scale molecular data sets. *Nucleic Acids Res.* **2012**, *40*, D109–D114. [CrossRef] [PubMed]

31. Schomburg, I.; Chang, A.; Placzek, S.; Sohngen, C.; Rother, M.; Lang, M.; Munaretto, C.; Ulas, S.; Stelzer, M.; Grote, A.; et al. BRENDA in 2013: Integrated reactions, kinetic data, enzyme function data, improved disease classification: New options and contents in BRENDA. *Nucleic Acids Res.* **2013**, *41*, D764–D772. [CrossRef] [PubMed]

32. Neidhardt, F.C.; Ingraham, J.L.; Schaechter, M. *Physiology of the Bacterial Cell: A Molecular Approach*; Sinauer Associates: Sunderland, MA, USA, 1990.

33. De Rosa, M.; Gambacorta, A.; Minale, L.; Bu'Lock, J. Cyclohexane fatty acids from a thermophilic bacterium. *J. Chem. Soc. D Chem. Commun.* **1971**, *21*, 1334a. [CrossRef]

34. Goto, K.; Tanaka, T.; Yamamoto, R.; Suzuki, T.; Tokuda, H. Characteristics of *Alicyclobacillus*. In *Alicyclobacillus*; Springer: Berlin, Germany, 2007; pp. 9–48.

35. Trinh, C.T.; Wlaschin, A.; Srienc, F. Elementary mode analysis: A useful metabolic pathway analysis tool for characterizing cellular metabolism. *Appl. Microbiol. Biotechnol.* **2009**, *81*, 813–826. [CrossRef] [PubMed]

36. Terzer, M.; Stelling, J. Large-scale computation of elementary flux modes with bit pattern trees. *Bioinformatics* **2008**, *24*, 2229–2235. [CrossRef] [PubMed]

37. Farrand, S.G.; Jones, C.W.; Linton, J.D.; Stephenson, R.J. The effect of temperature and pH on the growth efficiency of the thermoacidophilic bacterium *Bacillus acidocaldarius* in continuous culture. *Arch. Microbiol.* **1983**, *135*, 276–283. [CrossRef]

38. Lloyd, L. *HPLC Determination of Carbohydrates in Food and Drink*; Agilent Technologies: Santa Clara, CA, USA, 2011; Vol. SI-01407.

39. ASTM E1758-01. Standard method for the determination of carbohydrates by HPLC. In *2003 Annual Book of ASTM Standards*; American Society for Testing and Materials, International: Philadelphia, PA, USA, 2003.

40. Chaplin, M.F. *Carbohydrate Analysis: A Practical Approach*; IRL Press: Washington, DC, USA, 1986; p. 228.

41. Dubois, M.; Gilles, K.A.; Hamilton, J.K.; Rebers, P.A.; Smith, F. Colorimetric method for determination of sugars and related substances. *Anal. Chem.* **1956**, *28*, 350–356. [CrossRef]

42. Taylor, K. A modification of the phenol sulfuric acid assay for total carbohydrates giving more comparable absorbances. *Appl. Biochem. Biotechnol.* **1995**, *53*, 207–214. [CrossRef]

43. Trevelyan, W.E.; Harrison, J.S. Studies on yeast metabolism. 1. Fractionation and microdetermination of cell carbohydrates. *Biochem. J.* **1952**, *50*, 298–303. [CrossRef] [PubMed]

44. Beck, C.; Knoop, H.; Axmann, I.M.; Steuer, R. The diversity of cyanobacterial metabolism: Genome analysis of multiple phototrophic microorganisms. *BMC Genom.* **2012**, *13*, 56. [CrossRef] [PubMed]

45. Folsom, J.P.; Carlson, R.P. Physiological, biomass elemental composition and proteomic analyses of *Escherichia coli* ammonium-limited chemostat growth, and comparison with iron- and glucose-limited chemostat growth. *Microbiology* **2015**, *161*, 1659–1670. [CrossRef] [PubMed]

46. Vu, T.T.; Hill, E.A.; Kucek, L.A.; Konopka, A.E.; Beliaev, A.S.; Reed, J.L. Computational evaluation of *Synechococcus* sp. PCC 7002 metabolism for chemical production. *Biotechnol. J.* **2013**, *8*, 619–630. [CrossRef] [PubMed]

47. Van Handel, E. Rapid determination of glycogen and sugars in mosquitos. *J. Am. Mosq. Control Assoc.* **1985**, *1*, 299–301. [PubMed]

48. Del Don, C.; Hanselmann, K.W.; Peduzzi, R.; Bachofen, R. Biomass composition and methods for the determination of metabolic reserve polymers in phototrophic sulfur bacteria. *Aquat. Sci.* **1994**, *56*, 1–15. [CrossRef]

49. Sattler, L.; Zerban, F.W. The Dreywood anthrone reaction as affected by carbohydrate structure. *Science* **1948**, *108*, 207. [CrossRef] [PubMed]

50. De Mey, M.; Lequeux, G.; Maertens, J.; De Maeseneire, S.; Soetaert, W.; Vandamme, E. Comparison of DNA and RNA quantification methods suitable for parameter estimation in metabolic modeling of microorganisms. *Anal. Biochem.* **2006**, *353*, 198–203. [CrossRef] [PubMed]

51. Burton, K. Study of the conditions and mechanism of the diphenylamine reaction for the colorimetric estimation of deoxyribonucleic acid. *Biochem. J.* **1956**, *62*, 315–323. [CrossRef] [PubMed]

52. Gorokhova, E.; Kyle, M. Analysis of nucleic acids in *Daphnia*: Development of methods and ontogenetic variations in RNA-DNA content. *J. Plankton Res.* **2002**, *24*, 511–522. [CrossRef]

53. Downs, T.R.; Wilfinger, W.W. Fluorometric quantification of DNA in cells and tissue. *Anal. Biochem.* **1983**, *131*, 538–547. [CrossRef]

54. Beck, A.E.; Bernstein, H.C.; Carlson, R.P. Stoichiometric network analysis of cyanobacterial acclimation to photosynthesis-associated stresses identifies heterotrophic niches. *Processes* **2017**, *5*, 32. [CrossRef]

55. Sigma-Aldrich. *DNA Quantitation Kit, Fluorescence Assay: Technical Bulletin*; Sigma-Aldrich: St. Louis, MO, USA, 2009.

56. Hoiczyk, E.; Hansel, A. Cyanobacterial cell walls: News from an unusual prokaryotic envelope. *J. Bacteriol.* **2000**, *182*, 1191–1199. [CrossRef] [PubMed]

57. Izard, J.; Limberger, R.J. Rapid screening method for quantitation of bacterial cell lipids from whole cells. *J. Microbiol. Methods* **2003**, *55*, 411–418. [CrossRef]

58. Bligh, E.G.; Dyer, W.J. A rapid method of total lipid extraction and purification. *Can. J. Biochem. Physiol.* **1959**, *37*, 911–917. [CrossRef] [PubMed]

59. Sheng, J.; Vannela, R.; Rittrnann, B.E. Evaluation of methods to extract and quantify lipids from *Synechocystis* PCC 6803. *Bioresour. Technol.* **2011**, *102*, 1697–1703. [CrossRef] [PubMed]

60. Sheng, J.; Vannela, R.; Rittmann, B.E. Disruption of *Synechocystis* PCC 6803 for lipid extraction. *Water Sci. Technol.* **2012**, *65*, 567–573. [CrossRef] [PubMed]

61. Archanaa, S.; Moise, S.; Suraishkumar, G.K. Chlorophyll interference in microalgal lipid quantification through the Bligh and Dyer method. *Biomass Bioenergy* **2012**, *46*, 805–808. [CrossRef]

62. Vier, B.; Rogge, G.; Voigt, B. Production of lipids from a thermoacidophilic *Bacillus* strain. 1. Lipids from *Bacillus acidocaldarius* ZIMET 11274. *Acta Biotechnol.* **1992**, *12*, 37–40. [CrossRef]

63. Noble, J.E.; Knight, A.E.; Reason, A.J.; Di Matola, A.; Bailey, M.J.A. A comparison of protein quantitation assays for biopharmaceutical applications. *Mol. Biotechnol.* **2007**, *37*, 99–111. [CrossRef] [PubMed]

64. Noble, J.E.; Bailey, M.J.A. Quantitation of protein. In *Guide to Protein Purification*, 2nd ed.; Burgess, R.R., Deutscher, M.P., Eds.; Elsevier Science & Technology Books: New York, NY, USA, 2009; Vol. 463, pp. 73–95.

65. Fountoulakis, M.; Lahm, H.W. Hydrolysis and amino acid composition analysis of proteins. *J. Chromatogr. A* **1998**, *826*, 109–134. [CrossRef]

66. Henderson, J.W.; Ricker, R.D.; Bidlingmeyer, B.A.; Woodward, C. *Rapid, Accurate, Sensitive, and Reproducible HPLC Analysis of Amino Acids*; Agilent Technologies: Santa Clara, CA, USA, 2000.

67. Benthin, S.; Nielsen, J.; Villadsen, J. A simple and reliable method for the determination of cellular RNA content. *Biotechnol. Tech.* **1991**, *5*, 39–42. [CrossRef]

68. Imam, S.; Yilmaz, S.; Sohmen, U.; Gorzalski, A.S.; Reed, J.L.; Noguera, D.R.; Donohue, T.J. iRsp1095: A genome-scale reconstruction of the *Rhodobacter sphaeroides* metabolic network. *BMC Syst. Biol.* **2011**, *5*, 116. [CrossRef] [PubMed]

69. Liao, Y.C.; Huang, T.W.; Chen, F.C.; Charusanti, P.; Hong, J.S.J.; Chang, H.Y.; Tsai, S.F.; Palsson, B.O.; Hsiung, C.A. An experimentally validated genome-scale metabolic reconstruction of *Klebsiella pneumoniae* MGH 78578, iYL1228. *J. Bacteriol.* **2011**, *193*, 1710–1717. [CrossRef] [PubMed]

70. Liu, Z.J.; Gao, Y.; Chen, J.; Imanaka, T.; Bao, J.; Hua, Q. Analysis of metabolic fluxes for better understanding of mechanisms related to lipid accumulation in oleaginous yeast *Trichosporon cutaneum*. *Bioresour. Technol.* **2013**, *130*, 144–151. [CrossRef] [PubMed]

71. Carnicer, M.; Baumann, K.; Toplitz, I.; Sanchez-Ferrando, F.; Mattanovich, D.; Ferrer, P.; Albiol, J. Macromolecular and elemental composition analysis and extracellular metabolite balances of *Pichia pastoris* growing at different oxygen levels. *Microb. Cell Fact.* **2009**, *8*, 65. [CrossRef] [PubMed]

72. Van Veen, J.A.; Paul, E.A. Conversion of bio-volume measurements of soil organisms, grown under various moisture tensions, to biomass and their nutrient content. *Appl. Environ. Microbiol.* **1979**, *37*, 686–692. [PubMed]

73. Whyte, J.N.C. Biochemical composition and energy content of 6 species of phytoplankton used in mariculture of bivalves. *Aquaculture* **1987**, *60*, 231–241. [CrossRef]

74. Carlson, R.; Srienc, F. Fundamental *Escherichia coli* biochemical pathways for biomass and energy production: Creation of overall flux states. *Biotechnol. Bioeng.* **2004**, *86*, 149–162. [CrossRef] [PubMed]

75. Vu, T.T.; Stolyar, S.M.; Pinchuk, G.E.; Hill, E.A.; Kucek, L.A.; Brown, R.N.; Lipton, M.S.; Osterman, A.; Fredrickson, J.K.; Konopka, A.E.; et al. Genome-scale modeling of light-driven reductant partitioning and carbon fluxes in diazotrophic unicellular cyanobacterium *Cyanothece* sp. ATCC 51142. *PLoS Comput. Biol.* **2012**, *8*, e1002460. [CrossRef] [PubMed]

76. Fields, M.W.; Hise, A.; Lohman, E.J.; Bell, T.; Gardner, R.D.; Corredor, L.; Moll, K.; Peyton, B.M.; Characklis, G.W.; Gerlach, R. Sources and resources: Importance of nutrients, resource allocation, and ecology in microalgal cultivation for lipid accumulation. *Appl. Microbiol. Biotechnol.* **2014**, *98*, 4805–4816. [CrossRef] [PubMed]

77. Mus, F.; Toussaint, J.-P.; Cooksey, K.E.; Fields, M.W.; Gerlach, R.; Peyton, B.M.; Carlson, R.P. Physiological and molecular analysis of carbon source supplementation and pH stress-induced lipid accumulation in the marine diatom *Phaeodactylum tricornutum. Appl. Microbiol. Biotechnol.* **2013**, *97*, 3625–3642. [CrossRef] [PubMed]

78. Le Novere, N.; Finney, A.; Hucka, M.; Bhalla, U.S.; Campagne, F.; Collado-Vides, J.; Crampin, E.J.; Halstead, M.; Klipp, E.; Mendes, P.; et al. Minimum information requested in the annotation of biochemical models (MIRIAM). *Nat. Biotechnol.* **2005**, *23*, 1509–1515. [CrossRef] [PubMed]

79. Ravikrishnan, A.; Raman, K. Critical assessment of genome-scale metabolic networks: The need for a unified standard. *Brief. Bioinform.* **2015**, *16*, 1057–1068. [CrossRef] [PubMed]

80. Roels, J.A. Application of macroscopic principles to microbial metabolism. *Biotechnol. Bioeng.* **1980**, *22*, 2457–2514. [CrossRef]

81. Carlson, R.; Srienc, F. Fundamental *Escherichia coli* biochemical pathways for biomass and energy production: Identification of reactions. *Biotechnol. Bioeng.* **2004**, *85*, 1–19. [CrossRef] [PubMed]

processes

MDPI

Review

The Spectrum of Mechanism-Oriented Models and Methods for Explanations of Biological Phenomena

C. Anthony Hunt [1,*], Ahmet Erdemir [2], William W. Lytton [3], Feilim Mac Gabhann [4], Edward A. Sander [5], Mark K. Transtrum [6] and Lealem Mulugeta [7]

[1]   Department of Bioengineering and Therapeutic Sciences, University of California, San Francisco, CA 94143, USA
[2]   Department of Biomedical Engineering and Computational Biomodeling Core, Lerner Research Institute, Cleveland Clinic, Cleveland, OH 44195, USA; erdemira@ccf.org
[3]   Departments of Neurology and Physiology and Pharmacology, SUNY Downstate Medical Center, Department Neurology, Kings County Hospital Center, Brooklyn, NY 11203, USA; billl@neurosim.downstate.edu
[4]   Institute for Computational Medicine and Department of Biomedical Engineering, Johns Hopkins University, Baltimore, MD 21218, USA; feilim@jhu.edu
[5]   Department of Biomedical Engineering, University of Iowa, Iowa City, IA 52242, USA; edward-sander@uiowa.edu
[6]   Department of Physics and Astronomy, Brigham Young University, Provo, UT 84602, USA; mktranstrum@byu.edu
[7]   InSilico Labs LLC, Houston, TX 77002, USA; lealem.mulugeta@gmail.com
*   Correspondence: a.hunt@ucsf.edu; Tel.: +1-415-476-2455

check for updates

Received: 15 April 2018; Accepted: 6 May 2018; Published: 14 May 2018

Abstract: Developing and improving mechanism-oriented computational models to better explain biological phenomena is a dynamic and expanding frontier. As the complexity of targeted phenomena has increased, so too has the diversity in methods and terminologies, often at the expense of clarity, which can make reproduction challenging, even problematic. To encourage improved semantic and methodological clarity, we describe the spectrum of Mechanism-oriented Models being used to develop explanations of biological phenomena. We cluster explanations of phenomena into three broad groups. We then expand them into seven workflow-related model types having distinguishable features. We name each type and illustrate with examples drawn from the literature. These model types may contribute to the foundation of an ontology of mechanism-based biomedical simulation research. We show that the different model types manifest and exert their scientific usefulness by enhancing and extending different forms and degrees of explanation. The process starts with knowledge about the phenomenon and continues with explanatory and mathematical descriptions. Those descriptions are transformed into software and used to perform experimental explorations by running and examining simulation output. The credibility of inferences is thus linked to having easy access to the scientific and technical provenance from each workflow stage.

Keywords: computational model; explanatory model; hybrid model; mechanism; mechanistic model; modeling methods; provenance; workflow; systems modeling; simulation

1. Introduction

Within the large context of biological systems modeling and analysis, developing and improving mechanism-oriented computational models to better explain complex biological phenomena are expanding. As the complexity of the phenomena to be explained has increased, the diversity in methods and terminologies has also increased, often at the expense of clarity, which can enhance the impression of inaccessibility and make reproduction challenging, even problematic. Those characteristics can limit the credibility and acceptance of evidence and insights being presented within computational biology reports. This overview illustrates specific ways in which methodological and semantic clarity regarding mechanisms, explanations of phenomena and methods can be refined to improve accessibility and strengthen methodological and scientific credibility.

In the context of mechanism-oriented models intended to better explain a biological phenomenon, lack of clarity often involves the use of the terms "mechanistic" and "mechanistic model." There is considerable diversity in what is being implied when discussing mechanisms and/or describing models as mechanistic. Mechanistic model is a convenient yet ambiguous phrase typically used as an abbreviation for more accurate, more informative descriptors. Use of the term "mechanism" is often similarly ambiguous. Clarity within research reports and credibility of claims made are generally viewed as being correlated and computational biology is not an exception. Usage of ambiguous phrases within research reports can limit the credibility and acceptance of the evidence and insights being presented. This overview is motivated by ongoing collaborative efforts to improve credibility (Supporting Material provides background) and the belief that improvements in semantic and methodological clarity will strengthen the credibility of results leveraging simulation research.

The phrase "mechanistic model" has a variety of meanings ascribed to it that differ across biological domains. There is an increasing tendency to utilize "mechanistic model" both specifically and as an umbrella term. Herein, we define, characterize and cluster seven mechanistic model types and suggest specific terms for each. To insure clarity, we narrow the scope of discussions that follow by first limiting attention to reports seeking mechanism-oriented explanations of biological phenomena. We further restrict focus to research for which a scientific objective is to (1) provide deeper, more explanatory insight into the generation of biological phenomena; and/or (2) better predict, mimic, or emulate one or more biological phenomena.

We clarify various uses of "mechanistic model" and how they are represented computationally for explaining biological phenomena. We describe the spectrum of Mechanism-oriented Models and methods being used to develop explanations of biological phenomena. We cluster explanations of phenomena into three broad groups and then expand them into a total of seven model and simulation types. We name each type and illustrate with diverse examples drawn from the literature. We begin by framing the context and offering definitions. In "Methodological Complexity," we contrast how infrastructure and management of complexity influence clarity differently between wet-lab and simulation research. In the section that follows, we describe three spectra that are useful in describing, characterizing and distinguishing explanations of phenomena. Next, in "Three Groups of Models of Explanation," we use similarities and differences (with reference to the spectra) to guide characterizations that distinguish semantically among seven workflow-centered models of explanation, including four different types of computational models of explanation. The names used to identify each characterization are not intended as semantic standards; rather they are offered as suggestions to encourage movement in that direction and serve as a working foundation for an ontology to use in explanatory simulation research in the life sciences. In "Relevant Information, Multiple Sources," we illustrate why providing sufficient methodological information is essential to enhance the credibility of an explanatory simulation, whereas brevity weakens credibility at the expense of clarity. We characterize five different sources and types of information from which relevant details are needed to clearly distinguish among the four types of computational models of explanation. In "Workflow, Provenance and Hybrid Models," we comment on connections between workflows, methods, and semantics and on new technical issues that further increase the need for semantic and methodological clarity.

2. Background

Framing the Context: Mechanisms as Explanations of Phenomena

A prerequisite for discussing mechanism-oriented biological models is adopting a definition for "mechanism." Over the past two decades, within the philosophy of science literature, mechanism has emerged as a framework for thinking about fundamental issues in biology [1,2].

Braillard and Malaterre recently defined a biological mechanism [2]: "*A mechanism can be thought of as being composed of parts that interact causally (usually through chemical and mechanical interactions) and that are organized in a specific way. This organization determines largely the behavior of the mechanism and hence the phenomena that it produces. ... Mechanisms can be formalized in different ways, including with the help of diagrams and schemas and are usually supplemented by causal narratives that describe how the mechanisms produce the very phenomena to be accounted for.*"

Authors often augment their diagrams, schemas and causal narratives with a computational "narrative" (algorithm and implementation) that enables explicit predictions. We use the definitions listed under Working Definitions (Box 1) and specify that a mechanism is a real thing; it is concrete. A description is required for the term "mechanistic model." Kaplan and Craver state [3]: "*[That] the line that demarcates [mechanistic] explanations from merely empirically adequate models seems to correspond to whether the model describes the relevant causal structures that produce, underlie, or maintain the explanandum phenomenon. This demarcation line is especially significant as it also corresponds to whether the model in question reveals (however dimly) knobs and levers that potentially afford control over whether and precisely how the phenomenon manifests.*"

Thus, we see that there is a difference between a model that reproduces a phenomenon and a model that does so using a mechanism that recapitulates the actual underlying mechanism.

Box 1. Working Definitions.

- **mechanism**: (1) a structure, system (e.g., biological, mechanical, chemical, electrical and so on), or process performing a function in virtue of its component parts, component operations and their organization (adapted from [4]), where the function is responsible for the phenomenon to be explained; (2) entities and activities organized in such a way that they are responsible for the phenomenon to be explained (adapted from [5,6])
- **phenomenon**: (1) an observable fact or event: an item of experience or reality; (2) a fact or event of scientific interest susceptible to scientific description and explanation [7]
- **mechanistic**: (1) determined by, for example, a mechanical, chemical, and/or electrical mechanism, or executing software; (2) like, for example, a mechanical, chemical, or electrical mechanism in one or more ways; (3) of or relating to using a mechanism as an approach to explaining a biological phenomenon; (4) mechanism-oriented

Craver posits that mechanistic models are explanatory, but he notes [8]: "*Some models sketch explanations but leave crucial details unspecified or hidden behind filler terms. Some models are used to conjecture a how-possibly explanation without regard to whether it is a how-actually explanation.*"

The increasing variety and sophistication of published mechanism-oriented and mechanism-based explanatory models reflect that biological mechanisms exhibit features that are not expressed in the above definition of a mechanism. Darden discusses how features of mechanisms often become necessary parts of adequate descriptions and representations of a mechanism [9]. She identifies five features of biological mechanism, listed in Table 1, that often characterize mechanisms that adequately explain biological phenomena. These features will be useful in broadly distinguishing among model types and may provide a basis for further developing an ontology to support mechanism-oriented simulation research. The phenomenon to be explained is the first feature because the search for a mechanism-based model of explanation requires that the phenomenon is identified. Also, in biology, it is often the case that phenomena at a finer biological scale constitute the explanatory mechanism of the phenomenon of interest observed at coarser biological scale. The underlying finer details are the entities and activities responsible for observable coarser behavior.

Table 1. Five features of a biological mechanism (adapted from [9]): a biological mechanism exhibits all five. A computational mechanism-based model may strive to do the same.

Mechanism Features	Examples	Explanations
Phenomenon		A clearly identified phenomenon is the requisite for specifying the other four features of mechanism and for developing a credible explanation of that phenomenon.
Components	entities, activities, modules, processes, underlying finer details	Working entities act in the mechanism. Activities are producers of change. Some entities and activities can be organized into a module. Inner layer phenomena can be the entities and activities responsible for the outer layer phenomenon.
Spatial arrangement of components	localization, structure orientation, connectivity, compartmentalization	Components are typically localized and organized into a structure. A component's orientation can be a prerequisite for an activity. Producing change requires connectivity. Compartmentalization facilitates spatial arrangement within a structure.
Temporal aspects of components	order, rate, duration, frequency	Entities may play their role is a particular order. Some activities have characteristic rates. Activities can occur in stages and/or exhibit temporal organization. An activity and/or stage can repeat or exhibit frequencies. Stages can unfold in a particular order and have duration.
Contextual locations	location within a hierarchy and/or within a series	A mechanism is situated in wider context, such as within a hierarchy of mechanism levels or within a temporal series of mechanisms not directly influencing the phenomenon of interest.

3. Methodological Complexity

Methodological complexity has been increasing in wet-lab research for decades. Striving for clarity in descriptions of experiments remains an ingrained best practice. Although it is possible to document every aspect of software used, such clarity is not yet the norm in the computational biology research domain. Clarity in reports of wet-lab methods is facilitated and enabled by a large, trusted commercial infrastructure. Research reports can achieve clarity in part by including statements like the following within Methods sections, e.g., from [10]: *"Dulbecco's phosphate buffered saline (PBS), liver perfusion medium, hepatocyte wash medium . . . were purchased from Life Technologies (Carlsbad, CA) . . . Wild-type C57BL/6J, male mice (9 weeks of age), purchased from The Jackson Laboratory (Bar Harbor, ME), were acclimated . . . The resulting supernatant was injected into the high-performance liquid chromatography column using a Model 582 solvent delivery system and a Model 5600A CoulArray detector (ESA, Chelmsford, MA) . . . Protein content was determined using the Nanodrop 2000 Spectrophotometer (Thermo Scientific, Waltham, MA)."*

For each item, additional details are available on the manufacturer or supplier's websites. Also, many portions of wet-lab protocols are replicated from previous publications in which each step was explicated, e.g., *"cell toxicity was measured as in* [hypothetical reference]." There are even entire journals devoted to the distribution of standardized and generalizable protocols, e.g., "Journal of Visualized Experiments" and "Nature Protocols." A product of such infrastructure is a rich, evolving, consensus-supported nomenclature that facilitates methodological and semantic clarity.

By contrast, in biology simulation research, particular computational methods are often borrowed and repurposed but are rarely implemented and executed identically. Proprietary and open source simulation tools and packages are available, but we do not yet have commercial infrastructure specifically intended to facilitate biology simulation research.

Growth and diversification of the commercial infrastructure supporting biology research have been fueled in part by the requirement that, when needed, experiments can be independently reproduced and extended in a different laboratory. That requirement also drives the need for semantic and workflow clarity in wet-lab methods. Interest in independently reproducing results of simulation experiments and in reusing and repurposing simulation components is expected to increase as the healthcare implications and benefits of simulation experiments increase. Improved clarity at all workflow stages will facilitate those developments, and in the sections that follow, we present specific ways to improve and strengthen methodological and semantic clarity regarding mechanism-oriented explanations of phenomena.

4. Mechanism-Oriented Models of Explanation

Based on our sampling of the research literature, all explanations of phenomena that draw on features of mechanisms can be broadly described as being mechanism-oriented models of explanation. They differ from other models of explanation in that they try to organize knowledge about both phenomenon and its explanation around mechanisms [2]. The explanations are models because, even when there is considerable knowledge about a phenomenon, there is still uncertainty about details of the actual causal process and those details always exhibit biological variability. They range from being mechanism-oriented to fully mechanism-based models of explanation, as illustrated by the spectrum in Figure 1a and can be grouped under one of three broad characterizations (Roman numerals **I–VII** refer to the names of model Types characterized below in Group A, B and C subsections). **I:** The details of the explanation are mechanism-oriented but fall short of the definition of mechanism under Working Definitions (Box 1). **II:** The explanation is mechanism-based in that it builds on a description of a mechanism that meets the definition of a mechanism under Working Definitions (Box 1). However, the mechanism is an analogy based most often on a corresponding real or hypothetical engineering, physical, mechanical, chemical, or electronic mechanism. **III:** The details of the mechanism-based explanation strive to be biomimetic: some entities and activities map directly to biological counterparts. In a subsequent section, we explain and elaborate these three characterizations, extend them to include four types of computational models of explanation (**IV–VII**) and present examples.

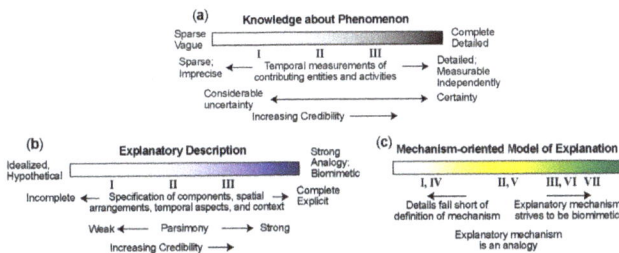

Figure 1. Three spectra for characterizing the explanation of a phenomenon. (**a**) This spectrum illustrates relative relationships among the three Model of explanation types (**I–III**) described in Figure 2. (**b**) Specifying an approximate location on this spectrum provides a clear, relativistic assessment about the strength of knowledge and information that is available to characterize the phenomenon. Independent of location, credibility is increased by making explicit information on (1) how the phenomenon has been measured, along with (2) details about temporal measurements of entities and activities thought to be contributing to its generation. Assessments of uncertainties further increase credibility. (**c**) Specifying an approximate location on this spectrum characterizes what is currently known or hypothesized about (1) how the phenomenon may be (or is) generated, (2) information about actual mechanism features listed in Table 1 and their orchestration, plus (3) simulation details illustrated that characterize the four types of computational models of explanation (**IV–VII**). Making that information explicit is essential for increasing credibility. There is often a correlation between characterization and locations on this spectrum and location on spectra **b** and **c**. For example, having locations on **b** and **c** that are right of center enables an Analogous-mechanism Model to be more biomimetic. Explanations that use mechanism analogies often have more centric locations on **b** and **c**.

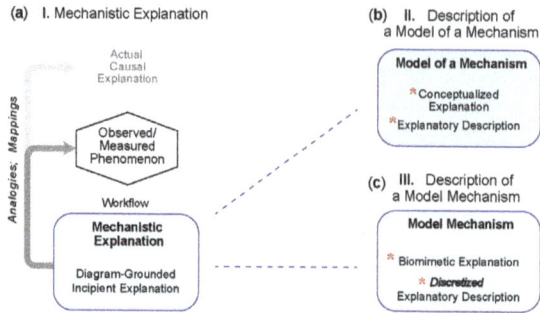

Figure 2. Three types of Mechanism-oriented Models of Explanation. There are three broad types of Mechanism-oriented Models of explanations of phenomena. They overlap to some extent. These illustrations highlight features that differentiate the three types. Credibility improves by making clear which type best characterizes a specific model of explanation. (**a**) A Mechanistic Explanation has three features and is located left of center in Figure 1a. All three differentiae are part of the model. The muted oval at top, which is repeated in Figures 3 and 4, reminds us that the actual causal explanation is yet to be discovered. The hexagon depicts the target phenomenon and represents the organized relevant information about the phenomenon that is being explained. Each phenomenon can be characterized by its relative (to other phenomena) location on the Figure 1b spectrum. The process (the workflow) of identifying and organizing information and features into a description of how the phenomenon might be generated is represented by the blue box. Part of the workflow involves establishing mappings and drawing analogies between features of the explanation and particular measurements; the darker gray arrow indicates that activity. The lighter gray arrow indicates a working hypothesis, in which those mappings and analogies will eventually extend to the actual causal explanation. (**b**) This model of explanation includes a detailed description and explanation along with the other elements in **a**. Information about possible generators is sufficient to conceptualize and describe an explanation that meets the definition of mechanism under Working Definitions (Box 1) by drawing on analogies to, for example, engineering, mechanical, chemical and electronic mechanisms. The result is an Analogous-Mechanism Model of explanation. The red asterisks designate characteristics that distinguish **II** from **I**. (**c**) Further right on both the Figure 1b,c spectra, knowledge about the phenomenon is sufficient to conceptualize a model of explanation that includes several of the Table 1 explanatory biomimetic features. The resulting detailed description and explanation is fundamentally different from **II**: it is a description of a Model Mechanism explanation. The red asterisks designate characteristics that distinguish **III** from **II**.

II and **III** have two requisites: there must be a clear mapping between the representation of entities and activities, and the target (referent) phenomenon; and the phenomenon must be specified clearly. Phenomena are grounded to the particular experiments or clinical trials in which they were observed and measured. In research, knowledge of a phenomenon can vary dramatically, yet there is a direct relationship between what is known about the phenomenon and the extent to which a mechanism-oriented model of explanation can become sufficiently accurate. The scope and depth of knowledge about a phenomenon can be characterized by an approximate location along the spectrum in Figure 1b. Phenomena that are the focus of more basic research tend to have central or left-of-center locations. A mechanism-oriented or mechanism-based explanation of how a phenomenon is thought to be—or might be—generated can be characterized by an approximate location along the spectrum in Figure 1c. Photosynthesis provides an example where the explanation of the phenomenon is located right of center on the Figure 1c spectrum. The depth of knowledge is such that explanatory mechanisms described in review articles and textbooks are broadly accepted as accurate, even though they fall far short of a complete account of what occurs in a particular plant under particular conditions. As such, it is accurate to describe such explanations as Model Mechanisms.

Autoprotection is described as resistance to toxicant re-exposure following acute, mild injury with the same toxicant, such as acetaminophen [11,12]. It is an example of a phenomenon that can be characterized as located on the far left of the spectrum 1b. Knowledge of the phenomenon is sparse and imprecise. Although there is considerable information about particular molecular details, only incomplete speculative explanations of the phenomenon are currently feasible, and it would be difficult to distinguish causes from effects. Such explanations would fall short of the definition of mechanism and so would be located considerably left of center on the spectrum 1c. As such, weak Mechanistic Explanation is an accurate descriptor and any possible mechanism-based account would be at best conjecture.

5. Three Groups of Models of Explanation

A huge variety of explanatory model types populates the Mechanism-oriented Models spectrum in Figure 1a. Having characterizations and descriptors that make it easier to distinguish among classes and types is essential to support clarity and credibility, aid in distinguishing among computational model types and provide a foundation for an ontology. We identify and describe seven broad types and cluster them into three groups. Group A includes the three characterizations illustrated in Figure 2. One of those characterizations is an essential core component of each of the four computational Mechanism-oriented Models illustrated in Figure 3 (elaborations of **I** and **II**) and Figure 4 (elaborations of **III**). As the descriptors and names for different models of explanation gain traction, attention can turn to discussions of finer grain model types, possibly drawing on features listed in Table 1.

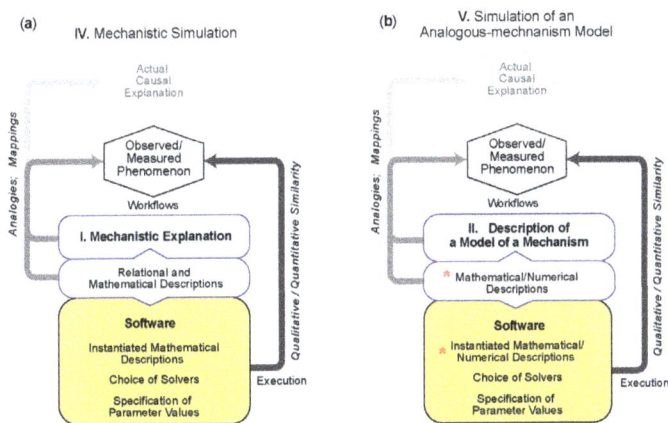

Figure 3. Characterizations of two types of simulation. Illustrated are work activities built upon explanations carried forward from **I** and **II**. Simulation operation is not illustrated. A requirement for both types of simulation is that output (specific computed solutions) match target phenomenon measurements within some tolerance. (**a**) Starting with a Mechanistic Explanation (**I**), the modeler completes two tasks. (1) Develop relational and continuum mathematical descriptions of the mechanistic explanation's salient information. (2) Faithfully instantiate in software all mathematical descriptions such that computed solutions simulate the output envisioned by those mathematical descriptions. The resulting system provides a Simulation of a Mechanistic Explanation. Before publication, the system has typically undergone several rounds of refinement and revision. (**b**) Starting with **II**, the modeler develops the mathematical descriptions needed to provide faithful characterizations of the analogous mechanism's salient features during operation. The requirements for software instantiation are the same as for **a**. The resulting system simulates output from **II** as if it were real. Red asterisks: characteristics that distinguish **b** from **a**.

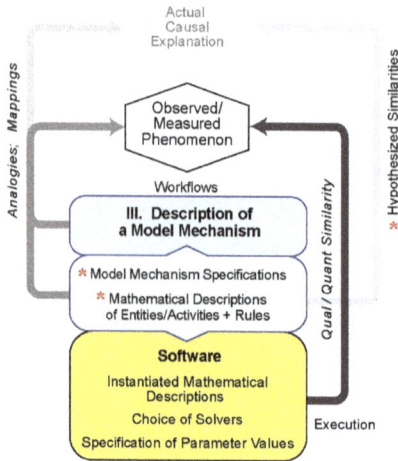

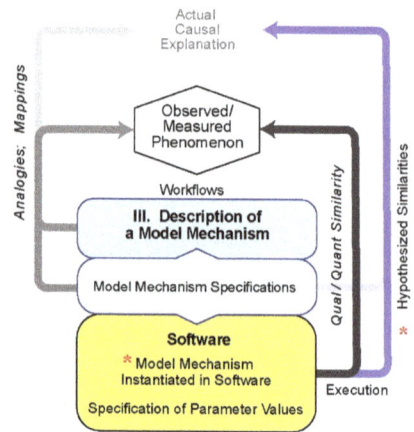

Figure 4. Model Mechanism: from simulation to instantiation. Snapshots of two different work activities built upon the detailed description of a Model Mechanism in **III** are illustrated. Simulation operation is not illustrated. A requirement for both is that output matches target phenomenon measurements within some tolerance. (**a**) Red asterisks identify characteristics that distinguish **VI** from **V**. Agent-based simulation methods are often utilized. To the extent feasible, envisioned entity activities are described using probabilistic and/or deterministic rules. Often, however, to simplify technical implementation challenges, behaviors of all or some Model Mechanism activities during execution are described using continuous mathematics, as in **V**, using physically grounded parameterizations; this prevents some or all of the software mechanisms during execution from meeting the definition of a mechanism. (**b**) The red asterisk identifies a characteristic that distinguishes **VII** from **VI**. Authors strive to use Model Mechanism specifications to instantiate an analog of the entire Model Mechanism in software. The product is a Computational Model Mechanism. To build credibility, authors demonstrate that a parameterized variant of **VII** has met the five requirements listed in the text. A distinguishing element is that features of the software mechanism during execution are observable, measurable and hypothesized to have biological counterparts (blue arrow).

5.1. Group A: Three Types of Mechanism-Oriented Models of Explanation

5.1.1. I—Mechanistic Explanation

Mechanistic explanations are pervasive in the life sciences research literature. In their simplicity, they are analogous to a cartoon; they are static and reflect observations. Knowledge about the phenomenon is characterized by a location considerably left of center in spectrum in Figure 1b and is insufficient to meet the definition of a mechanism under Working Definitions (Box 1). Nevertheless, there is often sufficient information to support an incipient coarse grain causal story that accounts reasonably well for the available evidence and explains how the phenomenon might have been generated. The blue box in Figure 2a represents the workflow required to identify and organize relevant information into a description of how the phenomenon might be generated. Such descriptions typically rely heavily on explanatory diagrams. They may also include mathematical descriptions, but they fall short of the definition of mechanism, which is clear in the three examples that follow. It is understood, but often not stated, that many somewhat different, yet equally possible explanatory models can be presented. An accurate descriptor is a Mechanism-oriented Model of Explanation. However, because we use that phrase as an umbrella expression, we prefer the abridged phrase, Mechanistic Explanation, which we use hereafter.

Example I.1: The well-known Hodgkin and Huxley model is a Mechanistic Explanation. It is an incomplete how-possibly story that provides preliminary insights into mechanisms responsible for generating and propagating action potentials along axons [13]. The authors make clear that their account is merely an explanatory model, not an actual explanation.

" . . . certain features of our equations were capable of a physical interpretation but the success of the equations is no evidence in favour of the mechanism of permeability change that we tentatively had in mind when formulating them."

Example I.2: Russmann et al. [14] offer a three-step Mechanistic Explanation of how hepatocyte death may be caused by drug-induced liver injury. 1) The initial injury results in direct cell stress possibly including mitochondrial impairment. 2) Death receptor-mediated pathways are triggered leading to mitochondrial permeability transition. 3) The result is apoptotic or necrotic cell death.

Example I.3: Bassler et al. [15] sought Mechanistic Explanations for unanticipated clinical side effects and efficacy limitations of integrin $\alpha IIb\beta 3$ antagonists. They posited a three-stage Mechanistic Explanation involving paradoxical platelet activation by $\alpha IIb\beta 3$ antagonists: a ligand-bound conformation change; receptor clustering; and pre-stimulation of platelets.

5.1.2. **II**—Analogous-Mechanism Model

It is common to encounter a mechanism-oriented explanation of a biological phenomenon that is framed as a mechanism analogy based on engineering principles, continuum mechanics, chemistry, electronics, computer science and so forth. When the analogical explanation meets the definition of mechanism, it can be accurately identified as an Analogous-Mechanism Model of Explanation (simply Analogous-Mechanism Model hereafter). It too is supported by diagrammatic depictions and often includes mathematical descriptions. Like **I**, it is still cartoonish. There are more cause and effect than in **I**. Although the mechanism's phenomenon is expected to be biomimetic, features of mechanism's components, their spatial arrangement, and/or temporal aspects are typically not biomimetic. The following are examples.

Example II.1: The three-element Hill muscle model for estimation of muscle force generation [16] is an idealized Analogous-Mechanism Model (Figure 2b). Such models do not have direct biological counterparts and any contextual location is hypothetical. However, measurements of the idealized mechanism during operation—if it were made real, concrete—are expected to adequately match measurements of the target phenomenon qualitatively and quantitatively.

Example II.2: Some therapeutic proteins such as trastuzumab, which is a monoclonal antibody, bind to pharmacological targets on cells. Efficacy is disrupted when the therapeutic protein binds instead to soluble targets shed from cells. Li et al. [17] describe a minimal physiologically based pharmacokinetic Analogous-mechanism Model intended to represent key features of a plausible mechanism hypothesized to be responsible for reduced efficacy. A computational description of their model in operation was used to simulate efficacy changes.

Example II.3: A demographic collapse of freshwater fish species, such as brown trout, can occur when rates of environmental change exceed the population's capacity to adapt. Ayllón et al. [18] describe a spatially explicit, multi-attribute, eco-genetic individual-based Analogous-mechanism Model that was used to study possible trout dynamics under three scenarios: (1) climate change-induced warming, (2) warming plus flow reduction resulting from climate and land use change and (3) a baseline of no environmental change.

A phenomenon that is explained using an Analogous-mechanism Model will be to the right of **I** in Figure 1b. As explanatory insight improves and the research workflow advances, one encounters research reports in which an earlier Mechanistic Explanation is replaced by an Analogous-mechanism Model. At that stage, authors typically assign names to some or all of the components of their model that are identical to real components and features of the referent biological system, that is, they draw directly from vocabularies of anatomical, biochemical, and physiological ontologies. While conceptually useful, such labeling may encourage conflating model explanation features with reality, which reduces both clarity and scientific credibility.

5.1.3. **III**—Model Mechanism

As explanatory knowledge about a phenomenon increases (moving further right on the Figure 1b spectrum), researchers begin conceptualizing and describing (hypothesizing about) a particular mechanism-based explanation of the phenomenon (Figure 2c) that is biomimetic; it is not an analogy of something else. Researchers strive to specify and characterize some or all of the explanatory features in Table 1. Model Mechanism is an accurate descriptor of a product of that process. Model Mechanisms are less cartoonish than **II** and more structured. An early stage model of explanation of this type would likely be assigned a central location on the Figure 1a spectrum. As the description matures, its location on all three Figure 1 spectra shifts rightward. Mappings exist between the Model Mechanism's discrete entities and activities and biological counterparts. The expectation is that, measurements of a phenomenon generated during simulation of a Model Mechanism would adequately match measurements of the actual target phenomenon qualitatively and quantitatively.

> Example **III**.1: An illustrative example is the two-dimensional model mechanism developed by Norton et al. [19] to facilitate achieving two related goals: (1) improve explanatory insight into the generation of the four distinguishable morphologies of ductal carcinoma in situ of the breast. (2) Disentangle the mechanisms involved in tumor progression. Additional examples are included with those provided below under Group C.

5.2. Group B: Using Simulation to Support and Enhance I and II

5.2.1. **IV**—Simulation of a Mechanistic Explanation

A frequent simulation research goal is to translate a Mechanistic Explanation (**I**) into simulation output that is (or is expected to be) qualitatively or quantitatively similar to reported measurements of the target phenomenon. An additional goal may be providing predictions and/or further improving insight into how the target phenomenon (and possibly other phenomena) may be generated. A Simulation of a Mechanistic Explanation (Figure 3a) builds upon **I** during three workflow activities. (1) Relational and continuum mathematical descriptions are developed of the salient explanatory information within the Mechanistic Explanation. (2) Those descriptions are instantiated in software; features to facilitate exploratory simulations are added; solvers are selected and the implementation undergoes verification. (3) An iterative workflow process achieves the desired qualitative and quantitative similarity between simulation output and measurements of the target phenomenon. During that process, the model and mathematical descriptions may be revised. To enable another modeler to independently reproduce reported simulation results, details of those workflow decisions should be made available when results are published [20]. For the second and third activities, it is increasingly common for researchers to rely on mathematical modeling tools, such as Matlab (The MathWorks, Inc., Natick, MA, USA), and/or proprietary or open source systems, including, for example, physiologically based simulation or emulation packages (e.g., see [21]). Use of standardized software increases credibility, reliability, and reproducibility by providing some assurance that the underlying numerical techniques are handled correctly. Use of open source software further improves reproducibility by making the simulation widely available while also opening the underlying techniques to later examination for correctness.

Technically, the simulation output is a model of solutions to the relational and mathematical descriptions under particular conditions; and the mathematical descriptions are a model of the mechanistic explanation in **I** given particular assumptions and constraints. Consequently, when "mechanistic model" or "mechanistic simulation model" is used to describe the work product, it can be difficult for a reader to know which model is being identified. To avoid misinterpretations, this type of work product can be identified accurately as a Simulation of a Mechanistic Explanation. The following are two related examples.

Example **IV**.1: The gamma rhythm is one of several characterized oscillations of activity in the brain (brain waves). The alpha rhythm of about 8 Hz is powerful enough that it can be readily detected outside of the head, something discovered in the 1920s by Hans Berger. In contrast to alpha, gamma oscillations are faster (~40 Hz) and more spatially localized, best detected by electrodes placed directly on the brain surface or into the brain parenchyma. A Simulation of a Mechanistic Explanation [22] helped explore how these gamma oscillations could be generated through inhibitory inputs, which were classically thought of as delaying or eliminating neural activity. Wang and Buzsaki demonstrated a mechanistic explanation wherein inhibitory inputs could in some cases paradoxically facilitate activity [23]. The dual roles of inhibition and facilitation allow it to entrain cell activity to a signal originating in inhibitory cells.

A relatively fine-grained, multi-formalism model is required to represent an entrainment mechanism by a simulated cell's inputs, at one scale, and the synchronization of multiple cells to plausibly generate gamma waves at a network scale. These simulations comprised local systems of ODEs, combined with a coarse PDE approximation to represent the single neuron, with event-driven techniques to connect cells into networks. To illustrate where this example fits into the spectrum of types (Figure 2a), it is useful to focus on the way the authors modeled ion channels, as systems of ODEs. Two cross-model alternatives were used, a coarse 3-channel and a fine 11-channel representation, both ultimately derived from the underlying Hodgkin-Huxley framework. Practically, using these alternatives\helped allow for cross-model validation in the face of the greater computational complexity of the 11-channel simulations. However, from a model of explanation perspective, it is important to note that the 11-channel parameterization maps more closely to ion channel biophysics. So, while both alternatives are simulations of mechanistic models, in that they are numerical solutions to systems of ODEs, the finer grained 11-channel representation is further to the right on the Figure 2a spectrum, toward an Analogous-mechanism Model and, ultimately, a Model Mechanism. Hence, this example exhibits different locations along the spectrum of types. It also demonstrates the use of methods for moving back and forth along that spectrum.

Example **IV**.2: More recent mechanistic explorations of gamma oscillations have focused on their possible role in the genesis of schizophrenia, where abnormalities in gamma oscillations have been demonstrated. Other clues to the biological explanation of schizophrenia have come from analogies with psychotomimetic drugs, such as ketamine. More recently, possible roles of particular molecular abnormalities have been suggested by a genome-wide association study. These many scales of causality were assessed by Neymotin et al. [24] using multiscale simulations of a mechanistic explanations to explore how alterations in one of the neural receptors at molecular scale might produce alterations in gamma oscillations in neuronal receptors at the molecular scale. By using both dynamical and information theoretic measures, simulation suggested how anomalies in neuronal activity might produce disturbances in function—disturbances in information flow. Thus, the model illustrates several levels of mechanistic explanation, connecting molecular anomalies with cellular anomalies, network anomalies and information transmission disturbance. Neurons were modeled with piecewise integrated difference equations, including inputs on the soma and dendrites, representing transmitted as well as background molecules and their receptors. Networks of simulated neurons were composed according to a fixed relationship between three different neuron types. Simulated current injections were used to drive the network to a baseline activity and then tuned to generate baseline theta, gamma and theta-modulated gamma oscillations in a Local Field Potential (LFP) spanning the simulated pyramidal neurons. The LFP oscillations provide the distinguishing phenomena. The simulated intervention mechanism consisted of turning on and off the NMDA (N-methyl-D-aspartate) inputs across 16 different cellular locations. Because the interventions are below the network scale, instantiated by the underlying software and mapped to the derived properties of the LFP oscillations, this model provides an excellent example of Simulation of a Mechanistic Explanation. Further, each neuron is, itself, an example of **IV**, in that it is a collection of sections (soma and dendrites), each of which is a system of difference equations propagating the inputs. However, the neuronal network is designed using random connectivity, since there are no data on actual cell-to-cell connectivity. Therefore, at this level, the model is only structurally evocative of the referent and thus approaches a Simulation of an Analogous-mechanism model (**V**). By using the information theoretic measures to relate the external inputs to spike outputs, the authors were able to demonstrate an inverse relation between gamma activity and the ability of the network to transmit information, to demonstrate how gamma oscillation might underlie information processing and how gamma oscillation anomalies could underlie the abnormal information processing in schizophrenia.

5.2.2. **V**—Simulation of an Analogous-mechanism Model

When starting with a description of an Analogous-mechanism Model (**II**), the simulation research goal is often to translate the knowledge contained within its description into simulation output that is qualitatively and quantitatively similar to measurements of the target phenomenon. When successful, an accurate descriptor of the work product is Simulation of an Analogous-mechanism Model. An increasing fraction of computational explanations of phenomena reported in the literature, including some "mechanistic models" described as being "multiscale" [25], fit reasonably well under that descriptor (e.g., see [26–31]).

Figure 3b is a snapshot of the process of building upon descriptions in **II** during two workflow activities that differ from those for **IV** in important ways. (1) The scientist creates mathematical descriptions of the Analogous-mechanism Model in operation. Continuum equations are adapted from descriptions of engineering, physical, mechanical, chemical, and/or electronic mechanisms. An important subset of those mathematical descriptions, for example, finite element analysis, goes beyond continuum mathematical descriptions because they also require numerical analysis techniques. (2) The mathematics is instantiated in software; features to support users are added; and solvers are selected. Computational solutions involve solving equations subject to boundary conditions and/or initial conditions and the implementation undergoes verification. (3) Authors undertake the iterative process of achieving qualitative and quantitative similarity between simulation output and measurements of the target phenomenon within some tolerance. The product of that process is output from selected parameterizations of a Simulation of an Analogous-mechanism Model. The following are examples.

Example **V**.1: Based on epidemiological studies, high-density lipoprotein (HDL) is believed to play an important role in lowering the risk of cardiovascular disease by mediating reverse cholesterol transport. Therapies that raise HDL-cholesterol, however, have been unable to confirm this hypothesis and demand a re-examination of the proposed mechanism. It is known that lipid-poor ApoA-I plays a role in initiating reverse cholesterol transport and that the drug RG7232 increases HDL-cholesterol. However, the influence of RG7232 on lipid-poor ApoA-I and reverse cholesterol transport is unclear because their direct measurement during dosing intervals is problematic. Lu et al. [27] developed an Analogous-mechanism Model and corresponding simulation to explore this response. The model is based on two other Analogous-mechanism Models, (1) a model of lipoprotein metabolism and kinetics and (2) a model of RG7232 pharmacokinetics. They are combined into a single simulation. The linked simulation goes further by additionally representing the hypothesis that the affinity of low-density lipoprotein (LDL) particles to LDL receptors are dependent on particle size or density. This hypothesis is implemented as a modified elimination rate. The resulting model describes temporal concentrations in two-compartments as coupled ordinary differential equations that are solved using the SimBiology toolbox of MathWorks. The simulation model is "analogous" in the sense that the proposed density-dependent elimination rate and compartmentalization is an analogy to chemical kinetics and chemical engineering. Parameters are estimated using a Bayesian approach that updates the parameter values from model components using the Matlab Global Optimization toolbox of MathWorks. The implementations simulate output from the linked Analogous-mechanism Model as if it were real.

Example **V.2**: More than 40% of astronauts who participate in long-duration missions return with ophthalmic changes similar to idiopathic intracranial hypertension. Experts posited that a microgravity-induced cephalic fluid shift elevates intracranial pressure (ICP). Feola et al. [32] hypothesized that elevated ICP would alter the peak strain environment in the optic nerve head (ONH) to cause tissue remodeling that may be contributing to the observed ophthalmic changes. They also suspected that variations in intraocular pressure (IOP) and mean arterial pressure (MAP) would affect the biomechanical strain in the OHN tissues. To explore that explanation, they implemented a finite element Analogous-mechanism Model in which a simulated structural mechanism is strongly analogous to (functions as an analog of) the ocular structure. The geometry of the analog was based on established ocular biomechanics research and it included representing coarse grain features of tissue structures known to play a significant role in the observed ophthalmic changes: sclera, preliminary neural tissue, lamina cribrosa, central retinal vessel, dura mater and pia mater of the optic nerve sheath. Furthermore, an annular ring was incorporated around the scleral canal to account for the circumferential alignment of the scleral collagen fibers around the ONH. The open source package Gmsh (V2.8.3) was used to generate the 3D finite element geometry and mesh and open source FE solver FEBio (V2.0) was used to solve for all simulations. The authors used Latin hypercube sampling of biologically plausible regions of parameter space to simulate biomechanical responses of their analog eye structure to various combinations of simulated ICPs, as well as varying IOP, MAP, and simulated tissue mechanical property conditions. Execution results showed that chronically elevated ICP coupled with interindividual differences in simulated optic nerve head mechanical properties could influence the risk for experiencing extreme optic nerve strains. The authors inferred that individuals with both soft optic nerve or pia mater and elevated ICP would be especially at risk.

Example **V.3**: Rosiglitazone is a PPARγ agonist, one of several approved insulin sensitizers used to treat diabetes. Despite being on the market for over a decade, the drug continues to be studied in the lab to understand the mechanism of action of this class of molecule. In Goto-Kakizaki rats, which are a rodent model of early-developing, non-obese type-2 diabetes, Gao and Jusko [30] show that rosiglitazone decreases glucose levels. To simulate how the insulin/glucose regulation might work, they built a feedback model—glucose stimulating insulin production and insulin increasing glucose consumption. The model is analogous to other simple feedback systems, without specifying the actual, detailed, biological mechanism (e.g., intermediate steps) for glucose/insulin co-regulation. The model also incorporates two pharmacodynamic effects of rosiglitazone that impact this feedback system: enhancing insulin sensitivity (i.e., increasing the rate of insulin-dependent loss of glucose) and inhibiting glucose production. As with many models of pharmacology, the pharmacokinetic part uses an idealized one-compartment model to fit observed drug absorption and loss. The simulation is implemented using coupled ODEs, plus analytical expressions for some of the molecules. Given its importance to diabetes and the system under study, the component representing the time-dependent body weight of the rats was a key variable being simulated along with the molecular components. Guided by experimental measurements (of drug, glucose and insulin levels over time), the model was parameterized for control, low dose and high dose rosiglitazone cases. The match between simulation output and experiment measurements showed that the Analogous-mechanism Model explained the observations sufficiently well. Using that model, the authors identified drug regimen design principles: specifically, to enhance insulin sensitivity in the long term (>6 weeks), a high-dose drug is needed continuously; neither lower-dose nor shorter-term treatment succeeded in elevating the sensitivity.

Example V.4: Attempts to design and build synthetic cellular memory systems using recombinases have thus far been hindered by a lack of validated computational models of a plausible mechanism representing DNA recombination. The predictive capabilities of such models are needed to reduce the number of iterative cycles required to align experimental results with design performance requirements. Bowyer et al. [31] developed and validated the first Simulation of an Analogous-mechanism Model for how DNA recombination might occur. The models were constructed by extracting verified biological details from an extensive review of the experimental literature and made use of a model analogy with well-established reactions networks common to chemistry and chemical engineering. Three essential biological details for which a consensus was lacking were included/excluded from the simulations. The computational model consisted of a system of ODEs, each representing the concentration of a distinct biological entity and model parameters that were optimized via the use of genetic algorithms to refine parameter values but no details on how the model was implemented were provided. Model predictions were compared to experimental data to determine which set of details might represent the most plausible mechanisms and thus serve as analogs of actual structural details by which DNA recombination works. They found that including unidirectional (versus bidirectional) excision, limiting recombinase directionality factor to monomeric form in solution (versus dimer or tetramer) and integrase monomer (versus dimer) binding to DNA produced the best model match to the data. Referring to Table 1, the contextual location this Analogous-mechanism Model is implied but is not part of the implemented computational model.

5.3. Group C: Using Computation to Support and Enhance Model Mechanisms

5.3.1. **VI**—Simulation of a Model Mechanism

The computational mechanisms used during simulation of an Analogous-mechanism Model have nothing in common with referent mechanism's spatiotemporal entities and activities within the biological context. When a description of a Model Mechanism is available (**III**), it is feasible to change that reality by striving to simulate an operating, concretized software (virtual) version of the Model Mechanism. The research goal becomes twofold. (1) Create a discretized specification of the operating Model Mechanism to guide development and instantiation of a virtual mechanism. Doing so requires meeting this requirement: key portions of the virtual Model Mechanism operate during execution as described in **III** and contribute to the simulation of Model Mechanism features. (2) Output and measurements taken during simulations are qualitatively and quantitatively similar to measurements of the target phenomenon.

The workflow characterization in Figure 4a is similar to that for **IV**, except that the Model Mechanism descriptions (light blue box) are distinct in three ways. (1) Descriptions of entities and activities are discretized sufficiently to specify in software a virtual analog of the Model Mechanism that is faithful to details in **III** (e.g., see [33,34]). (2) Evidence is presented that the entities and activities of the virtual analog are biomimetic. (3) The working hypothesis is that organized operation of software entities and activities will be capable of generating a biomimetic phenomenon.

To achieve computational efficiencies and/or fine grain details, such as receptor trafficking and molecular diffusion, influences of some entities and activities within the larger Model Mechanism are often described using a combination of rules and continuous mathematics, as in **V**, rather than being implemented as discrete biomimetic entities and activities. Doing so causes the software mechanisms during execution to fall short of the definition of mechanism [35]. Nevertheless, an accurate descriptor of the work product is a Simulation of a Model Mechanism. The following are examples.

Example **VI**.1: Simulations of Model Mechanisms are being used to help design and improve therapeutic interventions in disease [36–38]. For example, they are providing improved insight into possible failure modes of current treatments strategies for Tuberculosis (TB). Building on their multilevel, multi-attribute Model Mechanism of an immune response to TB, Linderman et al. [34] explored simulations of consequences of potential new pharmacological interventions on six different model entities and activities, including simulating immunomodulation by a cytokine; the consequences of oral and inhaled antibiotics; and the effect of vaccination. In line with the features of a biological mechanism (Table 1), their Model Mechanism identifies a phenomenon, the immune response of TB as indicated by granuloma formation and function. Components are represented at different spatial and temporal scales describe, starting with an agent-based analogy of cell behavior (macrophages and T cells) across a cross-section of lung tissue. Through rule-based probabilistic interactions, cell behavior is simulated in response to a bacterial environment. At the lowest levels of simulation hierarchy, ordinary differential equations were solved within each cell agent to simulate receptor/ligand binding, trafficking, and intracellular signaling. Partial differential equations were solved to simulate consequences of molecular diffusion. By linking their Simulation of a Model Mechanism for TB to ordinary differential equation-based pharmacokinetic and pharmacodynamic models, the authors simulated plausible consequences of the Model Mechanism's behavior during exposure to antibiotics. While simulations rely on some model compartments that are Analogous-mechanism Models, the whole system is arguably a Model Mechanism. It is biomimetic and represents an interconnected biological mechanism of granuloma formation and immune response that extends from molecular to organ levels.

Example **VI**.2: A decade ago, several laboratories sought improved models of explanation for vascular patterning defects observed in diabetic retinopathy and tumor angiogenesis. Evidence suggested that an explanatory mechanism would involve disruption of (1) Notch-driven specialization of endothelial cells into leading tip cells and following stalk cells and (2) a feedback loop that links VEGF-A tip cell induction with delta-like 4 (Dll4)-notch-mediated lateral inhibition. Bentley et al. [39] constructed a hierarchical Simulation of a Model Mechanism to explore the phenomenon of angiogenesis by connecting Analogous-mechanism Models of these processes into a large biomimetic system. The components included endothelial cell agents and membrane agents with multiple cell agents arranged as a cylindrical capillary with each cell having membrane agents distributed at the periphery. The study explored how different simulated VEGF environments and filopodia dynamics would affect simulations of Notch-mediated selection of tip cells. A staged simulation (temporally and spatially) first relied on a rule-based evaluation of membrane processes for filopodium retraction or extension or notch response to VEGF.

In following, the spatial sum of protein levels was calculated and redistributed within the endothelial cells and membrane agents. The modeling paradigm closely follows that of a Model Mechanism, where features reflect those of a biological mechanism (Table 1). An important observation of the simulations was that, by removing information that could influence simulated cell biasing, the simulated Dll4-notch lateral inhibition mechanism could generate an alternating pattern of cell fates characteristic of normal tip cell selection. The authors inferred from simulation results that abnormal patterning could be attributed to the dynamics of this particular sub-system, rather than any uncontrolled bias.

5.3.2. **VII**—A Computational Model Mechanism

This characterization differs from that in **VI** in one important way. All features of a Model Mechanism instantiated in software meet the definition of a mechanism during operation and may include all of the features in Table 1. To do so, five requirements are specified early in the workflow to guide software engineering, mechanism instantiation and simulation refinements. (1) Evidence is presented that entities, and activities of the virtual mechanism are biomimetic in prespecified ways. (2) Features of the Model Mechanism during execution are measurable. (3) Measurement of features of one or more simulation solutions match or mimic measurements of the target phenomenon within some tolerance (e.g., see [33,40]). (4) Arguments can be presented that, during execution, the Model Mechanism will have a biological counterpart (blue arrow in Figure 4b). (5) Biomimetic phenomena are generated during execution. Here are three examples.

Example **VII**.1: Enhanced mechanism-based explanations are needed to anticipate, prevent and reverse the liver injury caused by acetaminophen and other drugs. A characteristic acetaminophen phenomenon—the target phenomenon for this example—is that hepatic necrosis begins adjacent to central veins in hepatic lobules and progresses upstream. The prevailing (mechanism-oriented spatiotemporal) explanation (PE) is that location dependent differences in reactive metabolite formation within hepatic lobules (called zonation) are necessary and sufficient requisites to account for the phenomenon. Progress has been stymied because challenging that hypothesis in mice would require sequential intracellular measurements at different lobular locations within the same mouse, which is infeasible. Smith et al. [33] circumvent that impediment by performing experiments on virtual Mouse Analogs, where each is equipped with an in-silico liver that achieved multiple validation targets. Components and spaces at all levels of granularity are written in Java, utilizing the MASON multi-agent simulation toolkit. An accurate causal model of the PE that exhibits all Table 1 features was instantiated and parameterized so that, upon dosing with objects representing acetaminophen, metabolism and pharmacokinetic validation targets were achieved. However, the authors demonstrated that the PE failed to achieve the target phenomenon. Two parsimoniously more complex variants also failed to achieve the target phenomenon but a fourth variant met stringent tests of sufficiency. Execution of that forth Computational Model Mechanism provided a multilevel biomimetic causal explanation of key temporal features of acetaminophen hepatotoxicity in mice including the target phenomenon. The authors argue that the causal explanation provided during execution is strongly analogous to the actual causal mechanism in mice.

Example **VII**.2: Inflammation is not the result of one cell or molecule acting alone. It is a multicellular process that can be highly localized and yet also have diffuse actions. One of the keys to understanding tissue-level morphogenesis and spatially localized or heterogeneous processes such as inflammation is to explicitly study the spatial component—how the cells are arranged in the tissue and the influences that they have on each other. Thus, to gain insight into the pathogenesis of gastrointestinal inflammatory diseases, Cockrell et al. [41] developed a multi-level, discrete-event Model Mechanism that is used to study scenarios of how simulated cellular and molecular pathways may govern morphogenesis and inflammation in healthy and disease ileal mucosal dynamics. The system includes individual agents representing five different cell types, each with multiple independently acting instantiations at different physical locations. Cell agents have specific behaviors (proliferation, death, anoikis, etc.) and can influence each other's decision-making process. Inside each agent, there is also a simulated signaling network. The system uses algebraic rules to simulate most of the different components, including a representation of extracellular paracrine signaling between cells (with the addition of a grid-based partial differential equation to simulate consequences of diffusion), the dynamics of the simulated intracellular signaling networks and (using the current values of key intracellular signaling components as a basis) the likelihood of cell agents exhibiting each possible behavior. By simulating cell behavior in a virtual world that is analogous to biological microenvironments, the system can generate measurable phenomena (predictions) at multiple levels. Simulations provide insight into plausible pathological processes, including crosstalk between morphogenesis and inflammation and the effects of cell death on tissue health.

Example **VII**.3: Changes to savanna ecosystems related to climate change and land use practices are linked to fluctuations in savanna bird community structures, functional traits, and risk of extinction. Better, more insightful models of explanation are needed to support policy changes. However, detailed species-specific data for a given ecosystem are often limited. As a method test case for overcoming such limitations, Scherer et al. [42] used an agent-oriented approach (implemented in NetLogo) that merged trait-based and individual-based simulation methods to predict how different bird functional types might change in response to concurrent alterations to savanna rangeland from a combination of climate change and land use. The entire simulated ecosystem operates during execution as a Model Mechanism. Contained within are all of the features listed in Table 1. The system includes a spatial and stochastically varying set of entities representative of the type of individual, home range, vegetation, landscape, and environment. Each entity was characterized by a set of state variables, examples of which include age and reproductive status, or grasses, shrubs, or trees. Executions advance in uniform steps that map to an interval of up to 100 years and progress by randomly selecting, calculating and updating properties that control the spatial composition and configuration of simulated habitat and animals. Simulation results provided possible explanations for why simulated extinction risks for simulated larger- bodied insectivores, omnivores and small-bodied species were impacted differently by changes in simulated shrub-grass ratio and clumping intensity of shrub patches. Such predictions could prove essential for identifying better policies for conservation management.

6. Relevant Information, Multiple Sources

Essential relevant information from a variety of sources is needed to establish and enhance the credibility of improved insights derived from **IV–VII**. The Figure 1b,c spectra characterize two important sources. The three Figure 5 spectra identify additional information sources and types. The Figure 5 spectra are more closely linked to methodology than are the workflow characterizations in Figures 3–5. Having available sufficient information enables authors and readers to identify approximate locations on all six spectra, which improves clarity and brings into focus the characteristics that distinguish among **IV–VII**.

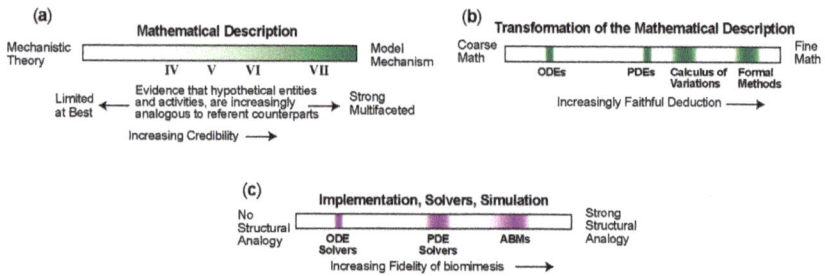

Figure 5. Characteristics three sources and types of relevant information. These three spectra are distinct from those in Figure 1. They bring into focus characteristics of methods and approach that distinguish among **IV–VII**. (**a**) The relationship within **I**, **II**, or **III** and the corresponding mathematical description must be clear. (**b**) Expanding a model or combining it with other models [43,44] is a strategy used to improve explanatory descriptions. The choice of mathematical description can influence faithfulness of deductive transformations. Four examples of commonly used mathematical model types illustrate that different types occupy different relative locations. Some mathematical model types cannot be easily modified and remain faithful to the target phenomenon while also preserving the original meaning(s) of the model's terms and model-to-target mappings provided in the explanatory descriptions. (**c**) This spectrum illustrates that implementation decisions (primarily within the yellow boxes in Figures 4 and 5) influence the fidelity of the biomimesis that can be built into the simulations during execution. Stronger analogies between the biology and model mechanisms during execution are expected to improve clarity, credibility and scientific usefulness.

The Figure 5a spectrum characterizes the mathematical descriptions used in **IV–VI**. Information is lost during derivation from the primarily prosaic description (including induction from data) in **II** and **III** to mathematical descriptions. Clarity about what is and is not lost can influence credibility. For example, the assumption behind Simulation of an Analogous-mechanism Model is that, if the model were made real, then some version of the phenomenon generated during operation would mimic the referent phenomenon. In most reports, the focus is primarily on mimicking the referent phenomenon and much less so on the model's entities, activities and organization during phenomenon generation. Consequently, it is often the case that mathematical descriptions are imbalanced, which can limit clarity and credibility.

The Figure 5b spectrum is about (primarily deductive) transformations of the descriptions in **I–III**. The research goal of improving mechanism-oriented explanations often involves inferring plausible biological details from explorations of the model's behavior and then seeking transformations (ways to change computational features) that provide improvement. Formal Methods refer to the computer science (and mathematics) that allows such transformations to be rigorous enough to reason over, i.e., to make them purely deductive. Particular types of mathematical models (e.g., ODEs) cannot be easily modified without breaking the extent to which the model represents the description in **II** or **III** and maps to the target phenomenon. Faithful deduction over a simulation, including modifications that are faithful to the target phenomenon, are those that preserve the original meaning(s) of the model's terms and model-to-target phenomenon mappings (for example [44]). The expectation is that credibility of **IV–VII** will increase as faithfulness to deductive transformations from mathematical descriptions increases.

The Figure 5c spectrum illustrates the influence of implementation decisions on the fidelity of biomimesis built into a simulation during execution. We anticipate that the deeper the insight, the stronger the analogy between the biology's mechanisms and simulation's mechanisms. Thus, credibility will increase by increasing structural analogies between implementations simulating the target phenomenon and the biological system generating the target phenomenon.

Moving rightward in Figure 1 on spectra 1b and 1c involves incorporating deeper (validated) insight into an expanding variety of interconnected biological processes and phenomena. Mechanism-oriented models that are developing deeper (validated) insight into an expanding variety of phenomena will be moving rightward on the Figure 5 spectra. As a consequence, implementations must change during each move to the right. During those changes, information that can influence—bias—simulation output can be lost and/or added. Documenting those influences enhances credibility. The absence of such documentation risks creating a barrier to credibility, thus limiting scientific usefulness.

7. Workflow, Provenance and Hybrid Models

Most biological scientists and clinicians have a general appreciation for and understanding of, the workflow, the systems utilized and methods employed in wet-lab research. When they read a research article reporting results of experiments, that knowledge influences their assessment of credibility. Biological scientists and clinicians outside of the simulation field may be drawn to (and may consider reading) a simulation-focused research report due to the prospect for improved explanatory insight or practical utility. However, they do not have a corresponding appreciation for, or understanding of, the workflow, the systems utilized, or the methods employed. Thus, there is a significant risk that missing information and lack of clarity will erode the reader's assessment of the credibility of arguments presented and of simulation approaches in general.

The credibility of inferences about a phenomenon based on results of wet-lab experiments depends on having easy access to the experiment's provenance [45], i.e., the full context of the experiment along with adequate descriptions of methods, materials and other important workflow details. Removing or distancing observations and/or data from the experiment's provenance abstracts away both information and knowledge, thus weakening justifications for their application or use elsewhere. By analogy, the credibility of explanations provided by simulations for how a phenomenon may be generated depends on use context and includes having easy access to the provenance of **IV–VII** [46]. Provenance begins with **I–III** and includes the full context of the simulation activities. Also, by analogy, unlinking an element (e.g., mathematical descriptions or software implementation details) from the information and knowledge provided by the original use context and provenance for application or reuse elsewhere can weaken or eliminate justifications for the intended application or reuse, thus eroding credibility and limiting scientific usefulness.

It is now common to encounter biology simulation research reports that seek merged explanations of two or more phenomena or a description of phenomena across multiple biological levels or scales. The software instantiations, commonly referred to as hybrid models, require means for the different, originally separate and independent mechanism-oriented models to interact during execution. Those means include adding software features and making changes to the previously independent implementations. Describing the product of that process as hybrid alerts readers to expect the merged system to behave in new ways. Some behaviors will be intended but others may be unintended. The situation is somewhat analogous to combining two reagents during a wet-lab protocol when, under some conditions, doing so risks an adverse interaction. The importance of providing clear details is obvious.

8. Concluding Remarks

Although credibility and clarity are often correlated, other factors can have an even greater influence on explanatory credibility. Each element in the **I–VII** characterizations will "resonate" differently with different scientists, clinicians, and stakeholders. Here are three examples: (1) The evidence selected to support a description of an Analogous-mechanism Model (**II**) may resonate well with engineers and system biologists but less so with oncologists. (2) For a particular characterization, the interpretations offered by authors in the context of selected simulation results will likely resonate differently with scientists approaching the problem from basic science and clinical perspectives. (3) The extent to which a particular set of mathematical expressions or software engineering methods resonates with a simulation researcher will likely have a significant impact on that person's determination of whether a particular computational mechanism-oriented model is sufficiently mechanistic or not, which, in turn, may impact that person's assessment of credibility. There are, of course, other influences and even larger issues to consider. For example, the interpretation of what is happening within all the above workflows is part of the philosophy of science. We put these important influences aside for now as they are beyond the scope of this overview.

Increasing complexity in pursuit of mechanism-oriented models that improve explanatory credibility is an explicit strategy within biology simulation research (e.g., see [26,44,46]). For the larger community of biologists, a priority is achieving deeper, more useful explanations of phenomena that facilitate advancing both science and health. The scientific usefulness of biology simulation as a discipline will become more evident to the larger community as credible multi-phenomena explanations become available. Achieving credible multi-phenomena explanations requires moving rightward on all spectra in Figures 1 and 5. But doing so requires increasing support from the larger biology community. Improving clarity, semantic and otherwise, is a necessary and essential small step to achieving that increased support.

By characterizing **I–III** and **IV–VII** we demonstrate how semantic clarity can be improved even as the complexity of those models of explanation increases. These categories of types of models and simulations may serve as a foundation for a clear ontology of mechanism-oriented simulation research in biology.

In summary, "mechanistic model" is used specifically and as an umbrella term within the computational biology community. Unclear, vague labeling of a computational model as "mechanistic" risks providing readers an ungrounded perception of its credibility, intentionally or unintentionally. We provide clear descriptions and illustrations of broad categories of explanatory models. We suggest terminology and language that modelers can use to more accurately explain how diverse mechanism-oriented computational models are—or are not—"mechanistic." The language is also intended to enable the audience of those models, which can be rather diverse, to more easily understand what it is about the model that is mechanistic.

Supplementary Materials: The following is available online at http://www.mdpi.com/2227-9717/6/5/56/s1, Supporting text: Role of the Committee on Credible Practice of Modeling and Simulation in Healthcare.

Author Contributions: C.A.H. managed manuscript preparation, content organization, development and editing; A.E., W.W.L., F.M.G, E.A.S., M.K.T., L.M. and C.A.H. contributed to the development of the presented ideas; C.A.H. created the Figures.

Acknowledgments: We thank Glen E.P. Ropella for providing content suggestions; Andrew Smith and Ryan Kennedy for constructive criticism during manuscript development; and Mitzi Baker editorial input. This work was supported in part by the National Institutes of Health: R01GM104139 (A.E.), R01HL101200 (F.M.G.), R01EB022903 (W.W.L.); The National Science Foundation: NSF:CAREER 1452728 (E.A.S.); InSilico Labs LLC (L.M.); and the UCSF BioSystems group (C.A.H.).

Conflicts of Interest: The authors declare no conflict of interest. InSilico Labs LLC supported L.M.'s effort but imposed no commercial restrictions or constraints. The funding sponsors had no role in the design of the study; in the collection, analyses, or interpretation of data; in the writing of the manuscript and in the decision to publish the results.

References

1. Craver, C.; Tabery, J. Mechanisms in Science. In *The Stanford Encyclopedia of Philosophy*; Springer: Berlin/Heidelberg, Germany, 2017.
2. Braillard, P.A.; Malaterre, C. *Explanation in Biology: An Enquiry into the Diversity of Explanatory Patterns in the Life Sciences*; Braillard, P.A., Malaterre, C., Eds.; Springer: Dordrecht, The Netherland, 2015.
3. Kaplan, D.M.; Craver, C.F. The explanatory force of dynamical and mathematical models in neuroscience: A mechanistic perspective. *Philos. Sci.* **2011**, *78*, 601–627. [CrossRef]

4. Bechtel, W.; Abrahamsen, A. Explanation: A mechanist alternative. *Stud. Hist. Philos. Sci. Part C* **2005**, *36*, 421–441. [CrossRef]
5. Craver, C.F. *Explaining the Brain: Mechanisms and the Mosaic Unity of Neuroscience*; Oxford University Press: Oxford, UK, 2007.
6. Illari, P.M.; Williamson, J. What is a mechanism? Thinking about mechanisms across the sciences. European. *J. Philos. Sci.* **2012**, *2*, 119–135. [CrossRef]
7. Merriam-Webster Unabridged. Available online: http://unabridged.merriam-webster.com/ (accessed on 2 October 2016).
8. Craver, C.F. When mechanistic models explain. *Synthese* **2006**, *153*, 355–376. [CrossRef]
9. Darden, L. Thinking again about biological mechanisms. *Philos. Sci.* **2009**, *75*, 958–969. [CrossRef]
10. Miyakawa, K.; Albee, R.; Letzig, L.G.; Lehner, A.F.; Scott, M.A.; Buchweitz, J.P.; James, L.P.; Ganey, P.E.; Roth, R.A. A Cytochrome P450–independent mechanism of acetaminophen-induced injury in cultured mouse hepatocytes. *J. Pharmacol. Exp. Therap.* **2015**, *354*, 230–237. [CrossRef] [PubMed]
11. Thakore, K.N.; Mehendale, H.M. Role of hepatocellular regeneration in CCl4 autoprotection. *Toxicol. Path* **1991**, *19*, 47–58. [CrossRef] [PubMed]
12. Rudraiah, S.; Rohrer, P.R.; Gurevich, I.; Goedken, M.J.; Rasmussen, T.; Hines, R.N.; Manautou, J.E. Tolerance to acetaminophen hepatotoxicity in the mouse model of autoprotection is associated with induction of flavin-containing monooxygenase-3 (FMO3) in hepatocytes. *Toxicol. Sci.* **2014**, *27*, kfu124. [CrossRef] [PubMed]
13. Hodgkin, A.L.; Huxley, A.F. A quantitative description of membrane current and its application to conduction and excitation in nerve. *J. Physiol.* **1952**, *117*, 500–544. [CrossRef] [PubMed]
14. Russmann, S.; Kullak-Ublick, G.A.; Grattagliano, I. Current concepts of mechanisms in drug-induced hepatotoxicity. *Curr. Med. Chem.* **2009**, *16*, 3041–3053. [CrossRef] [PubMed]
15. Bassler, N.; Loeffler, C.; Mangin, P.; Yuan, Y.; Schwarz, M.; Hagemeyer, C.E.; Eisenhardt, S.U.; Ahrens, I.; Bode, C.; Jackson, S.P.; et al. A mechanistic model for paradoxical platelet activation by ligand-mimetic αIIbβ3 (GPIIb/IIIa) antagonists. *Arterioscl. Thromb. Vasc. Biol.* **2007**, *27*, E9–E15. [CrossRef] [PubMed]
16. Zajac, F.E. Muscle and tendon Properties models scaling and application to biomechanics and motor. *Crit. Rev. Biomed. Eng.* **1989**, *17*, 359–411. [PubMed]
17. Li, L.; Gardner, I.; Rose, R.; Jamei, M. Incorporating target shedding into a minimal PBPK–TMDD model for monoclonal antibodies. *CPT Pharmacomet. Syst. Pharmacol.* **2014**, *3*, e96. [CrossRef] [PubMed]
18. Ayllón, D.; Railsback, S.F.; Vincenzi, S.; Groeneveld, J.; Almodóvar, A.; Grimm, V. InSTREAM-Gen: Modelling eco-evolutionary dynamics of trout populations under anthropogenic environmental change. *Ecol. Mod.* **2016**, *326*, 36–53. [CrossRef]
19. Norton, K.A.; Wininger, M.; Bhanot, G.; Ganesan, S.; Barnard, N.; Shinbrot, T. A 2D mechanistic model of breast ductal carcinoma in situ (DCIS) morphology and progression. *J. Theor. Biol.* **2010**, *263*, 393–406. [CrossRef] [PubMed]
20. Clegg, L.E.; Mac Gabhann, F. Molecular mechanism matters: Benefits of mechanistic computational models for drug development. *Pharmacol. Res.* **2015**, *99*, 149–154. [CrossRef] [PubMed]
21. Bangs, A.; Bowling, K.; Paterson, T. Simulating Patient-Specific Outcomes. US Patent 10/961,523, 7 October 2004.
22. Lytton, W.W.; Sejnowski, T.J. Simulations of cortical pyramidal neurons synchronized by inhibitory interneurons. *J. Neurophysiol.* **1991**, *66*, 1059–1079. [CrossRef] [PubMed]
23. Wang, X.J.; Buzsáki, G. Gamma oscillation by synaptic inhibition in a hippocampal interneuronal network model. *J. Neurosci.* **1996**, *16*, 6402–6413. [CrossRef] [PubMed]
24. Neymotin, S.A.; Lazarewicz, M.T.; Sherif, M.; Contreras, D.; Finkel, L.H.; Lytton, W.W. Ketamine disrupts theta modulation of gamma in a computer model of hippocampus. *J. Neurosci.* **2011**, *31*, 11733–11743. [CrossRef] [PubMed]
25. Xue, C.; Shtylla, B.; Brown, A. A stochastic multiscale model that explains the segregation of axonal microtubules and neurofilaments in neurological diseases. *PLoS Comput. Biol.* **2015**, *11*, e1004406. [CrossRef] [PubMed]

26. Meier-Schellersheim, M.; Fraser, I.D.; Klauschen, F. Multiscale modeling for biologists. *WIREs: Syst. Biol. Med.* **2009**, *1*, 4–14. [CrossRef] [PubMed]
27. Lu, J.; Cleary, Y.; Maugeais, C.; Kiu Weber, C.I.; Mazer, N.A. Analysis of "on/off" kinetics of a CETP inhibitor using a mechanistic model of lipoprotein metabolism and kinetics. *CPT Pharmacomet. Syst. Pharmacol.* **2015**, *4*, 465–473. [CrossRef] [PubMed]
28. Bassingthwaighte, J.B.; Chizeck, H.J.; Atlas, L.E. Strategies and tactics in multiscale modeling of cell-to-organ systems. *Proc. IEEE* **2006**, *94*, 819–831. [CrossRef] [PubMed]
29. Ménochet, K.; Kenworthy, K.E.; Houston, J.B.; Galetin, A. Simultaneous assessment of uptake and metabolism in rat hepatocytes: A comprehensive mechanistic model. *J. Pharmacol. Exp. Therap.* **2012**, *341*, 2–15. [CrossRef] [PubMed]
30. Gao, W.; Jusko, W.J. Modeling disease progression and rosiglitazone intervention in type 2 diabetic Goto-Kakizaki rats. *J. Pharmacol. Exp. Therap.* **2012**, *341*, 617–625. [CrossRef] [PubMed]
31. Bowyer, J.; Zhao, J.; Rosser, S.; Colloms, S.; Bates, D. Development and experimental validation of a mechanistic model of in vitro DNA recombination. In Proceedings of the 2015 37th Annual International Conference of the IEEE Engineering in Medicine and Biology Society (EMBC), Milano, Italy, 25–29 August 2015.
32. Feola, A.J.; Myers, J.G.; Raykin, J.; Mulugeta, L.; Nelson, E.S.; Samuels, B.C.; Ethier, C.R. Finite element modeling of factors influencing optic nerve head deformation due to intracranial pressure. *Investig. Ophthal. Vis. Sci.* **2016**, *57*, 1901–1911. [CrossRef] [PubMed]
33. Smith, A.K.; Petersen, B.K.; Ropella, G.E.P.; Kennedy, R.C.; Kaplowitz, N.; Ookhtens, M.; Hunt, C.A. Competing Mechanistic hypotheses of acetaminophen-induced hepatotoxicity challenged by virtual experiments. *PLoS Comput. Biol.* **2016**. [CrossRef] [PubMed]
34. Linderman, J.J.; Cilfone, N.A.; Pienaar, E.; Gong, C.; Kirschner, D.E. A multi-scale approach to designing therapeutics for tuberculosis. *Integr. Biol.* **2015**, *7*, 591–609. [CrossRef] [PubMed]
35. Issad, T.; Malaterre, C. Are dynamic mechanistic explanations still mechanistic? In *Explanation in Biology: An Enquiry into the Diversity of Explanatory Patterns in the Life Sciences*; Braillard, P.A., Malaterre, C., Eds.; Dordrecht, The Netherland, 2015; pp. 265–292.
36. Schoeberl, B.; Pace, E.A.; Fitzgerald, J.B.; Harms, B.D.; Xu, L.; Nie, L.; Linggi, B.; Kalra, A.; Paragas, V.; Bukhalid, R.; et al. Therapeutically targeting ErbB3: A key node in ligand-induced activation of the ErbB receptor-PI3K axis. *Sci. Signal* **2009**, *2*, ra31. [CrossRef] [PubMed]
37. Hosseini, I.; Mac Gabhann, F. Mechanistic Models Predict Efficacy of CCR5-Deficient Stem Cell Transplants in HIV Patient Populations. *CPT Pharmacomet. Syst. Pharmacol.* **2016**, *5*, 82–90. [CrossRef] [PubMed]
38. Moreno, J.D.; Zhu, Z.I.; Yang, P.-C.; Bankston, J.R.; Jeng, M.-T.; Kang, C.; Wang, L.; Bayer, J.D.; Christini, D.J.; Trayanova, N.A.; et al. A computational model to predict the effects of class I anti-arrhythmic drugs on ventricular rhythms. *Sci. Transl. Med.* **2011**, *3*, ra83. [CrossRef] [PubMed]
39. Bentley, K.; Gerhardt, H.; Bates, P.A. Agent-based simulation of notch-mediated tip cell selection in angiogenic sprout initialisation. *J. Theor. Biol.* **2009**, *250*, 25–36. [CrossRef] [PubMed]
40. Tang, J.; Hunt, C.A. Identifying the rules of engagement enabling leukocyte rolling, activation, and adhesion. *PLoS Comput. Biol.* **2010**, *6*, e1000681. [CrossRef] [PubMed]
41. Cockrell, C.; Christley, S.; An, G. Investigation of inflammation and tissue patterning in the gut using a spatially explicit general-purpose model of enteric tissue (SEGMEnT). *PLOS Comput. Biol.* **2014**, *10*, e1003507. [CrossRef] [PubMed]
42. Scherer, C.; Jeltsch, F.; Grimm, V.; Blaum, N. Merging trait-based and individual-based modelling: An animal functional type approach to explore the responses of birds to climatic and land use changes in semi-arid African savannas. *Ecol. Mod.* **2016**, *24*, 75–89. [CrossRef]
43. Weisel, E.W.; Petty, M.D.; Mielke, R.R. Validity of models and classes of models in semantic composability. In Proceedings of the Fall 2003 Simulation Interoperability Workshop, Orlando, FL, USA, 14–19 September 2003.
44. Yankeelov, T.E.; An, G.; Saut, O.; Luebeck, E.G.; Popel, A.S.; Ribba, B.; Vicini, P.; Zhou, X.; Weis, J.A.; Ye, K.; Genin, G.M. Multi-scale Modeling in Clinical Oncology: Opportunities and Barriers to Success. *Ann. Biomed. Eng.* **2016**, *44*, 2626–2641. [CrossRef] [PubMed]

45. Kazic, T. Ten Simple Rules for Experiments' Provenance. *PLoS Comput. Biol.* **2015**, *11*, e1004384. [CrossRef] [PubMed]
46. McDougal, R.; Bulanova, A.; Lytton, W. Reproducibility in computational neuroscience models and simulations. *IEEE Trans. Biomed. Eng.* **2016**, *63*, 2021–2035. [CrossRef] [PubMed]

MDPI

St. Alban-Anlage 66

4052 Basel

Switzerland

Tel. +41 61 683 77 34

Fax +41 61 302 89 18

www.mdpi.com

Processes Editorial Office

E-mail: processes@mdpi.com

www.mdpi.com/journal/processes

www.ingramcontent.com/pod-product-compliance
Lightning Source LLC
Chambersburg PA
CBHW051846210326
41597CB00033B/5794